THE FOOD-ENERGY-WATER TRIANGLE FOR SUSTAINABLE DEVELOPMENT

Integrative and Synergic Solutions

THE FOOD-ENERGY-WATER TRIANGLE FOR SUSTAINABLE DEVELOPMENT

Integrative and Synergic Solutions

Pedro Aguilar-Zárate, PhD
Mónica L. Chávez-González, PhD
Ann Rose Abraham, PhD
A. K. Haghi, PhD
Editors

AAP | APPLE ACADEMIC PRESS

First edition published 2025

Apple Academic Press Inc.
1265 Goldenrod Circle, NE,
Palm Bay, FL 32905 USA

760 Laurentian Drive, Unit 19,
Burlington, ON L7N 0A4, CANADA

CRC Press
2385 NW Executive Center Drive,
Suite 320, Boca Raton FL 33431

4 Park Square, Milton Park,
Abingdon, Oxon, OX14 4RN UK

Apple Academic Press exclusively co-publishes with CRC Press, an imprint of Taylor & Francis Group, LLC

Library and Archives Canada Cataloguing in Publication

Title: The food-energy-water triangle for sustainable development : integrative and synergic solutions / Pedro Aguilar-Zárate, PhD, Mónica Chávez-González, PhD, Ann Rose Abraham, PhD, A.K. Haghi, PhD, editor.
Names: Aguilar-Zárate, Pedro, editor. | Chávez-González, Mónica Lizeth, 1987- editor | Abraham, Ann Rose, editor | Haghi, A. K., editor
Description: First edition. | Includes bibliographical references and index.
Identifiers: Canadiana (print) 20240486641 | Canadiana (ebook) 20240487567 | ISBN 9781774917855 (hardcover) | ISBN 9781774917862 (softcover) | ISBN 9781003561453 (ebook)
Subjects: LCSH: Sustainable development. | LCSH: Water security. | LCSH: Food security. | LCSH: Energy security. | LCSH: Climatic changes.
Classification: LCC HD75.6 .F66 2025 | DDC 333.7—dc23

Library of Congress Cataloging-in-Publication Data

CIP data on file with US Library of Congress

ISBN: 978-1-77491-785-5 (hbk)
ISBN: 978-1-77491-786-2 (pbk)
ISBN: 978-1-00356-145-3 (ebk)

About the Editors

Pedro Aguilar-Zárate, PhD
Professor, Laboratorio Nacional CONAHCYT de Apoyo a la Evaluación de Productos Bióticos (LaNAEPBi), Unidad de Servicio Tecnológico Nacional de México/ Instituto Tecnológico de Ciudad Valles

Pedro Aguilar-Zárate, PhD, is a Professor at the National Technological Institute of Mexico, Technological Institute of Ciudad Valles, Mexico. His research focuses on the development of bioprocesses for the extraction and purification of secondary plant metabolites, the development of microbial fermentations in solid and liquid media to produce metabolites of industrial interest, and the study of protein–polyphenols interactions. Dr. Aguilar has published over 40 articles in indexed journals, over 20 book chapters, and nine articles in refereed journals. He has directed and codirected 24 undergraduate theses and two MSc theses and has made several presentations at scientific and academic events. He has been in charge of four research projects funded by the National Technological Institute of Mexico and collaborated in two projects funded by the sectoral funds SEP-CONACYT and CONAFOR-CONACYT. He is a guest editor of *Frontiers in Food Science and Technology, Polymers, Plant Science Today,* and the *International Journal of Research and Technological Innovation* (RIIT). This has earned him recognition as a member of the National System of Researchers level 1 (Mexico). His research and scientific career have been recognized by the Mexican Association of Food Science (2014), the Mexican Society of Biotechnology and Bioengineering (2015), the Mexican Association of Food Engineering and Biotechnology (2020), and the Potosino Council of Science and Technology (2021). He earned his PhD degree in Food Science and Technology at Autonomous University of Coahuila, Mexico.

Mónica L. Chávez-González, PhD
Professor, School of Chemistry, Universidad Autónoma de Coahuila, Mexico

Mónica Chávez-González, PhD, is a Research Professor at the School of Chemical Sciences of the Universidad Autónoma de Coahuila, Mexico. She is a pharmacobiological chemist. Her master's and doctoral program was in Food Science and Technology. She is the General Coordinator and founder of the Bioprocessing Mx-LATAM Network, whose objective is to promote R&D in the area of bioprocessing through interaction between universities, institutes, research centers, and relevant industries in Mexico and Latin America. She has been awarded the Latin American Women in Chemistry Award in the category of "Emerging Leader," granted by the American Chemical Society and the Latin American Federation of Chemical Associations (FLAQ). Her expertise is in the areas of fermentation processes, microbial biotransformation, enzyme production, industrial food waste valorization, extraction of bioactive compounds (antioxidants), and chemical characterization.

Ann Rose Abraham, PhD
Assistant Professor, Sacred Heart College (Autonomous), Thevara, Kochi, Kerala, India

Ann Rose Abraham, PhD, is currently an Assistant Professor in the Department of Physics, Sacred Heart College (Autonomous), Thevara, Kochi, Kerala, India. She has expertise in the field of condensed matter physics, nanomagnetism, multiferroics, polymeric nanocomposites, etc. She has research experience at various reputed national institutes, including Bose Institute, Kolkata, India; SAHA Institute of Nuclear Physics, Kolkata, India; and the UGC-DAE CSR Centre, Kolkata, India. She has collaborated with various international laboratories. She is a recipient of a young researcher award in physics and best paper awards—2020 and 2021. She has served as an Assistant Professor and Examiner at the Department of Basic Sciences, Amal Jyothi College of Engineering, under APJ Abdul Kalam Technological University, Kerala, India. Dr. Abraham is a frequent speaker at national and international conferences. She has a good number of publications to her credit in many peer-reviewed high-impact journals of international repute. She has authored many book chapters and edited more than 10 books with Taylor and Francis, Elsevier, etc.

A. K. Haghi, PhD
Research Associate, Department of Chemistry,
University of Coimbra, Portugal

A. K. Haghi, PhD, is a retired professor and has written, cowritten, edited or coedited more than 1000 publications, including books, book chapters, and papers in refereed journals with over 3800 citations and h-index of 32, according to the Google Scholar database. He is currently a research associate at University of Coimbra, Portugal. Professor Haghi has received several grants, consulted for several major corporations, and is a frequent speaker to national and international audiences. He is founder and former Editor-in-Chief of the *International Journal of Chemoinformatics and Chemical Engineering* and *Polymers Research Journal.* Professor Haghi has acted as an editorial board member of many international journals. He has served as a member of the Canadian Research and Development Center of Sciences & Cultures. He has supervised several PhD and MSc theses at the University of Guilan (UG) and cosupervised international doctoral projects. Professor Haghi holds a BSc in urban and environmental engineering from the University of North Carolina (USA) and holds two MSc degrees, one in mechanical engineering from North Carolina State University (USA) and another one in applied mechanics, acoustics, and materials from the Université de Technologie de Compiègne (France). He was awarded a PhD in engineering sciences at the Université de *Franche-Comté* (France). He is a regular reviewer of leading international journals.

Contents

Contributors

Pedro Aguilar-Zarate
Engineering Department. Tecnológico Nacional de México/I. T. de Ciudad Valles, Ciudad Valles, San Luis Potosí, México

Carlos Alonso
Grupo de Investigación en Alimentos Funcionales y Nutracéuticos, TecNM/Instituto Tecnológico de Durango, Felipe Pescador Ote. Durango, Dgo., México

Brenda Alvarado
Autonomous University of San Luis Potosi. Faculty of Professional Studies Huasteca Zone. Master in Biochemical Sciences, Ciudad Valles, San Luis Potosí, México

Bivitha
Post-Graduate and Research Department of Physics, Catholicate College, Pathanamthitta, Kerala, India

Agness Chacha
Affiliated to School of Sciences, Jain University, Bangalore, India

Stephany Nefertari Chávez-García
Food Research Department, School of Chemistry, Saltillo, Coahuila, México

Itza Nallely Cordero-Soto
Tecnológico Nacional de México/Instituto Tecnológico de Durango Departamento de Ingenierías Química y Bioquímica, Felipe Pescador Ote, Colonia Nueva Vizcaya, C.P. Durango, Durango, México

María Gorety Contreras-Hernández
Tecnológico Nacional de México/Instituto Tecnológico de Durango Departamento de Ingenierías Química y Bioquímica, Felipe Pescador Ote, Colonia Nueva Vizcaya, C.P. Durango, Durango, México

Dora Elisa Cruz-Casas
Food Research Department, School of Chemistry, Saltillo, Coahuila, México

Jincemon Cyriac
Marian College (Autonomous) Kuttikkanam, Kuttikkanam, Peermade, Idukki, India

Adriana Carolina Flores-Gallegos
Food Research Department, School of Chemistry, Saltillo, Coahuila, México

José Alberto Gallegos-Infante
Grupo de Investigación en Alimentos Funcionales y Nutracéuticos, TecNM/Instituto Tecnológico de Durango, Felipe Pescador Ote. Durango, Dgo., México

Priyanka Ghow
Affiliated to School of Sciences, Jain University, Bangalore, India

Manuel Efraín González-Mercado
Grupo de Investigación en Ciencias del Movimiento Humano, Universidad Juárez del Estado de Durango, Constitución Sur, Durango, Dgo., México

Rubén Francisco González-Laredo
Laboratorio Nacional CONAHCYT de Apoyo a la Evaluación de Productos Bióticos (LaNAEPBi), Unidad de Servicio, Tecnológico Nacional de México/ I.T. de Durango (TecNM/ITD), Durango, Dgo., México

Silvia Marina González-Herrera
Tecnológico Nacional de México/Instituto Tecnológico de Durango Departamento de Ingenierías Química y Bioquímica, Felipe Pescador Ote, Colonia Nueva Vizcaya, C.P. Durango, Durango, México
Research Group on Functional Foods and Nutraceuticals. TecNM/Instituto Tecnológico de Durango, Felipe Pescador Ote., Durango, Dgo., México

José Alberto Gallegos-Infante
Research Group on Functional Foods and Nutraceuticals. TecNM/Instituto Tecnológico de Durango, Felipe Pescador Ote., Durango, Dgo., México

Akhina H.
School of Energy Materials, Mahatma Gandhi University, Kottayam, India

Karen Marlenne Herrera-Rocha
Grupo de Investigación en Alimentos Funcionales y Nutracéuticos, TecNM/Instituto Tecnológico de Durango, Felipe Pescador Ote, Durango, Dgo., México

Parvathi Jayasankar
Department of Chemistry and Biochemistry, JAIN (Deemed-to-Be) University, Bangalore, India

Saritha K. Jose
Department of Physics, Deva Matha College, Kuravilangad, Kottayam, Kerala, India

Linta P. Joseph
Marian College (Autonomous) Kuttikkanam, Kuttikkanam, Peermade, Idukki, India

Satheesh Kumar K. V.
Postgraduate and Research Department of Chemistry, Government College, Kasaragod, India

Archana L. S.
Government College for Women, Vazhuthacaud, Thiruvananthapuram, Kerala, India

Erik De León-Juárez
Food Research Department, School of Chemistry, Saltillo, Coahuila, México

Keerthana M.
Affiliated to School of Sciences, Jain University, Bangalore, India

Soosen Samuel M.
Post-Graduate and Research Department of Physics, Catholicate College, Pathanamthitta, Kerala, India

Alfredo de Jesús Martínez-Roldán
CONAHCYT-Tecnológico Nacional de México/Instituto Tecnológico de Durango. Maestría en Sistemas Ambientales. Departamento de Ingenierías Química y Bioquímica, Felipe Pescador Ote, Colonia Nueva Vizcaya, C.P. Durango, Durango, México

Sunny Mathew
Marian College (Autonomous) Kuttikkanam, Kuttikkanam, Peermade, Idukki, India

Rekha Mirle
Affiliated to School of Sciences, Jain University, Bangalore, India

Manjusha. M. V.
Department of Physics, The Cochin College, Kochi, Ernakulum, Kerala, India

Santhosh M. V.
SCMS School of Engineering and Technology, Karukutty, Kerala, India

Martha Rocío Moreno-Jiménez
Laboratorio Nacional CONAHCYT de Apoyo a la Evaluación de Productos Bióticos (LaNAEPBi), Unidad de Servicio, Tecnológico Nacional de México/ I.T. de Durango (TecNM/ITD), Blvd. Felipe Pescador 1830 Ote., C.P. 34080, Durango, Dgo., México.

Vedashree N.
Affiliated to School of Sciences, Jain University, Bangalore, India

Álvaro Javier Pastrana Pastrana
Food Research Department, School of Chemistry, Saltillo, Coahuila, México

Tressia Alias Princy Paulose
Post Graduate and Research Department of Chemistry, Bishop Moore College, Mavelikara, Kerala, India

Vaisakh. P. H.
Post-Graduate and Research Department of Chemistry, Bishop Moore College, Mavelikara, Kerala, India

Eapen Philip
Post-Graduate and Research Department of Chemistry, Bishop Moore College, Mavelikara, Kerala, India

Steffanny Sánchez Portillo
Food Research Department, School of Chemistry, Saltillo, Coahuila, México

Geethu R.
SCMS School of Engineering and Technology, Karukutty, Kerala, India

Dhanya Raj
Department of Physics, The Cochin College, Kochi, Ernakulam, Kerala, India

Abigail Reyes
Autonomous University of San Luis Potosi. Faculty of Professional Studies Huasteca Zone. Master in Biochemical Sciences. Ciudad Valles, San Luis Potosí, México

Veena R. K.
Government College for Women, Vazhuthacaud, Thiruvananthapuram, Kerala, India

MaríaJosé Rivas-Arreola
University Iberoamericana de Puebla. Blvd. del Niño Poblano, Reserva Territorial Atlixcáyotl, Centro Comercial Puebla, San Andrés Cholula

Nuria Elizabeth Rocha-Guzmán
Laboratorio Nacional CONAHCYT de Apoyo a la Evaluación de Productos Bióticos (LaNAEPBi), Unidad de Servicio, Tecnológico Nacional de México/ I.T. de Durango (TecNM/ITD), Durango, Dgo., México

Olga Miriam Rutiaga-Quiñones
Tecnológico Nacional de México/Instituto Tecnológico de Durango Departamento de Ingenierías Química y Bioquímica, Felipe Pescador Ote, Colonia Nueva Vizcaya, C.P. Durango, Durango, México

María Dolores Josefina Rodríguez-Rosales
Tecnológico Nacional de México/Instituto Tecnológico de Durango Departamento de Ingenierías Química y Bioquímica, Felipe Pescador Ote, Colonia Nueva Vizcaya, C. P. Durango, Durango, México

Christian Sánchez
Autonomous University of San Luis Potosi. Faculty of Professional Studies Huasteca Zone. Master in Biochemical Sciences. Ciudad Valles, San Luis Potosí, México

Pedro Hassiel Sariñana-Núñez
Research Group on Functional Foods and Nutraceuticals. TecNM/Instituto Tecnológico de Durango, Felipe Pescador Ote., Durango, Dgo., México

Tina Sebastian
Department of Physics, Deva Matha College, Kuravilangad, Kottayam, Kerala, India

Liyana Sherin
Marian College (Autonomous), Kuttikkanam, Peermade, Idukki, India

K. Sreelatha
Department of Physics, Sree Narayana College, Chengannur, Alappuzha, Kerala, India

Ridha Zain T.
Marian College (Autonomous) Kuttikkanam, Kuttikkanam, Peermade, Idukki, India

Binoy T. Thomas
Post-Graduate and Research Department of Botany, Catholicate College, Pathanamthitta, Kerala, India

Meghana U.
JAIN (Deemed-to-be University, Bangalore, India

Vedavalli V.
JAIN (Deemed-to-be University, Bangalore, India

Sergio Valle-Cervantes
Tecnológico Nacional de México/Instituto Tecnológico de Durango Departamento de Ingenierías Química y Bioquímica, Felipe Pescador Ote, Colonia Nueva Vizcaya, C. P. Durango, Durango, México

Bincy Varghese
Post-Graduate and Research Department of Physics, Catholicate College, Pathanamthitta, Kerala, India

Sneha Sara Varghese
Post-Graduate and Research Department of Chemistry, Bishop Moore College, Mavelikara, Kerala, India

Ariel Vazquez
CIATEJ Research and Assistance Center in Technology and Design of the State of Jalisco A. C. Guadalajara, Jalisco, Mexico

Agampodi Sandali Inupama De Zoysa
JAIN (Deemed-to-be University, Bangalore, India

Preface

Water, energy, and food are the main resources to sustain life. Water, energy, and food are interlinked in multiple ways, and the term "nexus" represents the interconnections.

This book provides a broad overview of both the nexus of water–food–energy and climate and sustainable development that will have huge impacts on water and food availability.

Solutions for food, clean water, and energy for the current world population of seven billion is one of the greatest challenges that require technology innovation and changes in public policy. As universities have seen a considerable increase in interdisciplinary energy and environmental courses, this research-oriented reference book could be a practical guide for scholars interested in gaining knowledge in the area of nexus studies without a strong background in all three fields and the different aspects of nexus studies, considering the fact that the population growth increases the demand on water, food, and energy.

This new volume is also a valuable guide for professionals with strong interests in interdisciplinary research involving environmental science, environmental management, and sustainability science.

This research-oriented volume includes:

- Introductory chapters for conceptual framework on how to better understand the water–energy–food nexus
- Case studies to illustrate synergies among water–energy–food sectors
- Introduction to engineering solutions that include both "soft path" and technology innovation methods
- Case studies for management of water, energy, agricultural growth, and food security using empirical analyses and technological interventions
- Case studies to minimize water–energy–food nexus pressures and achieve green growth

The book highlights fundamentals of synergy framework on water, food, energy, and the future of sustainability. It illustrates practical concepts for higher level undergraduate and graduate level students.

CHAPTER 1

The Sustainable Use of Cactaceae as Functional Foods: The Hypoglycemic Effect

ABIGAIL REYES[1], CHRISTIAN SÁNCHEZ[1], BRENDA ALVARADO[1], ARIEL VAZQUEZ[2], and PEDRO AGUILAR-ZARATE[3]

[1]*Autonomous University of San Luis Potosi, Faculty of Professional Studies Huasteca Zone, Master in Biochemical Sciences, San Luis Potosí, México*

[2]*CIATEJ Research and Assistance Center in Technology and Design of the State of Jalisco A. C. Guadalajara, Jalisco, Mexico*

[3]*Laboratorio Nacional CONAHCYT de Apoyo a la Evaluación de Productos Bióticos (LaNAEPBi), Unidad de Servicio Tecnológico Nacional de México/ Instituto Tecnológico de Ciudad Valles, Ciudad Valles, San Luis Potosí, México.*

ABSTRACT

The Cactaceae family can be easily found in arid and semi-arid regions, being undervalued in terms of their potential, since their use has generally been limited to forage, even though many research studies have demonstrated the antioxidant properties and bioactive compounds in cactus species, This chapter aims to review and publicize the biological properties of nopal (*Opuntia ficus-indica*) jacube, tuna (cactus fruit), pitahaya or dragon fruit (fruit of the genus *Hylocereus*), to be considered as an alternative preventive and/or functional food control against hyperglycemia that often becomes type-2 diabetes mellitus (T2DM). They possess important amounts of micro, macronutrients, and bioactive compounds, which allows them a wide variety

The Food-Energy-Water Triangle for Sustainable Development: Integrative and Synergic Solutions.
Pedro Aguilar-Zárate, Mónica Chávez-González, Ann Rose Abraham, and A. K. Haghi (Eds.)

of uses, fresh or processed, and for industrial purposes for the production and extraction of compounds of interest.

1.1 INTRODUCTION

Diabetes mellitus (DM) is a chronic metabolic disease characterized by hyperglycemia related to impaired carbohydrate, lipid, and protein metabolism, and a lack of insulin secretion or decreased sensitivity to the metabolic effects of insulin.[1] Type-2 diabetes mellitus (T2DM) is the result of the body's ineffective use of insulin. T2DM is a series of metabolic disorders that share the common characteristic of hyperglycemia, which generates reactive oxygen species (ROS), causing lipid peroxidation and damage to the membrane. These free radicals play an important role in the appearance of complications. secondary DM, damaging kidneys, eyes, blood vessels, and nerves.[2] The World Health Organization (WHO) in 2021, reports that 5% of people with diabetes have T2DM, which is the result of excess body weight and physical inactivity. The T2DM prevalence is increasing worldwide and is related to the increase in the prevalence of obesity. Globally, between 2000 and 2016, there was a 5% increase in premature mortality from diabetes. In the Americas in 2019, diabetes was the sixth leading cause of death, with an estimated 244,084 deaths directly caused by T2DM.[3,4] Due to the increasing number of patients with diabetes, the search for new food alternatives from natural resources,[5] such as supplementing diets with plants, has been considered a healthier option, since they are dense in nutrients, rich in fiber, phytochemicals, and micronutrients, low in calories and saturated fat. Phytochemicals and fiber from plant foods improve immune function and positively influence health. Due to their properties, some plants are considered as a functional food option, when properly administered, they are healthy and offer adequate nutrition for all age groups.[6] Science concentrates its efforts on natural compounds, promoting the use of nutraceuticals, bioactive compounds of vegetable origin with important nutritional values.[7] A large majority of plants with beneficial properties are known in traditional medicine, this has prompted a growing interest in bioactive compounds of plant origin with important nutritional values.[8] One of the families of plants that have been shown to have beneficial functional properties for health, such as a hypoglycemic effect, is that belonging to the cacti.[9] This family is endemic to the American continent, distributed from North America to South America and in the Antilles. In Mexico, 669 species of cacti are Mexican, which are distributed in the central zone of the country, such as in the states of Tamaulipas, Veracruz, Querétaro and San Luis Potosí.[10] This chapter aims to disseminate and assess cacti as a nutraceutical and functional food option with a hypoglycemic effect on T2DM.

1.2 CACTUS FAMILY

This family is endemic to the American continent, distributed from North America to South America and in the Antilles. In this region, there are various types of habitats in which cacti can develop, such as deserts, semi-arid areas, dry forests, and coastal areas. In North America, cacti are found primarily in the deserts of the southwestern United States and northern Mexico. In Central America and the Caribbean, they can be found in the arid areas of Mexico and on the Caribbean islands. In South America, cacti are most common in the deserts of the Peruvian coast, northern Chile, and Argentina. In Mexico, 669 species of cacti are Mexican, which are distributed throughout the central zone of the country, such as in the states of Tamaulipas, Veracruz, Querétaro, and San Luis Potosí.

The species of the cactus family are characterized by the production of mucilage that, together with different tissues, represents a source of functional metabolites for humans. Within the functional metabolites, phenolics have gained strong attention since they affect the quality of plant foods, both fresh and processed and their contribution to the maintenance of human health has been described.[12]

Among the food groups, cacti can be used as raw material to develop new functional foods.[68] In Mexico there are several widely known plants, such as the nopal-tuna (*Opuntia ficus-indica*) and the pitaya that belongs to *Hylocereus undatus* and *Hylocereus polyrhizus*. Bravo-Hollis (1978) points out that pitayos are columnar cacti native to subtropical regions of Mexico that produce attractively colored edible fruits.[69] This crop has been a very relevant activity in Mexico for many years. Likewise, efforts and agricultural practices have been made in the selection of better phenotypes and biomass productivity in crops with high quality to increase the value in the market.[70,71] On the contrary, Esquivel (2004) points out the importance of continuing with the agronomic aspects of columnar and climbing cacti" mainly on the fruit (prickly pear) of the cactus (*Opuntia* sp.).[72]

1.2.1 RELATIONSHIP OF CACATACEAS WITH T2DM

These foods represent an important source of nutrients for the population that is in conditions of prevalence of diabetes in Mexico. To contextualize the importance of increasing access to these products in the Mexican diet, it is important to highlight that, according to the report of the Ministry of Health (2022), in Mexico there were approximately 12.4 million people with diabetes according to the National Survey of Health and Nutrition (Ensanut) 2021, representing a significant risk because "causing irreversible damage to the eyes, kidneys or skin, indicated the head of Internal Medicine of the National Homeopathic Hospital (HNH) of the Ministry of Health" Secretariat of Health (2022).

Of the 12.4 million that suffer from diabetes in the country, there are 8.9 million are diagnosed by a doctor, distributed in the federal entities in order of importance as follows: the State of Mexico has 12.79% of the total Mexico City, 8.69%; Veracruz, 7.53%; Jalisco, 6.32%; Nuevo León, 5.79%; Guanajuato, 4.63%; Puebla, 4.37%; Oaxaca, 4.03%; Tamaulipas, 3.47%; Michoacán, 3.13%; Guerrero, 2.96%; Chihuahua, 2.75%; San Luis Potosí, 2.74%; Baja California, 2.68%; Hidalgo, 2.65%; Coahuila, 2.57%; Chiapas, 2.34%; Sinaloa, 2.05%; Querétaro, 1.85%; Sonora, 1.84%; Tabasco, 1.80%; Yucatan, 1.77%; Quintana Roo, 1.73%; Morelos, 1.73%; Zacatecas, 1.40%; Durango, 1.32%; Aguascalientes, 0.96%; Tlaxcala, 0.93%; Campeche, 0.88%; Nayarit, 0.87%; Baja California Sur, 0.74%; and Colima, 0.69%.[73]

1.2.2 CONSUMERS OF CACTACEAE

Based on the analysis of the National Survey of Household Income and Expenses of the National Institute of Geography and Statistics—(ENIGH-INEGI, 2020), in Mexico there is a demand for the consumption of nopal, pitaya, or prickly pear, where approximately 4.7 million heads of households reported spending for the consumption of these products. Of this total, 84.39% is located in urban areas and 15.61% in rural territories. When multiplying households by family members, it is obtained that there are 17.8 million people who have been consuming these cacti, of which 73.92% correspond to men and 26.08 to women.[74]

When the characteristics of public health and food insecurity classified according to the Food and Agriculture Organization of the United Nations (FAO, 2012) are related, the Mexican households surveyed that indicated having consumed the aforementioned cacti present the following characteristics socioeconomic: 41.46% (1.9 million heads of households) of which 662 thousand were in poverty and 1.3 million in conditions of nonpoverty. In addition, 343,000 Mexican households that consumed the aforementioned cacti and bought medicines for the prevention of diabetes presented health deficiencies and 651,000 social security deficiencies.[74] On the contrary, 58.54% (2.8 million households consuming cacti) stated that they had not purchased diabetes medicine, of which 1.1 million heads of households were in poverty and 1.6 million in nonpoverty conditions, according to the measurement of CONEVAL (2020).[75] At the same time, the distribution of heads of households that consume cacti according to food insecurity, in 2020, ranged from 2.2 million heads of households in conditions of food security, 1.4 million with mild food insecurity,

It is important to note that the prices of the main cacti, such as the pitaya, oscillated between \$40 and \$70 pesos in the first quarter of 2023, on average, in the Mexican Republic; which shows that as it is an exotic plant, only those consumers with a medium-high income level will be able to have access to it. In the case of the tuna, it fluctuated on average at \$33 pesos, while the nopal in the supply centers was between \$15 pesos on average.[76] This relationship explains that 61.55% of the population that consumes these cacti is in nonpoverty conditions while 38.45% is in poverty conditions.

1.2.3 INCOME DISTRIBUTION OF CACTUS CONSUMERS

To better understand the income distribution of men and women consumers of cacti, the Gini coefficient and the Lorenz curve of the 4.7 million consumers of nopal, pitaya (pitahaya), and nopal were analyzed. The results are generated from the ENIGH-INEGI (2020)[74] referring to a quarter of current monetary income (income) from work (ing_lab); current monetary income from transfers (ing_tra); current monetary income from income (ing_ren); mostly. The Gini coefficient has a range of values between 0 and 1. A value of 1 represents income concentration and maximum inequality, and 0 shows maximum equality (income is distributed among individuals).

Table 1.1 shows the relationships of the main sources of income for cacti consumers in Mexico. To better contextualize the relevance of income, we examine the importance of the different sources in relation to total income (S_k), the condition of equitable or unequal distribution of the source of income (G_k), and the relationship between the source of income and the distribution of total income (R_k).

TABLE 1.1 Gini Breakdown in Cactus Consumers in Mexico.

Total income			**Variable: ing_mon**		
Income	S_k	G_k	R_k	**Stake**	**Change (%)**
Employment income	0.7854	0.5291	0.8857	0.8212	0.0358
Rental income	0.0159	0.9792	0.6427	0.0224	0.0064
Income from transfers	0.1987	0.7670	0.4602	0.1564	–0.0422
Average monthly income for programs for the elderly	0.0133	0.8872	–0.1560	–0.0041	–0.0174
Income total		0.4482			

Source: Own elaboration with data from Ref. [74].

Labor income shows that a 1% increase in this source of income, other things being equal, increases the Gini coefficient of total income by 0.0358%: with a slightly equalizing effect on income distribution. Labor income is unevenly distributed G_k (0.5291) and the Gini correlation between labor income and total income is R_k (0.8857), indicating that labor income favors the rich more than any other source. For its part, income from rentals shows that a 1% increase in this source of income, other things being equal, increases the Gini coefficient marginally by 0.0064%, however, G_k (0.9792) and G_k (0.9792) and are distributed unevenly. The Gini correlation between rental and total income is R_k (0.6427). Besides, income from transfers has a slight equalizing effect on the distribution of total income, generating a marginal reduction in the Gini of –0.0422% and the average monthly income for programs for the elderly (–0.0422%). The average monthly income from programs for the elderly is very unequally distributed G_k (0.8872) and the Gini correlation between this income and the sum of all types of income is negative R_k (–0.1560).

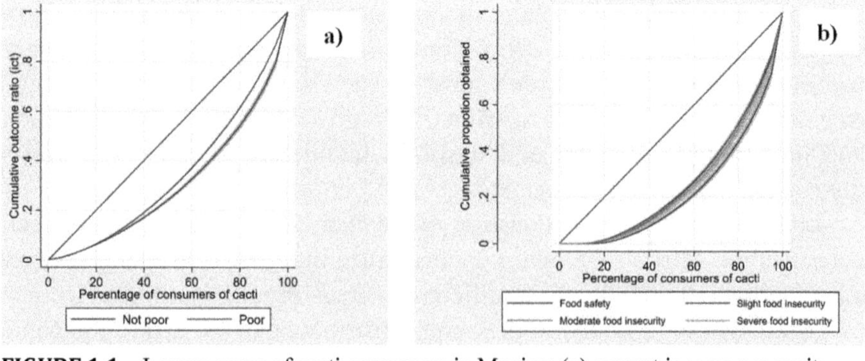

FIGURE 1.1 Lorenz curve of cacti consumers in Mexico: (a) current income per capita (b) labor income of the consumer population.

Source: Own elaboration with data generated from Ref. [74].

Inequality increases as the Lorenz curve moves away from the straight line of distribution toward the axes. On the contrary, when the curve coincides with the diagonal, the average distribution of income adopts an egalitarian form. Graph 1a shows greater inequality among the nonpoor consumer population in terms of per capita income. Figure 1.1b shows the distribution of current labor income of households that consume cacti in the Lorenz Curve according to food insecurity generated from the Latin American and Caribbean Scale of Food Insecurity—ELCSA.[77] It is observed that consumers with levels of food security move further toward an erratic equidistribution (diagonal) which reveals a less equitable relationship where approximately 80% of the population concentrates 43% of income; In this case, it is evident that the population in conditions of insecurity, in general, presents a better distribution of per capita income among people, in contrast to the population with food security.

In this way, it is essential to emphasize that the access to the main cacti produced in Mexico in the different territories of the country is found with a better reach in those inhabitants with better income levels. To measure the distribution that exists in Mexico regarding production, it is necessary to visualize the territories, for which reason, these issues are studied in the following sections.

1.2.4 CACTACEAE PRODUCING TERRITORIES

From the economic point of view, the national supply of nopal production (nopalitos), in the year 2021, 12,500 ha were planted, where the states of Morelos, Mexico City, the State of Mexico, and Tamaulipas stood out with an area planted with 4200 , 2,200, 1,000 , and 995 ha, respectively[78] (Figure 1.2). It is important to point out how much of it is distributed by the Mexican Republic but is also exported considering the high values that consumers from other countries are willing to pay over the price given the quality characteristics in the market.

Regarding the cultivation of tuna, in 2021 there were a total of 90.4 thousand ha distributed in at least 16 federal entities, with the State of Mexico, Zacatecas and Puebla standing out with 31.7 thousand ha, 23.2 thousand ha and 11.0 thousand ha respectively (Figure 1.3).

Regarding the cultivation of pitaya, in 2019 there were 3675 ha planted in the country, reaching a production of 14,836.18 tons. These are found in the following states in order of percentage importance, according to the official

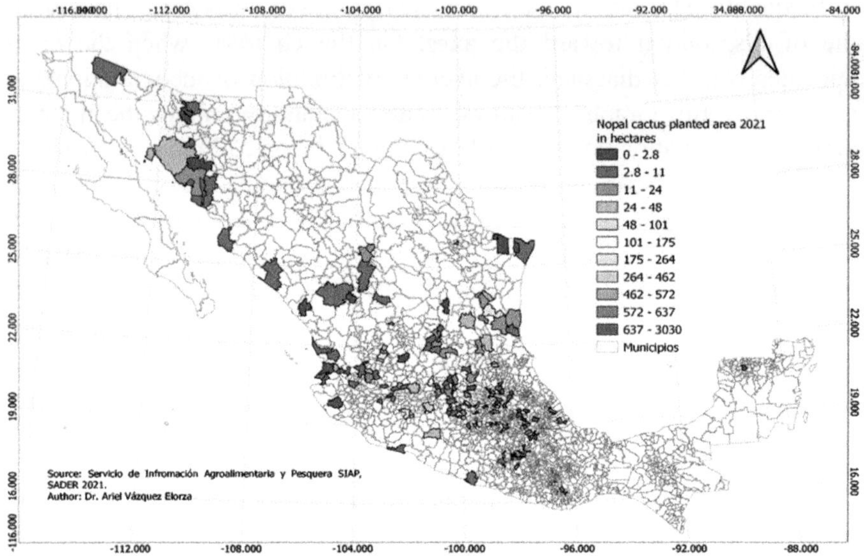

FIGURE 1.2 Distribution of nopal production in the year 2021.

Source: Own elaboration with data from Ref. [78].

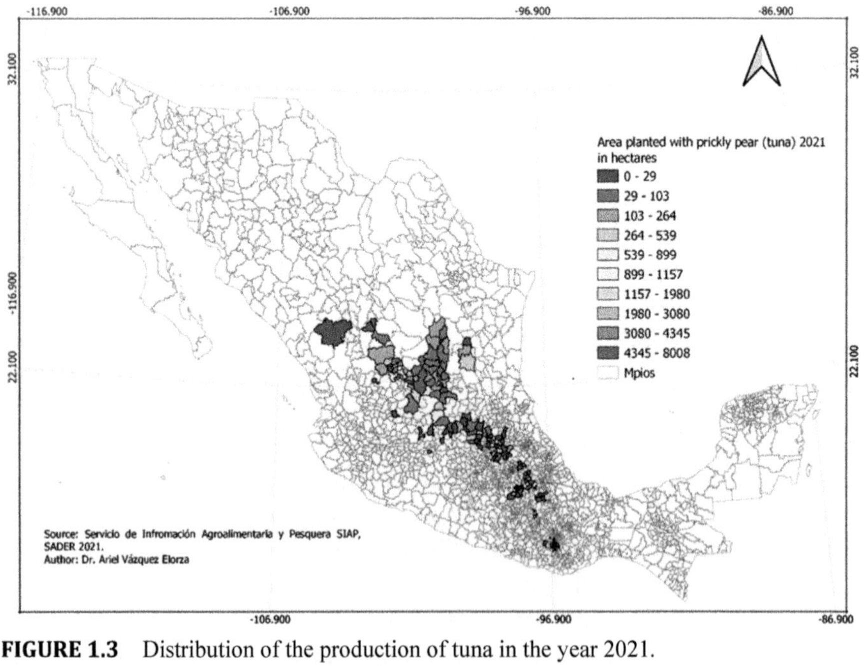

FIGURE 1.3 Distribution of the production of tuna in the year 2021.

Source: Own elaboration with data from Ref. [78].

records of SADER Quintana Roo 32.67%, Oaxaca 30.09%, Jalisco 13.67%, Yucatán 11.27%, Puebla 8.23%, Baja California Sur 0.98%, mainly (Figure 1.4).

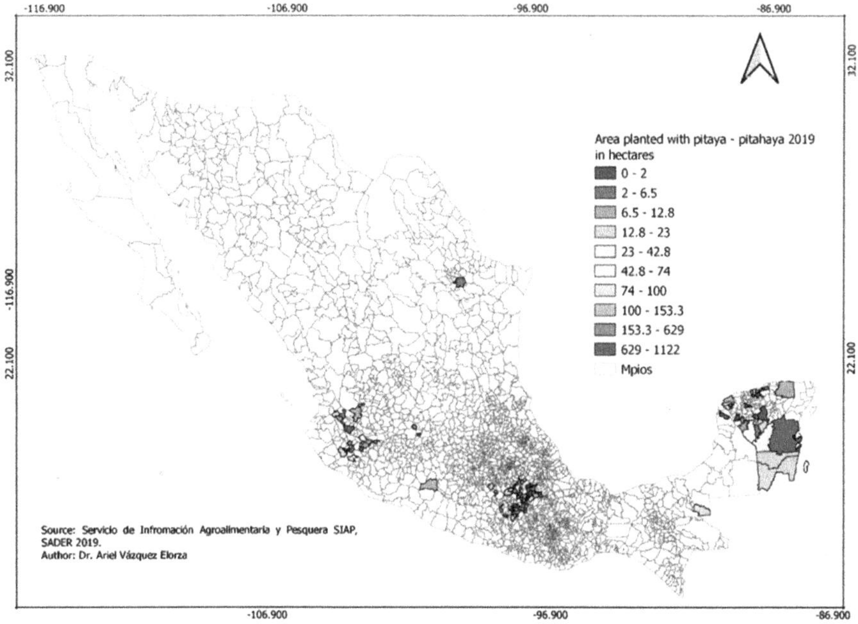

FIGURE 1.4 Distribution of pitaya production in 2019.

Source: Own elaboration with data from Ref. [78].

Undoubtedly, the pitaya represents an important crop for the Mexican southeast, mainly for the state of Yucatan and Quintana Roo. Likewise, it is found in the present west, among other territories. Undoubtedly, the production of fruits and the generation of derivatives from cacti is essential to enrich the nutritional needs of the population, especially those who have less access and are in prevalent conditions of diabetes. However, in some areas where they are produced, these products may be undervalued or poorly processed, preventing them from generating social and nutritional value.

It is important to emphasize the need to promote the production and transformation of cacti among the different territories of the country to increase supply and supply–demand with better and more competitive prices. It is necessary to increase their nutritional value and promote scientific and consumption innovations among the population with major diabetes problems. Encouraging local production of these fruits can also create jobs and improve local economies while promoting healthy and balanced diets.

It is necessary to increase society's awareness of their nutritional benefits. This requires implementing and designing public policies to disseminate and transfer knowledge, as well as launching awareness campaigns and food education programs to promote consumption and inclusion in the daily diet.

1.3 CACTI WITH FUNCTIONAL PROPERTIES

Cactaceae plants and their fruits are considered rich sources of functional metabolites, among which can be found phenolic compounds, betalains, vitamins, carotenoids, minerals, and soluble fiber. Some cactus matrices have been consumed and considered a source of food for a long time, as in Mexico, an example of this is the consumption of nopal, which is considered a functional food, since its consumption can confer benefits for health. health.[13] These plants could be a source of new relevant bioactive compounds for the field of phytomedicine.[9] Some of the best-known cacti include the following:

1.3.1 NOPAL

The nopal (*O. ficus-indica*) is an endemic product of Mexico. There are 200 species of nopales, of which 101 of them in our country. The nopales are divided into two groups; *Opuntia* and *Nopalea* are very similar in their stems, but their difference lies in the shape of their flowers. Those of the genus *Opuntia* look like cups with small petals and are pollinated by insects, and those of *Nopalea* are elongated with a pistil, which are pollinated by hummingbirds.[14] Within its composition it is a rich source of vitamins C, B1, B2, A, and E, minerals such as calcium, potassium, magnesium, iron and phosphorus, as well as bioactive compounds such as carotenoids, betalains and phenolic compounds. Among them, phenolic acids, betalains and flavonoids stand out, which are the main responsible for the healthy properties of this plant[26] where the extracts of *O. ficus-indica* cladodes can reduce the cholesterol level and transmit antiulcer and anti-inflammatory mechanisms and the aqueous extract significantly improves wound healing.[62,63] Investigations carried out in patients and animals diagnosed with diabetes demonstrate the hypoglycemic properties of nopal, where dietary intake of nopal (*O. ficus-indica*) improves the postprandial response of glucose, insulin, the insulinotropic peptide index dependent on glucose [gastric inhibitory polypeptide (GIP)] and peptide index is similar to glucagon [glucagon-like peptide 1

(GLP-1)] in patients with T2DM after consumption of a carbohydrate-rich or soy–protein breakfast.[27] The hypoglycemic effect could be increased by placing *Opuntia* extracts in heat, as reported by Frati-Munari et al.,[28] in patients who consumed grilled *Opuntia streptacantha* extracts. Similarly, the stems of *Opuntia humifusa* cause a decrease in blood glucose and cholesterol in rats treated with streptozotocin (STZ).[29] In another study, Andradc-Cetto et al.[30] showed that extracts of *O. streptacantha* do not reduce glycemia in rats treated with STZ compared to control, but show a hypoglycemic effect when administered before an oral glucose tolerance test (OGTT). Similarly, a similar result was observed in prediabetic obese patients (men and women) treated for 16 weeks with OpunDia™ (a mixture of extracts of cladodes and *O. ficus-indica* fruits).

1.3.2 TUNA (OPUNTIA FRUIT)

The fruits of *O. ficus-indica* known as tuna, which is a strawberry of the Cactaceae family, has a thick and thorny shell with a pulp abundant in seeds and constitutes a food of great nutritional value, it is endemic to Mexico and is available in almost all year round. In addition, it is considered one of the best fruits to consume for all the benefits it provides, some of the reasons are that it is a natural alternative to lower cholesterol and triglyceride levels, it reduces the production of gastric acid, and alleviates the discomfort caused by ulcers, it contains a large amount of fiber that helps to have better digestion, helps to stabilize and regulate the level of glucose in blood.[14] Studies have shown that the whole fruit contains 9–10% seeds.[15,16] Prickly pear seeds have been shown to contain polysaccharides, cellulose, and hemicelluloses.[17] Many studies have suggested that prickly pear seeds are a new source of fruit oils.[18,19] It has been reported that this fruit has several biological activities, such as antioxidant in vivo[20] and in vitro,[21,22] antidiabetic.[17,23] In an investigation, it was observed that glycemia and glycosylated hemoglobin decreased to normal values in rats induced with diabetes, after treatment with STZ and supplemented with an extract of prickly pear from *O. fuliginosa*.[32] Similarly, the stems of *O. humifusa* promote a decrease in blood glucose and cholesterol in rats treated with STZ.[31] Many studies have suggested that prickly pear seeds are a new source of fruit oils.[18,19] It has been reported that this fruit has several biological activities, such as antioxidant in vivo[20] and in vitro,[21,22] antidiabetic.[17,23] In an investigation, it was observed that glycemia and glycosylated hemoglobin decreased to normal values in rats induced with diabetes, after treatment with STZ and supplemented with an extract of prickly pear from *O. fuliginosa*.[32] Similarly, the stems of *O. humifusa* promote a decrease in blood glucose and cholesterol in rats treated with STZ.[31] Many studies have suggested that prickly pear seeds are a new source of fruit oils.[18,19] It has been reported that this fruit has several biological activities, such as antioxidant in vivo[20] and in vitro,[21,22] antidiabetic.[17,23] In an investigation, it was observed that glycemia and glycosylated hemoglobin decreased to normal values in rats induced with diabetes, after treatment with STZ and supplemented with an extract of prickly pear from *O. fuliginosa*.[32] Similarly, the stems of *O. humifusa* promote a decrease in blood glucose and cholesterol in rats treated with STZ.[31] It was observed that glycemia and glycosylated hemoglobin decreased to normal values in rats induced with diabetes, after treatment with STZ and supplemented with an extract of *O. fuliginosa* prickly pear.[32] Similarly, the stems of *O. humifusa* promote a decrease in blood glucose and cholesterol in rats treated with STZ.[31] It was observed that glycemia and glycosylated hemoglobin decreased to normal values in rats induced with diabetes, after treatment with STZ and supplemented with an extract of *O. fuliginosa* prickly pear.[32] Similarly, the stems of *O. humifusa* promote a decrease in blood glucose and cholesterol in rats treated with STZ.[31]

1.3.3 JACUBE

The jacube (*Acanthocereus tetragonus* (L.) Hummelinck) is a cactus also known as spider, cardón, is a cactus of the Cactaceae family. It is a creeping plant, although it can also be columnar and shrubby, its average height is 2–3 m, often forming thickets with multiple stems 3–8 cm thick, it has 3–5 longitudinal ribs, with gray and acicular spines. no obvious leaves. It has light green flowers with brown wool on the outside and creamy purple on the inside. It blooms at the end of summer and its flowers open at night. It produces 2.5-cm edible fruits that are red or purple in color when ripened. It is a species of columnar cactus native to Central and South America, it is an endemic species to Mexico. The jacube grows in well-drained soils and in sunny places, semi-desert or desert areas, it is resistant to drought and high temperatures, so it can be grown in hot and dry climates.[37] In a study, the composition of immature stems of *Acanthocereus* and *Hylocereus* cacti is analyzed. Where it is reported that the spider has large amounts of protein and fiber as well as minerals such as potassium, phosphorus, magnesium, sodium, copper, and a high amount of calcium.[11]

1.3.4 PITAYAS

The pitaya is an exotic fruit also known as dragon fruit, it is a fruit that belongs to two species of cacti of the Cactaceae family, mainly to the genus *Hylocereus* and *Selenicereus.* They are native to the tropical regions of Central and South America, but due to their good adaptability to different types of climates and soils, they can be found distributed in various parts of the world.[33–35] There are several types of pitayas, but *H. undatus* and *H. polyrhizus* are the two most produced and traded species due to their attractive sensory characteristics for consumers.[33] These species produce fruits 7–14 cm long and 5–9 cm wide, with bright red skin with overlapping green bracts, juicy white (*H. undatus*) or red–purple *(H. polyrhizus*), and small black seeds.[34–36]

1.4 MAIN NUTRITIONAL COMPOUNDS OF SOME CACTACEAE

The nutritional composition of cacti is relevant because they are rich in nutrients that can contribute to human health and can be considered as a food option. Nopales (*O. ficus-indica*) have been reported to contain substantial amounts of ascorbic acid, vitamin E, carotenoids, fibers, amino acids, and antioxidant compounds (phenols, flavonoids, betaxanthin, and betacyanin), all of which have been shown to possess functional properties. that improve health, among them hypoglycemia, hypolipidemia, and antioxidant properties can be mentioned.[56,58] Research has documented the abundance of vitamins and minerals in these cacti.[51] The fruit of *O. ficus-indica* is a valuable source of nutrients,[59] known as prickly pear, which has good amounts of protein, carbohydrates (arabinose, galactose, xylose, mannose, and glucose), lipids (oleic, palmitic, linoleic, and linolenic acids), fiber, minerals (Ca, Mg, K, P, Fe, Cu, Se, and Mn), vitamins (C, E, B1, B2, B3, B6, B9, and B12), pigments with bioactive properties (chlorophyll, carotenoids, phenolic compounds, and betalains) and a high antioxidant capacity.[40–42] *O. ficus-indica* flowers contain different flavonoids, in particular kaempferol and quercetin.[60] Regarding its peel and seed, these can be used to prepare cactus oil, since the lipids of the peel are rich in essential fatty acids and fat-soluble antioxidants.[61] This species has nutritive cladodes rich in carbohydrates, such as rhamnose (7.23 $\mu g\ mg^{-1}$), fucose (2.67 $\mu g\ mg^{-1}$), arabinose (101.86 $\mu g\ mg^{-1}$), xylose (23.42 $\mu g\ mg^{-1}$), mannose (12.59 $\mu g\ mg^{-1}$), galactose (26.49 $\mu g\ mg^{-1}$), glucose (53.81 $\mu g\ mg^{-1}$), and galacturonic acid (187.62 μg

mg^{-1}).[43] *H. undatus* produces very nutritious fruits, abundant in composition, especially organic acids (such as ascorbic, citric, isocitric, and malic acids), carbohydrates, amino acids (proline), minerals (mainly potassium and magnesium), and lipids (proline), minerals (mainly potassium and magnesium) and lipids.[44] Its peel can be considered a good source of dietary fiber and vitamin C.[45] Functional lipids can be extracted from its seeds, which can be a new source of essential oil, with a high content of saturated (myristic, palmitic, and stearic) and unsaturated (oleic, linoleic, and linolenic) acids.[46] On the contrary, in *H. polyrhizus* organic acids (citric, D-isocitric, lactic, and L-malic acids) have been identified, as well as some natural pigments (such as betacyanin and phenolic compounds), and a pulp rich in minerals, mainly K, Ca. The species *H. polyrhizus* has been reported to contain organic acids (citric acid, D-isocitric acid, lactic acid, and L-malic acid), as well as some natural pigments (betacyanin and phenolic compounds) and contains a pulp rich in minerals (K, Ca, and Mg).[47] In a study, a comparison was made with the pulp of the fruit, where it was shown that the skin has higher amounts of dietary fiber (6.81%), ash (1.04%), and vitamin C (22.51 mg 100 g^{-1}).[45] On the contrary, an investigation carried out by Lim et al. (2010), demonstrated that the seeds of *H. polyrhizus* contained a large amount of oil (18.33–28.37%), prevailing within them, linoleic, oleic and palmitic acids, being the main fatty acids found. Finally, in another investigation, it was found that the peel of the fruit has good amounts of pectin (26.38% dw), which consisted mainly of galacturonic acid (39.11%) mannose (17.78%), rhamnose (14 47%), galactose (11.91%), glucose (10.82%), xylose (2.41%), and arabinose (3.49%).[34]

1.5 HYPOGLYCEMIC EFFECT OF COMPOUNDS PRESENT IN SOME CACTACEAE

The composition of some edible cacti described in this chapter has been of interest in the field of research, not only for being a rich source of nutrients but also for the functional properties that the bioactive compounds they contain can generate. These can prevent metabolic disorders such as diabetes. An investigation carried out by Lagunes (2019), demonstrated that the intake of *A. tetragonus* extract and its mucilage to Winstar rats induced with metabolic syndrome have a beneficial effect on parameters of glucose, cholesterol, triglycerides, and blood pressure, for which it can be used as a therapeutic for this condition.[47] Similarly, it has been shown that the ingestion of the

TABLE 1.2 Nutritional Composition of Some Cactaceae.

Fruit/ vegetable	Part	Composition	Content	Reference
Prickly pear (*O. ficus-indica*)	Whole fruit	Humidity	7.65% to	[38]
		Ash	9.00	
		Protein	0.86	
		Total carbohydrates	88.85 g 100 g^{-1} [a]	
		Arabinose	10.83 g 100 g^{-1} [a]	
		Galactose	2.13 to	
		Xylose	4.26 to	
		Mannose	2.13 to	
		Glucose	69.30 a.m.	
		Calcium	14159.90 mg L^{-1} [a]	
		Magnesium	2759.63	
		Manganese	8.51	
		Potassium	2161.02	
		Match	321.98	
		Copper	177.53	
		Aluminum	4004.94	
		Nickel	6.86	
		Sodium	2807.92	
		Iron	94.32 μg g^{-1} [a]	
		Selenium	0.38	
	Pulp	Full acidity	0.06 b g citric acid 100^{-1}	[48]
		pH	5.90	
		Total soluble solids	13.30 b °Brix	
	Seed oil	Tridecylic acid (C13:0)	0.01	[40]
		Myristic acid (C14:0)	0.13	
		Pentadecyl acid (C15:0)	0.02	
		Palmitoleic acid (C16:1*n*–7)	1.77	
		Palmitelaidic acid (C16:1*n*–9)	0.04	
		Palmitic acid (C16:0)	20.34	
		Margaric acid (C17:0)	0.04	
		Oleic acid (C18:1*n*–9)	41.22	
		Vaccenic acid (C18:1*n*–7)	1.77	
		Stearic acid (C18:0)	1.66	
		Linoleic acid (C18:2*n*–6)	19.94	
		α-Linolenic acid (C18:3*n*–3)	0.43	

TABLE 1.2 *(Continued)*

Fruit/ vegetable	Part	Composition	Content	Reference
		Arachidic acid (C20:0)	0.15	
		Gondoic acid (C20:1*n*–9)	0.02	
		Paulinic acid (C20:1*n*–7)	0.19	
		Eicosadienoic acid (C20:2*n*–6)	0.03	
		Heneicosylic acid (C21:0)	0.01	
		Behenic acid (C22:0)	0.03	
		Erucic acid (C22:1n-9)	0.01	
		Tricosylic acid (C23:0)	0.01	
		Lignoceric acid (C24:0)	0.03	
		Pentacosylic acid (C25:0)	0.01	
		Cerotic acid (C26:0)	0.02	
	Whole fruit	Chlorophylls	16.66 mg kg^{-1}	[39,49]
		Total carotenoids	1480.00	
		Total phenols	25.15 mg GAE 100 g^{-1}	
		Total flavonoids	16.08 mg QE 100 g^{-1}	
		Total betacyanins	1830.00 µg IE g^{-1} [b]	
		Total betaxanthins	760.00 µg BE g^{-1} [b]	
		Total betalains	2580.00 µg g^{-1} [b]	
		Vitamin C	71.94 mg AE 100 g^{-1} [b]	
		Vitamin E	63.20 µg g^{-1} [a]	
		Vitamin B1	8.60	
		Vitamin B2	6.72	
		Vitamin B3	307.20	
		Vitamin B6	21.20	
		Vitamin B9	708.80	
		B12 vitamin	4.00	
		DPPH	126.49b µmol TE 100 g^{-1}	
		ORAC	38.66[b]	
Jacube (*A. tetragonus L.*) Hummelinck	Complete plant	Humidity	94.18 g	[36,37,47]
		Ashes	0.68 g	
		Fat (ethereal extract)	0.10 g	
		Protein	1.09 g	
		Dietary fiber	2.94 g	

TABLE 1.2 *(Continued)*

Fruit/ vegetable	Part	Composition	Content	Reference
		Total digestible carbohydrates dry basis 1.01	1.01g	
	Mucilage	Humidity	8.97 g	[47]
		Ashes	22.87 g	
		Fat (ethereal extract)	1.03 g	
		Protein	18.75g	
		Dietary fiber	28.60 g	
		Total digestible carbohydrates	19.78 g	
H. undatus	Pulp	Pulp moisture	85.39%	[15]
		Ashes	0.47 g $100g^{-1}$	
		Proteins	1.31	
	Juice	Juice proline	415.00 mg L^{-1}	[51]
	Pulp	Pulp lipids	0.47%	[32]
	Juice	Juice pH	5.32	
		Dietary fiber	2.14	
		Full acidity	3.30 g malic acid L^{-1}	
		Total soluble solids	9.40 °Brix	
	Pulp	Calorific value of pulp	55.99 calories	
		Total carbohydrates	12.95 g 100 g^{-1}	
	Juice	Glucose	46.60 g L^{-1}	
		Fructose	18.4	
		Potassium	3995.00 mg L^{-1}	
		Sodium	33.00	
		Calcium	30.60	
		Magnesium	265.50	
	Seed oil	Saturated fatty acids	17.99 g 100 g^{-1}	[54]
		Monounsaturated fatty acid	25.20	
		Polyunsaturated fatty acid	56.81	
		Myristic acid (C14:0)	0.19	
		Palmitic acid (C16:0)	12.78	
		Stearic acid (C18:0)	4.67	
		Arachidic acid (C20:0)	0.35	
		Palmitoleic acid (C16:1)	0.64	
		Oleic acid (C18:1)	24.43	
		Erucic acid (C22:1)	0.13	

TABLE 1.2 *(Continued)*

Fruit/ vegetable	Part	Composition	Content	Reference
		Linoleic acid (C18:2)	55.63	
		Linolenic acid (C18:3)	1.18	
	Pulp	Phenols total	358.42 mg GAE 100 g^{-1}	
		Flavonoids total	4.77 mg QE g^{-1}	
		Ascorbic acid	28.40 mg 100 g^{-1}	
	Juice	Citric acid	132.50 mg L^{-1}	
		D-isocitric acid	19.20	
		L-Lactic Acid	153.50	
		L-malic acid	7.20	
	Pulp	DPPH IC_{50}	0.84	
		FRAP	59.27 μmol Fe Eq. g^{-1}	
H. polyrhizus	Pulp	Humidity	89.46%	
		Protein	1.06	[32,33]
		Ashes	0.36	
		Lipids	0.36	
		Dietary fiber	3.26	
		Calorific value	44.87 calories	
		Full acidity	0.24 mg CAE 100 g^{-1}	
		pH	4.88	
		Soluble solids	11.00 °Brix	
		Carbohydrates total	7.93%	
	Peel	Glucose	55.4 g L^{-1}	[51]
		Fructose	19.2 g L^{-1}	
		Match	1.02 g kg^{-1}	
		Potassium	14.31	
		Calcium	0.41	
		Magnesium	1.47	
		Sulfur	0.14	
		Sodium	0.22	
		Copper	2.75	
		Iron	9.75	
		Zinc	8.00	
		Manganese	6.00	
	Seed oil	Saturated fatty acids	22.78 g 100 g^{-1}	[45]
		Monounsaturated fatty acid	27.92	

TABLE 1.2 *(Continued)*

Fruit/ vegetable	Part	Composition	Content	Reference
		Polyunsaturated fatty acid	49.30	
		Myristic acid (C14:0)	0.19	
		Palmitic acid (C16:0)	16.53	
		Stearic acid (C18:0)	5.57	
		Arachidic acid (C20:0)	0.49	
		Palmitoleic acid (C16:1)	0.94	
		Oleic acid (C18:1)	26.80	
		Erucic acid (C22:1)	0.18	
		Linoleic acid (C18:2)	48.00	
		Linolenic acid (C18:3)	1.30	
	Pulp	Polyphenols total	118.57 GAE 100 g^{-1}	[46]
		Betacyanin total	42.84 mg 100 g^{-1} [a]	
	Juice	Citric acid	579.00 mg L^{-1}	
		D-isocitric acid	10.00	
		L-lactic acid	18.50	
		L-malic acid	4.80	
	Pulp	Ascorbic acid	20.69 mg 100 g^{-1}	
	Peel	DPPH IC_{50}	0.83 mg mL^{-1}	

[a]Values expressed in dry weight. [b]Values expressed in fresh weight. GAE: Gallic Acid Equivalents; IE: Indicaxanthin equivalents; CAE: Citric Acid Equivalents; BE: Betanin Equivalents BE: Betanin Equivalents; AE: Ascorbic acid equivalents; TE: Trolox equivalents; QE: Quercetin equivalents.

TABLE 1.3 Bioactive Compounds Identified in Some Cactaceae Plants.

Gender	Part	Type	Compound	Content	Reference
Nopal (*O. ficus-indica*)	Fruit oil	Phytosterol	Campesterol	21.65 mg 100 g^{-1}	[40]
			Stigmasterol	11.26	
			β-Sitosterol	387.44	
			Stigmastanol	47.04	
	Full fruit	Betaxanthins	Portulacaxanthin I	12.9 μg g^{-1} [a]	[48]
			Portulacaxanthin III	84.1	
			Vulgaxanthin III	25.5	
			Vulgaxanthin I	49.3	
			Vulgaxanthin II	16.3	

TABLE 1.3 *(Continued)*

Gender	Part	Type	Compound	Content	Reference
			Betaxanthins -aminobutyric acid	13.4	
			Indicaxanthin	207.4	
			Betaxanthins-Tryptophan	17.9	
	Whole fruit	Betacyanins	Betanine	292.1	
			Isobetanine	21.4	
			Betanidine	22.4	
			Gomphrenin I	31.7	
			Neobetanine	16.8	
	Whole fruit	Carotenoids	(all-E)-Violaxanthin	23.78 μg 100 g^{-1} [b]	[49]
			(all-E)-neoxanthin	12.24	
			(9Z)-violaxanthin	8.41	
			(all E)-antheraxanthin	11.83	
			(all-E)-lutein	332.16	
			(all-E)-zeaxanthin	13.50	
			Lutein-5,6-epoxide	13.50	
			(all-E)-α-carotene	3.61	
			(all-E)-β-carotene	65.81	
			(9Z)-β-carotene	6.75	
			Lycopene	26.40 μg g^{-1} [a]	[49]
	Pulp	Phenolic acids	Protocatechuic acid-hexoside	0.02 mg g^{-1} [a]	[52]
			Ferulic acid derivative	0.08	
			Ferulic acid-hexoside	0.14	
			Sinapic acid-hexoside	0.10	
			Piscidic acid derivative	982.3 μg g^{-1} [a]	
			Piscidic acid	21331.80	
			Derivative of 4-hydroxybenzoic acid	816.80 mg g^{-1} a	
	Pulp	Flavonols	Quercetin-hexoside-pentoside	0.01	
			Quercetin-hexoside	0.01	
			Quercetin glycoside	71.50 μg g^{-1} [a]	

TABLE 1.3 *(Continued)*

Gender	Part	Type	Compound	Content	Reference
			Isorhamnetin Glucoxyl-rhamnosyl-rhamnoside	15.10	
			Glucoxyl-rhamnosyl-Isorhamnetin pentoside	8.30	
			Isorhamnetin hexosyl-hexosyl-pentoside	6.30	
			Isorhamnetin glucoxyl-pentoside	4.40	
			Routine	15.5	
			Kaempferol-glucosyl-rhamnoside	9.40	
			Isorhamnetin derivative	0.02 mg g^{-1} to	
			Isorhamnetin-rutinoside	0.02	
	Pulp	Flavanone	Naringin	0.04	
	Whole fruit	Terpenoids	β-pinene	0.20%	[53]
			Camphene	5.30	
			α-flandrene	0.10	
			ρ-cymene	1.40	
			Limonene	3.90	
			1,8-cineole	0.60	
			β-Ocimene	0.60	
			3-carene	0.80	
			γ-terpinene	0.50	
			Linalool	3.80	
			α-Farnesene	1.00	
H. undatus	Pulp oil	Tocopherol	α-Tocopherol	24.00 mg 100 g^{-1}	[45]
			γ-Tocopherol	12.70	
	Pulp oil	Phytosterol	Campesterol	198.00	
			Stigmasterol	72.00	
			β-Sitosterol	548.00	
	Peel		γ-Sitosterol	19.32	
	Peel	Phenolic acid	Gallic acid	0.20	
			Protocatechuic	0.93	

TABLE 1.3 *(Continued)*

Gender	Part	Type	Compound	Content	Reference
			p-Hydroxybenzoic	0.72	
			Vanillic acid	0.70	
			Caffeinated	0.71	
			Syringic acid	0.21	
			p-Cumaric	0.79	
			Salicylic acid	0.98 mg g^{-1} to	[54]
			Caffeotanic acid	0.27	
			Benzoic acid	0.19	
	Peel	Monomeric Flavan-3-ols	Catechin	0.29	
		Flavonols	Routine	0.30	
		Alkane	Eicosan	1.92%	[55]
			Heptacosane	5.52	
			Nonacosan	5.02	
			Octadecan	9.25	
	Peel	Triterpenoid	β-amyrin	23.39	
		Ergostanoid	Ergosta-4,6,8(14),22-tetraen-3-one	1.46	
H. polyrhizus	Pulp oil	Tocopherol	α-Tocopherol	31.90 mg 100 g^{-1}	[45]
			γ-Tocopherol	11.60	
	Pulp oil	Phytosterol	Campesterol	252.00	
			Stigmasterol	106.00	
			β-Sitosterol	676.00	
	Peel		γ-Sitosterol	9.35	
	Pulp oil	Phenolic acid	Gallic acid	0.25	
			Protocatechuic	0.93	
			p-Hydroxybenzoic	0.66	
			Vanillic acid	0.64	
			Caffeinated	0.08	
			Syringic acid	0.08	
			p-Cumaric	0.78	
	Peel	Alkane	Eicosan	3.64%	[55]
			Tetratriacontane	1.04	
			Heptacosane	0.44	
			Nonacosan	1.43	
			Octadecan	6.27	
			Docosan	3.19	

TABLE 1.3 *(Continued)*

Gender	Part	Type	Compound	Content	Reference
	Peel	Triterpenoids	α-Amyrin	13.90	
			β-Amyrin	15.87	
	Fruit	Betacyanin	Betanine	23.81 mg 100 g^{-1}b	[46]
			Isobetanine	3.11	
			Phyllocactin	27.17	
Jacube (*A. tetragonus* L.) Hummelinck	Raw immature stems	Acids	Threonic acid	135.0294	[11]
			2-Hydroxy-succinic acid (malic acid)	133.0139	
			2-Hydroxy-succinic acid (malic acid) isomer	133.0139	
	Cooked immature stems		Citric acid derivative	147.0293	
	Raw and cooked immature stems		Citric acid	191.01973	
	Raw and cooked immature stems		Derivative of Isomeric citric acid	147.02916	
	Raw immature stems		Piscidic acid	255,051	
			Citric acid derivative	175.0612	
	Raw and cooked immature stems		Eucomic acid	239.0561	
	Cooked immature stems		Eucomic acid isomer	239.0561	
			dihydroxybenzoic acid	153.0188	
	Raw and cooked immature stems		Luteolin Rutinoside	593.1518	

TABLE 1.3 *(Continued)*

Gender	Part	Type	Compound	Content	Reference
			Isomer luteolin rutinoside	593.1518	
			Diphenylphenol	245.0972	
			sinapic acid	223.0607	
	Raw immature stems		4-Hydroxycinnamic acid Isomer acid I	163.0396	
	Cooked immature stems		4-Hydroxycinnamic acid Isomer acid II	163.0396	
			4-Hydroxycinnamic acid Isomer acid III	163.0396	
			4-Hydroxycinnamic Acid Isomer Acid IV	163.0396	
			Gallic acid	169.0144	
			Di-butyl di-tert-butyl 4-amino-4-(3-(tert-butoxy)-3-oxopropyl) heptanedioate	414.2861	
			1,4-Benzenediol, 2,2'-(6-dodecyl-1,12-diyl) bis[3,6-dimethoxy-]	501.2493	
			Linolenic acid derivative	721.3663	
			Dirhamnosil linolenic acid	559.3124	
			N,*N*-dihexyl-4-hydroxy-3,5-dimethoxybenzamide	364.2493	
			Dodecyl 4-*O*-acetyl-2-*O*-(4-*O*-acetyl-6-deoxy-αL-mannopyranosyl)-6-deoxy-β-d-galactopyranoside	561.3280	
Cereus jamacaru	Cladodes	Alkaloids	*N*-methyltyramine	33.79 mg 100 g^{-1} to	[87]
			Hordenine	0.69	

[a]Values expressed in dry weight. [b]Values expressed in fresh weight.

extract of *C. jamacaru* (210 and 420 mg kg^{-1}) decreases appetite and stops the increase in body weight in male Wistar rats.[66] On the contrary, it has been shown that the intake of *Pilosocereus gounellei* cladodes (125, 250, and 500 mg kg–1) reduced body weight, total cholesterol, low-density lipoprotein (LDL) cholesterol, and triglyceride levels, which had a positive impact on the cholesterol [high-density lipoprotein (HDL)] levels and glucose tolerance, in turn reducing epididymal fat and steatosis in male Swiss mice.[65] In another study, it was shown that the use of *O. ficus-indica* can prevent diabetes induced by alloxan, where it is described that the administration of oil from the seeds of this species (2 mL kg^{-1}) in Swiss albino mice increased the rate of animal Survival, where hyperglycemia lowered which prevented the formation of free radicals, loss of body weight, and inhibited pancreatic tissue damage.[18] In another study, obesity and insulin resistance were evaluated in male C57BL/6J mice, the results obtained showed that the intake of betacyanins from the rind of *H. undatus* (50, 100, and 200 mg kg^{-1}) gradually decreased the body weight, improved glucose intolerance and insulin resistance, and also improved lipid metabolism (adiponectin receptor 2 (AdipoR2), carnitine palmitoyltransferase 1b (Cpt1b), acyl-coenzyme A peroxisomal oxidase 1 (Acox1), receptor peroxisome proliferator-activated gamma (PPARγ), insulin-induced gene 1 (Insig1) and insulin-induced gene 2 (Insig2).[66]

1.6 APPLICATION IN THE DEVELOPMENT OF FOOD PRODUCTS

1.6.1 PASTA

In the market there are a variety of pastas that are made from various vegetables. In this sense, it seeks to introduce new products that contain ingredients with proven biological potential, however, one of the limitations in the pasta-making process is that high temperatures are required, which degrade the active compounds. That is why research has been carried out on the incorporation of edible cacti in pasta. Oniszczuk et al. (2020)[25] studied the fruits (prickly pear) of *O. ficus-indica* as a potential ingredient for the preparation of a paste. Among the results obtained, they showed that the antioxidant activity of the paste was positively correlated with the amount of tuna added, the content of free phenolic acids, and the sum of polyphenols. The total content of free phenolic acids also increased with the increase in the content of functional additives in the paste samples, reaching values of 26,671, 28,359, 29,525, 32,503, 34,097, 36,153, and

38,876 μg/g of dry matter, respectively, for the control without addition and the samples enriched with the addition of 2.5, 5.0, 7.5, 10.0, 12.5, and 15.0% fruit. Regarding the temperature used (40°C–50°C), it is not high enough to deactivate the hydrolytic or oxidative enzymes that can influence the phenolic content of the analyzed materials. For the control without addition and the samples enriched with the addition of 2.5, 5.0, 7.5, 10.0, 12.5, and 15.0% fruit. Regarding the temperature used (40–50°C), it is not high enough to deactivate the hydrolytic or oxidative enzymes that can influence the phenolic content of the analyzed materials. For the control without addition and the samples enriched with the addition of 2.5, 5.0, 7.5, 10.0, 12.5, and 15.0% fruit. Regarding the temperature used (40°C–50°C), it is not high enough to deactivate the hydrolytic or oxidative enzymes that can influence the phenolic content of the analyzed materials.[25]

1.6.2 BREAD

Another of the applications that has been given to cacti is in the bakery. In a study, it was observed that breads formulated with pulverized *O. ficus-indica* cladodes instead of wheat flour (2.5, 5.0, 7.5, and 10%) showed low extensibility, strain energy, and changes in color parameters, in addition to a 25% increase in the specific volume of bread (supplementation with 5% powder), phenolic content and antioxidant capacity by 10 and 139 times in supplementation with 10%, respectively.[79] In another study, the loaf bread formulation used 10% flour from the skin of the fruit of *C. jamacaru* dehydrated at 50°C resulting in a product with better color and flavor.[80] On the contrary, CE Silva (2019), evaluated the production of cakes using *P. gounellei* in partial substitution of wheat flour, where he observed that formulations of 20, 40, and 60% of *P. gounellei* flour had greater acceptability and purchase intention, while the formulation with 80% resulted in a moister cake, slightly more acidic and less accepted by consumers. In addition, they found that this partial substitution increased the content of ash (20%), protein (20%), and ascorbic acid (60 and 80%) in the formulated cake.[81]

1.6.3 YOGURT

Within the fermented products, investigations have been carried out on the incorporation of some cacti in different types of yogurts. In a study conducted by Nobrega et al. in 2020, where goat yogurt was prepared with

15% *C. jamacaru* pulp gelatin and passion fruit with oligofructose (3.4 g L^{-1}), which assessed the stability during 28 days of refrigerated storage, the results obtained showed an increase in water at the end of storage and changes in the physicochemical properties, especially in acidity, pH and lactose, concluding that the addition of these ingredients had a good incorporation and could be an interesting approach for the formulation of a ncw dairy product.[82] In another investigation, work was done with pulp and peel of *C. jamacaru* and *O. ficus-indica*, where a natural yogurt was elaborated with pulp and peel of each fruit with a ratio (1: 1, w/v), in which it was determined that the yogurts formulated with *C. jamacaru* pulp and peel had a higher content of soluble (9.40° Brix), pH (3.53) and total acidity (0.84%), while the formulation containing *O. ficus-indica* contained a higher content of vitamin C (4.87 mg 100 g^{-1}).[83]

1.6.4 JUICE

Experiments have also been made in the preparation of formulations that involve juices or juices of some fruits added with some cacti. Maataoui et al.[84] demonstrated that the juice of the purple prickly pear of *O. ficus-indica* has higher antioxidant activity than that of the yellow-orange prickly pear juice. In another study, it was found that prickly pear showed a high DPPH free radical scavenging potential, equal to 99.78%.[25] On the contrary, they studied the influence of clarification of *O. ficus-indica* fruit juice. Similarly, it was found that although microfiltration and ultrafiltration were used to clarify the juice, it did not alter the content of total soluble solids, pH, and acidity, but impacted the content of phenolic compounds (6, 1%) as they decreased in the ultrafiltration permeate fraction and the protein content in the microfiltration and ultrafiltration permeate fractions (76.9 and 78.3%, respectively). On the contrary, the retained fraction can be considered a source of bioactive compounds (e.g., phenolics and vitamin C), betacyanins, and betaxanthins, with the potential to be mousses, ice creams, jellies, and functional foods.[85]

1.6.5 GUMMIES

Within a little explored field, there is a piece of research that studied the production of jellybeans, by evaluating the application of encapsulated betalains extracted from *O. ficus-indica*. The observed results of the investigation mention that the viscoelastic properties, gel strength, and stress relaxation depended on the type of gelatin and the ratio of betalain-rich capsules/gelatin used, in addition to the purple–red coloration of the chewing gum. (Characterized by the addition of betalain) had good storage stability at 4°C for 30 days, apparently absent of relevant changes in total color parameters.[86]

1.6.6 ICE CREAM

One way to preserve the biological properties of some foods is through a temperature drop, in this sense, there is a field of opportunity for the manufacture of ice creams made with edible cacti. In a study carried out by Pinto et al. (2017), they made ice creams with pulp of *C. jamacaru* (400–500 g), *P. gounellei* (400–500 g) and *C. jamacaru* + *P. gounellei* (200 gc/u) to which determined by physicochemical analysis: humidity (71, 89–79.13%), pH (5.71–6.19), total soluble solids (3.9–4.9%), total acidity (0.21–0.32%) and ashes (0.70–0.90%), in addition, sensory analyzes were carried out, where it was reported that good sensory acceptance was obtained, highlighting the formulation that contained 500 g of *C. jamacaru* pulp, which showed a greater sensory acceptance and consequently a greater purchase intention. The results they obtained indicated that these ice creams met the appropriate parameters for their industrialization and commercialization.[88] In another study carried out on ice cream, in the elaborated formulations it was obtained that they all showed an acid pH and that the ice cream made with *C. jamacaru*, which consisted of 100 g of ice cream and 100 g of pulp or shell, had a higher content of total soluble solids (21.66 Brix) and total acidity (0.41%), and in the ice cream made with *O. ficus-indica* (1:1. w/w) it was obtained that it contained a higher content of vitamin C (6.24 mg 100 g^{-1}). On the contrary, an attempt has been made to replace some ice cream ingredients with others that do not promote any disease. Given this problem, an investigation tested the application of *H. polyrhizus* skin flour, where it was shown to be an alternative to reducing fat of the ice cream (73.5%), in addition to having a good acceptability and improving the rheological behavior of the sample.[89] Due to the results obtained in the investigations, there is scientific evidence that the production of ice cream made with the fruits of *O. ficus-indica* and *C. jamacaru* can be an interesting strategy to expand its commercial exploitation.

1.7 CONCLUDING REMARKS

There are several research papers related to the nutritional benefits, medicinal use, and food value of the Cactaceae family. The species reviewed above are excellent sources of phytochemicals responsible for several biological properties, including the hypoglycemic effect. The species of the Cactaceae family have the potential to be used in the pharmaceutical and cosmetic industries. In addition, they can be an excellent alternative to crops in arid and semi-arid regions to meet food and fodder needs. However, they are still an underutilized and underexploited crop. Thus, future approaches are needed to target their cultivation, diversification of their uses, and commercialization. The latter is important and can contribute to the food and nutritional security and socioeconomic development of the population in the regions where these species naturally occur.

KEYWORDS

- **functional foods**
- **sustainability**
- **Cactaceae**
- **hypoglycemic effect**
- **antioxidant properties**

REFERENCES

1. Uusitupa, M.; Khan, T. A.; Viguiliouk, E.; Kahleova, H.; Rivellese, A. A.; Hermansen, K.; Pfeiffer, A.; Thanopoulou, A.; Salas-Salvadó, J.; Schwab, U.; Sievenpiper, J. L. Prevention of Type 2 Diabetes by Lifestyle Changes: A Systematic Review and Meta-Analysis. *Nutrients* **2019,** *11* (11), 2611. doi: 10.3390/nu11112611. PMID: 31683759; PMCID: PMC6893436.
2. World Health Organization. https://www.who.int/es/news-room/fact-sheets/detail/diabetes
3. Pan American Health Organization. https://www.paho.org/en/issues/diabetes
4. Zhu, D.; Yan, Q.; Liu, J.; Wu, X.; Jiang, Z. Can Functional Oligosaccharides Reduce the Risk of? *FASEB J.* **2019,** *33* (11), 11655–11667. doi: 10.1096/fj.201802802RRR. Epub 2019 Aug 15. PMID: 31415188; PMCID: PMC6902710.

5. Tuso, P. J.; Ismail, M. H.; Ha, B. P.; Bartolotto, C. Nutritional Update for Physicians: Plant-Based Diets. *Perm. J.* **2013,** *17* (2), 61–66, https://doi.org/10.7812/TPP/12-085.
6. Sinicropi, M. S.; Baldino, N.; Ceramella, J.; Iacopetta, D.; Scali, E.; Basile, G.; Saturnino, C.; Catalano, A. *Opuntia ficus-indica* (L.) Mill. an Ancient Plant Source of Nutraceuticals. *Curr. Top. Med. Chem.* **2022,** *22.* https://doi.org/10.2174/15680266226 66220803151814.
7. Aragona, M.; Lauriano, E. R.; Pergolizzi, S.; Faggio, C. *Opuntia ficus-indica* (L.) Miller as a Source of Bioactivity Compounds for Health and Nutrition. *Nat. Product Res.* **2018,** *32* (17), 2037–2049. https://doi.org/10.1080/14786419.2017.1365073
8. Aispuro-Hernández, E.; Vergara-Jiménez, M. J.; Cárdenas-Torres, F. I.; Martínez-Téllez, M. A.; Ontiveros, N. Cactaceae Plants as Sources of Active Bioavailable Phytochemicals. *Food Funct.* **2022,** 13 (19). https://doi.org/10.1039/D2FO01863B
9. Secretary of Agriculture and Rural Development [Online] **2022.** https://www.gob.mx/agricultura/es/articulos/un-alimento-milenario-y-nutritivo-el-nopal#:~:text=Entre%20sus%20propiedades%20destaca%20su,a%20d%C3%ADa%20work%20our%20%23H%C3%A9roesDeLaAlimentaci%C3%B3n.
10. Ramawat, K. G. Desert Plants. Biology and Biotechnology. In *ML Sukhadia University Department of Botany Udaipur* 313001. Publisher Springer: India; **2010.** ISBN: 978-3-642-02549-5 and-ISBN: 978-3-642-02550-1. DOI 10.1007/978-3-642-02550-1. Springer Heidelberg Dordrecht London New York
11. Cornejo-Campos, J.; Adriana Gómez-Aguirre, Y.; Rodolfo Velázquez-Martínez, J.; Javier Ramos-Herrera, O.; Estefanía Chávez-Murillo, C.; Cruz-Sosa, F.; Areche, C.; Cabañas-García, E. Citation: Molecules Impact of the Cooking Process on Metabolite Profiling of *Acanthocereus tetragonus*, a Plant Traditionally Consumed in Mexico, **2022.** https://doi.org/10.3390/molecules27123707
12. Kashif, R. R.; D'Cunha, N. M.; Mellor, D. D.; Alexopoulos, N. I.; Sergi, D.; Naumovski, N. Prickly Pear Cacti (Opuntia Spp.) Cladodes as a Functional Ingredient for Hyperglycemia Management: A Brief Narrative Review. *Medicine (Kaunas)* **2022,** *58* (2), 300. doi: 10.3390/medicine58020300. PMID Cardoso PDS, da Silva INB, Ferreira-Ribeiro
13. SADER. **2020.**
14. El Mannoubi, I.; Barrek, S.; Skanji, T.; Casabianca, H.; Zarrouk, H. Characterization of *Opuntia ficus-indica* Seed Oil from Tunisia. *Chem. Nat. Compd.* **2009,** *45,* 616.
15. Taoufik, F.; Zine, S.; El Hadek, M.; Idrissi Hassani, L. M.; Gharby, S.; Harhar, H.; Matthäus, B. Oil Content and Main Constituents of Cactus Seed Oils *Opuntia ficus-indica* of Different Origin in Morocco. *Mediterr. J. Nutr. Metab.* **2015,** *8,* 85–92.
16. Berraaouan, A.; Ziyyat, A.; Mekhfi, H.; Legssyer, A.; Sindic, M.; Aziz, M.; Bnouham, M. Evaluation of Antidiabetic Properties of Cactus Pear Seed Oil in Rats. *Pharm. Biol.* **2014,** *52,* 1286–1290.
17. Ramadan, M. F.; Morsel, J.-T. Oil Cactus Pear (Opuntia-Indica L.). *Food Chem.* **2003,** *82,* 339–345.
18. Berraaouan, A.; Abderrahim, Z.; Hassane, M.; Abdelkhaleq, L.; Mohammed, A.; Mohamed, B. Evaluation of Protective Effect of Cactus Pear Seed Oil (*Opuntia ficus-indica* L. MILL.) Against Alloxan-Induced Diabetes in Mice. *Asian Pac. J. Trop. Med.* **2015,** *8,* 532–537.

19. Bardaa, S.; Turki, M.; Ben Khedir, S.; Mzid, M.; Rebai, T.; Ayadi, F.; Sahnoun, Z. The Effect of Prickly Pear, Pumpkin, and Linseed Oils on Biological Mediators of Acute Inflammation and Oxidative Stress Markers. *BioMed Res. Int.* **2020,** *2020,* 5643465.
20. Khemiri, I.; Hedi, B. E.; Zouaoui, N. S.; Gdara, N. B.; Bitri, L. The Antimicrobial and Wound Healing Potential of *Opuntia ficus-indica* L. Inermis Extracted Oil from Tunisia. *Evid. Based Complement Alternate Med.* **2019,** *2019,* 9148782.
21. Ortega-Ortega, M. D.; Cruz-Cansino, N.D.; Alanís-García, E.; Delgado-Olivares, L.; Ariza-Ortega, J. A.; Ramirez-Moreno, E.; Manríquez-Torres, J. D. Optimization of Ultrasonic Extraction of Cactus Pear (*Opuntia ficus-indica*) Seed Oil Based on Antioxidant Activity and Evaluation of Its Antimicrobial Activity. *J. Food Qual.* **2017,** *2017,* 9315360.
22. Al-Naqeb, G. Effect of Prickly Pear Cactus Seeds Oil on the Blood Glucose Level of Streptozotocin-Induced Diabetic Rats and Its Molecular Mechanisms. *Int. J. Herb. Med.* **2015,** *3,* 29–34.
23. Benattia, F. K.; Arrar, Z.; Khabba, Y. Evaluation the Hypo and Antihyperglycemic Activity of Cactus Seeds Extracts (*Opuntia ficus-indica* L.). *PCBS J.* **2017,** *11.* [CrossRef]
24. Ennouri, M.; Fetoui, H.; Bourrett, E.; Zeghal, N.; Attia, H. Evaluation of Some Biological Parameters of *Opuntia ficus-indica.* 1. Influence of a Seed Oil Supplemented Diet on Rats. *Bioresour. Technol.* **2006,** *97,* 1382–1386.
25. Oniszczuk, A.; Wójtowicz, A.; Oniszczuk, T.; Matwijczuk, A.; Dib, A.; Markut-Miotła, E. Opuntia Fruits as Food Enriching Ingredient, the First Step towards New Functional Food Products. *Molecules (Basel, Switzerland)* **2020,** *25* (4). https://doi.org/10.3390/MOLECULES25040916.
26. López-Romero, P.; Pichardo-Ontiveros, E.; Avila-Nava, A., et al. The effect of Nopal (*Opuntia ficus-indica*) on Postprandial Blood Glucose, Incretins, and Antioxidant Activity in Mexican Patients with Type 2 Diabetes After Consumption of Two Different Composition Breakfasts. *J. Acad. Nutr. Dietet.* **2014,** *114* (11), 1811–1818.
27. Frati-Munari, A. C.; Gordillo, B. E.; Altamirano, P.; Ariza, C. R. Hypoglycemic Effect of *Opuntia streptacantha* Lemaire in NIDDM. *Diabetes Care* **1988,** *11* (1), 63–66.
28. Hahm, S.-W.; Park, J.; Son, Y.-S. *Opuntia humifusa* Stems Lower Blood Glucose and Cholesterol Levels in Streptozotocin-Induced Diabetic Rats. *Nutr. Res.* **2011,** *31* (6), 479–487.
29. Andrade-Cetto, A.; Wiedenfeld, H. Anti-Hyperglycemic Effect of *Opuntia streptacantha* Lem. *J. Ethnopharmacol.* **2011,** *133* (2), 940–943.
30. Godard, M. P.; Ewing, B. A.; Pischel, I.; Ziegler, A.; Benedek, B.; Feistel, B. Acute Blood Glucose Lowering Effects and Long-Term Safety of OpunDia Supplementation in Pre-Diabetic Males and Females. *J. Ethnopharmacol.* **2010,** *130* (3), 631–634.
31. Trejo-Gonzalez, A.; Gabriel-Ortiz, G.; Puebla-Perez, A. M., et al. A Purified Extract from Prickly Pear Cactus (Opuntia Fuliginosa) Controls Experimentally Induced Diabetes in Rats. *J. Ethnopharmacol.* **1996,** *55* (1), 27–33.
32. de Abreu, W. C.; de Lopes, C. O.; Pinto, K. M.; Oliveira, L. A.; De Carvalho, G. B. M.; De Barcelo, M. F. P. Physicochemical Characteristics and Total Antioxidant Activity of Red and White Pitaya. *Adolfo Lutz Inst. Mag.* **2012,** *71* (4), 656–661. http://periodicos.ses.sp.bvs.br/pdf/rial/v71n4/v71n4a07.pdf.

33. Muhammad, K.; Mohd. Zahari, N. I.; Gannasin, S. P.; Mohd. Adzahan, N.; Bakar, J. High Methoxyl Pectin from Dragon Fruit (*Hylocereus polyrhizus*) Peel. *Food Hydrocolloids* **2014,** *42,* 289–297. https://doi.org/10.1016/j.foodhyd.2014.03.021.
34. Ramírez-Rodríguez, Y.; Martínez-Hu´elamo, M.; Pedraza-Chaverri, J.; Ramírez, V.; Martínez-Tagüena, ˜N.; Trujillo, J. Ethnobotanical, Nutritional and Medicinal Properties of Mexican Drylands Cactaceae Fruits: Recent Findings and Research Opportunities. *Food Chem.* **2020,** *312,* 126073. https://doi.org/10.1016/j. foodchem.2019.126073.
35. Arul, V.; Edison, T. N. J. I.; Lee, Y. R.; Sethuraman, M. G. Biological and Catalytic Applications of Green Synthesized Fluorescent N-Doped Carbon Dots Using *Hylocereus undatus. J. Photochem. Photobiol. B: Biol.* **2017,** *168,* 142–148. https://doi.org/10.1016/j.jphotobiol.2017.02.007.
36. Peña-Montes Carolina; Lagunes-Vela Kalid Gabriela; Morales-Cano Karla Lizzeth; Ramírez-Higuera Abril1; Oliart-Ros Rosa María. Food Research and Development Unit (UNIDA), Technological Institute of Veracruz, National Technological Institute of Mexico, Calzada Miguel Ángel De Quevedo 2779. Col. Formando Hogar, Veracruz, Ver. MEXICO CP 91897. Tel: +52-229 934 15 00, #Correspondence: Carolina Peña Montes, Laboratory of Applied Genetics, E-mail: carolina.pm@veracruz.tecnm.mx
37. Juarez Cruz, A.; Livera Muñoz, M.; Sosa Montes, E.; Goytia Jiménez, M. A. A.; González Hernández, V. A.; Bárcena Gama, R. CHEMICAL COMPOSITION OF IMMATURE STEMS OF Acanthocereus Spp. and *Hylocereus undatus* (Haw.) Britton & Rose. *Mexican Phytotechn. Mag.* **2012,** *35,* 171–175.
38. Salehi, E.; Emam-Djomeh, Z.; Askari, G.; Fathi, M. *Opuntia ficus-indica* Fruit Gum: Extraction, Characterization, Antioxidant Activity and Functional Properties. *Carbohydrate Polymers* **2019,** *206,* 565–572. https://doi.org/10.1016/j.carbpol.2018.11.035.
39. Bakar, B.; Çakmak, M.; Ibrahim, M. S.; Ozer, ¨D.; Saydam, S.; Karatas, F. Investigation of Amounts of Vitamins, Lycopene, and Elements in the Fruits of *Opuntia ficus-indica* Subjected to Different Pretreatments. *Biol. Trace Element Res.* **2020,** *198* (1), 315–323. https://doi.org/10.1007/s12011-020-02050-w.
40. Brahmi, F.; Haddad, S.; Bouamara, K.; Yalaoui-Guellal, D.; Prost-Camus, E.; De Barros, J. P. P.; Prost, M.; Atanasov, A. G.; Madani, K.; Boulekbache-Makhlouf, L.; Lizard, G. Comparison of Chemical Composition and Biological Activities of Algerian Seed Oils of Pistacia Lentiscus L., *Opuntia ficus-indica* (L.) Mill. and Argania Spinosa L. Skeels. *Ind. Crops Products* **2020,** *151,* 112456. https://doi.org/10.1016/j.indcrop.2020.112456.
41. Salehi, E.; Emam-Djomeh, Z.; Askari, G.; Fathi, M. *Opuntia ficus-indica* Fruit Gum: Extraction, Characterization, Antioxidant Activity and Functional Properties. *Carbohydrate Polymers* **2019,** *206,* 565–572. https://doi.org/10.1016/j.carbpol.2018.11.035.
42. Blando, F.; Russo, R.; Negro, C.; De Bellis, L.; Frassinetti, S. Antimicrobial and Antibiofilm Activity Against Staphylococcus Aureus of *Opuntia ficus-indica* (L.) Mill. Cladode polyphenolic extracts. *Antioxidants* **2019,** *8* (5), 117. https://doi.org/10.3390/antiox8050117
43. Esgote Junior, J. D. Chemical Composition and Antioxidant Activity of Different Species of Pitaias [Universidade Tecnologica Federal do Parana] [Online]. **2017**. http://repositorio .utfpr.edu.br/jspui/bitstream/1/3410/1/composicaoquimicaatividadeantioxidantepitaias.pdf.

44. De Abreu, W. C.; De Lopes, C. O.; Pinto, K. M.; Oliveira, L. A.; De Carvalho, G. B. M.; De Barcelo, M. F. P. Physicochemical Characteristics and Total Antioxidant Activity of Red and White Pitaya. *Adolfo Lutz Inst. Mag.* **2012,** *71* (4), 656–661. http://periodicos.ses.sp.bvs.br/pdf/rial/v71n4/v71n4a07.pdf
45. Lim, H. K.; Tan, C. P.; Karim, R.; Ariffin, A. A.; Bakar, J. Chemical Composition and DSC Thermal Properties of Two Species of *Hylocereus cacti* Seed Oil: *Hylocereus undatus* and *Hylocereus polyrhizus*. *Food Chem.* **2010,** *119* (4), 1326–1331. https://doi.org/10.1016/j.foodchem.2009.09.002.
46. Yong, Y. Y.; Dykes, G.; Lee, S. M.; Choo, W. S. Comparative Study of Betacyanin Profile and Antimicrobial Activity of Red Pitahaya (*Hylocereus polyrhizus*) and Red Spinach (Amaranthus Dubius). *Plant Foods Hum. Nutr.* **2017,** *72* (1), 41–47. https://doi.org/10.1007/s11130-016-0586-x.
47. Lagunes-Vela, K. Characterization of Crosshead Mucilage and Effect of Its Consumption in a Mouse Model with Metabolic Syndrome. In *Thesis to Obtain the Degree of Master of Science in Biochemical Engineering*. National Technological of Mexico: Veracruz; **2019**.
48. García-Cayuela, T.; Gomez-Maqueo, A.; Guajardo-Flores, D.; Welti-Chanes, J.; Cano, M. P. Characterization and Quantification of Individual Betalain and Phenolic Compounds in Mexican and Spanish Prickly Pear (*Opuntia ficus-indica* L. Mill) Tissues: A Comparative Study. *J. Food Comp. Anal.* **2019,** *76,* 1–13. https://doi.org/10.1016/j.jfca.2018.11.002.
49. Cano, M. P.; Gomez-Maqueo, A.; García-Cayuela, T.; Welti-Chanes, J. Characterization of Carotenoid Profile of Spanish Sanguinos and Verdal Prickly Pear (*Opuntia ficus-indica*, Spp.) Tissues. *Food Chem.* **2017,** *237,* 612–622. https://doi.org/10.1016/j.foodchem.2017.05.135.
50. Esgote Junior, J. D. Chemical Composition and Antioxidant Activity of Different Species of Pitaias [Universidade Tecnologica Federal do Parana] [Online] **2017**. http://repositorio.utfpr.edu.br/jspui/bitstream/1/3410/1/composicaoquimicaatividadeantioxidantepitaias.pdf
51. Stintzing, F. C.; Schieber, A.; Carle, R. Evaluation of Color Properties and Chemical Quality Parameters of Cactus Juices. *Eur. Food Res. Technol.* **2003,** *216* (4), 303–311. https://doi.org/10.1007/s00217-002-0657-0.
52. Mena, P.; Tassotti, M.; Andreu, L.; Nuncio-Jauregui, N.; Legua, P.; Del Rio, D.; Hern´andez, F. Phytochemical Characterization of Different Prickly Pear (*Opuntia Ficus-Indica* (L.) Mill.) Cultivars and Botanical Parts: UHPLC-ESI-MSn Metabolomics Profiles and Their Chemometric Analysis. *Food Res. Int.* **2018,** *108,* 301–308. https://doi.org/10.1016/j.foodres.2018.03.062.
53. Oumato, J.; Zrira, S.; Petretto, G. L.; Saidi, B.; Salaris, M.; Pintore, G. Volatile Constituents and Polyphenol Composition of Opuntiaficus-Indica (L.) Mill from Morocco. *Rev. Mar. Sci. Agron. V'et.* **2016,** *4,* 5–11.
54. Zhuang, Y.; Zhang, Y.; Sun, L. Characteristics of Fiber-Rich Powder and Antioxidant Activity of Pitaya (*Hylocereus undatus*) Peels. *Int. J. Food Sci. Technol.* **2012,** *47* (6), 1279–1285. https://doi.org/10.1111/j.1365-2621.2012.02971.x.
55. Luo, H.; Cai, Y.; Peng, Z.; Liu, T.; Yang, S. Chemical Composition and in Vitro Evaluation of the Cytotoxic and Antioxidant Activities of Supercritical Carbon Dioxide Extracts of Pitaya (Dragon Fruit) Peel. *Chem. Central J.* **2014,** *8* (1), 1–7. https://doi.org/10.1186/1752-153X-8-1.

56. Osorio-Esquivel, O.; Alicia-Ortiz-Moreno; Alvarez, V. B.; Dorantes-Alvarez, L.; Giusti, M. M. Phenolics, Betacyanins and Antioxidant Activity in Opuntia Joconostle Fruits. *Food Res. Int.* **2011,** *44,* 2160–2168.
57. Paiz, R. C.; Juarez-Flores, B. I.; Aguirre, R. J. R.; Cardenas, O. C.; Reyes, A. H. A.; Garcia, C. E.; Álvarez, F. G. Glucose-Lowering Effect of Xoconostle (Opuntia Joconostle A. Web. Cactaceae) in Diabetic Rats. *J. Med. Plants Res.* **2010,** *4,* 2326–2333.
58. Schaffer, S.; Schmitt-Schillig, S.; Muller, W. E.; Eckert, G. P. Antioxidant Properties of Mediterranean Food Plant Extracts: Geographical Differences. *J. Physiol. Pharmacol.* **2005,** *56* (Suppl. S1), 115–124.
59. Stintzing, F. C.; Schieber, A.; Carle, R. Phytochemical and Nutritional Significance of Cactus Pear. *Eur. Food Res. Technol.* **2001,** *212,* 396–407.
60. DeLeo, M.; Abreu, M. B. D.; Pawlowska, A. M.; Cioni, P. L.; Braca, A. Profiling the Chemical Content of *Opuntia ficus-indica* Flowers by HPLC–PDA-ESI-MS and GC/EIMS Analyses. *Phytochem. Lett.* **2010,** *3,* 48–52.
61. Ramadan, M. F.; Morsel, J.-T. Oil Cactus Pear (*Opuntia ficus-indica* L.). *Food Chem.* **2003,** *82,* 339–345.
62. Galati, E. M.; Mondello, M. R.; Monforte, M. T.; Galluzzo, M.; Miceli, N.; Tripod, M. M. Effect of *Opuntia ficus-indica* (L.) Mill. Cladodes in the Wound-Healing Process. *J. Prof. Assoc. Cactus Dev.* **2003,** *5,* 1–16.
63. Laughton, M. J.; Evans, P. J.; Moroney, M. A.; Hoult, J. R.; Halliwell, B. Inhibition of Mammalian 5-Lipoxygenase and Cyclo-Oxygenase by Flavonoids and Phenolic Dietary Additives. Relationship to Antioxidant Activity and to Iron Ion-Reducing Ability. *Biochem. Pharmacol.* **1991,** *42,* 1673–1681.
64. Song, H.; Chu, Q.; Xu, D.; Xu, Y.; Zheng, X. Purified Betacyanins from *Hylocereus undatus* Peel Ameliorate Obesity and Insulin Resistance in High-Fat-Dietfed Mice. *J. Agric. Food Chem.* **2016,** *64* (1), 236–244. https://doi.org/10.1021/acs.jafc.5b0517710.1021/acs.jafc.5b05177.s001.
65. De Oliveira, A. M.; De Freitas, A. F. S.; De Costa, M. D. S.; Da Torres, M. K. S.; De Castro, Y. A. A.; Almeida, A. M. R.; Paiva, P. M. G.; Carvalho, B. M.; Napole˜ao, T. H. *Pilosocereus gounellei* (Cactaceae) Stem Extract Decreases Insulin Resistance, Inflammation, Oxidative Stress, and Cardio-Metabolic Risk in Diet-Induced Obese Mice. *J. Ethnopharmacol.* **2021,** *265,* 113327. https://doi.org/10.1016/j. jep.2020.113327.
66. De Medeiros, I. U.; De Medeiros, R. A.; Bortolin, R. H.; De Queiroz, F. M.; Silbiger, V. N.; Pflugmacher, S.; Schwarz, A. Genotoxicity and Pharmacokinetic Characterization of *Cereus iamacaru* Ethanolic Extract in Rats. *Biosci. Rep.* **2019,** *39* (1), 20180672. https://doi.org/10.1042/BSR20180672.
67. Perez-Gutierrez, R. M.; Enriquez-Alvirde, S. G. Ursane Derivatives Isolated from Leaves of *Hylocereus undatus* Inhibit Glycation at Multiple Stages. *Chinese J. Nat. Med.* **2018,** *16* (11), 856–865. https://doi.org/10.1016/S1875-5364(18) 30127-4.
68. Hernandez, H. M.; Godinez, H. Contribution to the Knowledge of the Threatened Mexican Cacti. *Mexican Bot. Act* **1994,** *26,* 33–52.
69. Bravo-Hollis, H. *The Cactaceae of Mexico: Vol.* 1; National University of Mexico, 1978.
70. Nobel, P. S. *Remarkable Agaves and Cacti*; Oxford University Press; 1994.
71. Nobel, P. S.; Pimienta-Barrios, E. Monthly Stem Elongation for Stenocereus Queretaroensis: Relationships to Environmental Conditions, Net CO2 Uptake and Seasonal Variations in Sugar Content. *Environ. Exp. Bot.* **1995,** *35* (1), 17–24.

72. Esquivel, P. The Fruits of Cacti and Their Potential as Raw Material. *Mesoamerican Agronomy* **2004,** *15* (2), 215–219.
73. National Institute of Public Health. National Health and Nutrition Survey (ENSANUT). Databases and Questionnaires for ENSANUT Continua 2021 [Online]; **2021**. https://ensanut.insp.mx/encuestas/ensanutcontinua2021/descargas.php.
74. National Institute of Geography and Statistics (INEGI). National Survey of Household Income and Expenses (ENIGH). **2020** New series. INEGI-ENIGH [Online] 2020. https://www.inegi.org.mx/programas/enigh/nc/2020/
75. CONEVAL. Evaluation of Poverty Lines by Income. Income Poverty Lines [Online] **2020**. https://www.coneval.org.mx/Medicion/MP/Paginas/Lineas-de-bienestar-y-canasta-basica.aspx.
76. National System of Information and Integration of Markets. Prices [Online] **2023**. http://www.economia-sniim.gob.mx/nuevo/Home.aspx
77. Food and Agriculture Organization of the United Nations (FAO). *Latin American and Caribbean Food Security Scale (ELCSA): Manual of Use and Applications*In Rome: FAO; **2012**.
78. SIAP. *Consultation Agrifood Information System (SIACON-NG)*; Food and Fisheries Information Service (SIAP); **2021**.
79. do Birth, M. A. G. Agro-Industrial Use of Cacti in the Brazilian Semi-Arid Region. In *Federal University of Campina Grande*; Nepomuceno, N.C., Santos, A.S.F., Oliveira, J.E., Glenn, G,M., Eds., **2014**.
80. Msaddak, L.; Abdelhedi, O.; Kridene, A.; Rateb, M.; Belbahri, L.; Ammar, E.; Nasri, M.; Zouari, N. *Opuntia ficus-indica* Cladodes as a Functional Ingredient: Bioactive Compounds Profile and their Effect on Antioxidant Quality of Bread. *Lipids Health Dis.* **2017,** *16* (1), 1–8. https://doi.org/10.1186/s12944-016-0397-.
81. Silva, C. E. *Development, Characterization and Sensory Analysis of Bolus from Farinha De Xiquexique (Pilosocereus gounellei)*; Federal Institute of Sertao˜. Pernambuco; **2019**.
82. De Nobrega, J. P. M.; Jeronimo, H. M. A.; De Ramos, J. A.; Soares, J. K. B.; De Oliveira, M. E. G.; Viera, V. B.; Martins, A. C. S. Physical and Physical Characterization of Prebiotic Caprine Yogurt Added in Mandacaru and Passion Fruit Jelly. *Res. Soc. Dev.* **2020,** *9* (6), e191963788. https://doi.org/10.33448/rsdv9i6.3788.
83. Fidelis, V. R. L.; Pereira, E. M.; Silva, W. P.; Gomes, J. P.; Silva, L. A. Production of Sorvetes and Yoghurts from Two Fruits: Figo Da India and Mandacaru. *Rev. Verde Agroecol. Desenvolvimento Sustentavel* **2015,** *10* (4), 17. https://doi.org/10.18378/rvads.v10i4.3698.
84. Maatoui, B. S.; Hmyene, A.; Hilali, S. Activités Anti-Radiculaires D'extracts De Jus De Fruits Du Figuier De Barbarie (*Opuntia ficus-indica*). *Leban. Sci. J.* **2006,** *7,* 3–8.
85. Cassano, A.; Conidi, C.; Drioli, E. Physical-Chemical Parameters of Cactus Pear (*Opuntia ficus-indica*) Juice Clarified by Microfiltration and Ultrafiltration Processes. *Desalination* **2010,** *250* (3), 1101–1104. https://doi.org/10.1016/j.desal.2009.09.117.
86. Otalora, M. C.; De Jesús Barbosa, H.; Perilla, J. E.; Osorio, C.; Nazareno, M. A. Encapsulated Betalains (*Opuntia ficus-indica*) as Natural Colorants. Case Study: Gummy Candies. *LWT* **2019,** *103,* 222–227. https://doi.org/10.1016/j.lwt.2018.12.074.
87. Davet, A.; Carvalho, J. L. S.; Dadalt, R. C.; Vituoso, S.; Dias, J. F. G.; Miguel, M. D.; Miguel, O. G. *Cereus jamacaru*: A Non-Buffered Lc Quantification Method to Nitrogen Compounds. *Chromatographia* **2009,** *69* (SUPPL. 2), 245–247. https://doi.org/10.1365/s10337-009-1130-z.

88. Pinto, M. P. Desenvolvimento de sorvete a ` base de polpa de mandacaru e xiquexique [Federal Institute of Education, Science and Technology of Piauí]. **2017**; http://bia.ifpi.edu.br:8080/jspui/handle/prefix/256.
89. Utpott, M.; Ramos De Araujo, R.; Galarza Vargas, C.; Nunes Paiva, A. R.; Tischer, B.; De Oliveira Rios, A.; Hickmann Flores, S. Characterization and Application of Red Pitaya (*Hylocereus polyrhizus*) Peel Powder as a Fat Replacer in Ice Cream. *J. Food Proc. Pres.* **2020,** *44* (5), 1–10. https://doi.org/10.1111/jfpp.14420.
90. Fidelis, V. R. L.; Pereira, E. M.; Silva, W. P.; Gomes, J. P.; Silva, L. A. Production of Sorvetes and Yoghurts from Two Fruits: Figo Da India and Mandacaru. *Green Mag. Agroecol. Sustainable Dev.* **2015,** *10* (4), 17. https://doi.org/10.18378/rv ads.v10i4.3698

CHAPTER 2

Microbial Technology to Produce Sustainable Food Bioproducts: Insights into Submerged Fermentation

ITZA NALLELY CORDERO-SOTO[1],
MARÍA GORETY CONTRERAS-HERNÁNDEZ[1],
SILVIA MARINA GONZÁLEZ-HERRERA[1],
ALFREDO DE JESÚS MARTÍNEZ-ROLDÁN[2], and
OLGA MIRIAM RUTIAGA-QUIÑONES[1]

[1]*Tecnológico Nacional de México/Instituto Tecnológico de Durango, Departamento de Ingenierías Química y Bioquímica, Felipe Pescador 1803 Ote, Colonia Nueva Vizcaya, Durango, Durango, México*

[2]*CONAHCYT-Tecnológico Nacional de México/Instituto Tecnológico de Durango. Maestría en Sistemas Ambientales, Departamento de Ingenierías Química y Bioquímica, Felipe Pescador 1803 Ote, Colonia Nueva Vizcaya, Durango, Durango, México*

ABSTRACT

Nowadays, consumers are interested in food bioproducts with health properties and nutritional attributes. In this sense, the market has been searching for natural bioproducts with biological activities due to the growing demand for organic, sustainable products, highlighting microbial bioproducts. Bacteria, yeasts, and fungi can produce natural biological active products used as food ingredients, enzymes with a wide application in the food industry, and proteins that can be consumed directly or used to enrich food formulations. However, key factors in obtaining these bioproducts are the microorganism's growth system strategy, substrate, and culture conditions. Solid-state fermentation and submerged

The Food-Energy-Water Triangle for Sustainable Development: Integrative and Synergic Solutions.
Pedro Aguilar-Zárate, Mónica Chávez-González, Ann Rose Abraham, and A. K. Haghi (Eds.)

fermentation (SmF) have been extensively explored, and SmF has shown advantages in stability and control parameters such as temperature, pH, nutrient distribution, and oxygen transfer. These parameters, as well as the system configuration, are crucial in process viability and industrial scale-up. Several companies have exploited microorganisms through SmF to obtain various products. Some of the most industrially explored bioproducts are proteins, which have demonstrated stability in food formulation and represent a viable, sustainable alternative to obtaining meat analogs. Nevertheless, one of the main challenges is to increase consumer acceptability and improve the food formulation to satisfy their flavor and sensorial necessities.

2.1 INTRODUCTION

Globalization has impacted the behavior of society, with an increased awareness of the consumption of safe products and high-quality food as well as leading a healthy lifestyle. Consumers, especially middle-aged adults, are interested in bioproducts with properties that play a role in healthy aging and holistic treatments for chronic diseases. Therefore, consumers have been searching for bioproducts with biological activities such as antitumor, anti-obesity, antihypocholesterolemia, anti-Alzheimer disease, antioxidant, and immunosuppressive properties. Thus, there is a growing demand for natural and organic products labeled "clean-sustainable".[1,2]

This approach becomes relevant to mitigate the ecological impacts of the industrial sector on the environment. Thus, there is a need to recover and reuse natural resources, focusing on the circular economy. These proposals have a perspective of future scenarios that lead to a sustainable era, solving the high demand of an increasing population and diminishing diseases due to the poor quality of commercial food products. To this end, microorganisms have been considered promising nutritional food sources and cell factories to obtain bioproducts and biomolecules that can be used in the food industry.[3–5]

Microbial processes stand out because of their low carbon footprint, sustainability, and the use of food waste and by-products to obtain high-valued natural products. Bacteria, yeast, and filamentous fungi have been widely studied to develop novel strategies to assess a circular economy. In this sense, submerged fermentation (SmF) presents several advantages compared to other growth culture strategies, such as solid-state fermentation, including the control of the parameters related to microbial growth, such as pH, temperature, aeration, oxygen, and mass transfer, leading to a stable bioprocess.[6]

Using SmF as a culture system, bacteria have been widely studied for the obtention of natural products and biomolecules. Some of these microorganisms have been used as cell models since their metabolism has been well characterized.[5] Lactic acid bacteria (LAB) are some of the most important bacteria used in the food industry for the fermentation of food and beverages and the synthesis of exopolysaccharides (EPS), flavors, and bioactive compounds.[7]

Conversely, yeasts have shown remarkable applications in the food industry.[8] *Saccharomyces cerevisiae* is the most known and studied yeast and has been used as a cell model factory to obtain proteins, enzymes, amino acids, and other food products.[8, 9] However, non-*Saccharomyces* and nonconventional yeast can be considered as a new yeast generation for the production of high-value bioproducts. Yeasts are promising flavor, cell–protein, and pigment producers that, besides their attributes, confer nutritional and antioxidant properties to food products.[8, 9]

Filamentous fungi have been investigated as readily available sources of various food products that can be integrated into the food formulation or consumed naturally and offer a high content of bioactive compounds. These microorganisms are used as natural dyes and functional compounds in bakery and beverage products. Therefore, different sustainable biotechnological tools are currently being sought to isolate new colors of fungal origin.[10] Other studies have investigated the production of enzymes and mycoprotein, resulting in economically viable food bioproducts.[11–13] Moreover, microalgae have produced important bioproducts such as EPS, pigments, proteins, and animal feed products. In addition, some types of microalgae are consumed as food since their nutritional characteristics and bioactive compounds have attracted the interest of consumers, leading to their commercialization.[14, 15]

In general, microbial food bioproducts are well accepted by the consumers when their formulations, properties, and health benefits are known.

This chapter aims to present the principal bioproducts obtained from microbial sources to lead in an economically and sustainable process by SmF. First, this chapter presents the microorganisms used in the food industry (Section 2.2). The sustainable production of high-value products by SmF is discussed in Section 2.3, and microbial protein is given as an example of a successful bioproduct (Section 2.4). Finally, Section 2.5 presents the conclusions, perspectives, and challenges.

2.2 MICROORGANISMS IN THE FOOD INDUSTRY

2.2.1 BACTERIA

Bacteria are microorganisms that present two growth modalities: unicellular (planktonic), which allows bacterial dispersion, and multicellular, where a biofilm favors bacterial proliferation.[16] Several studies have focused on bacteria as cell factories to obtain natural products and biomolecules. *Escherichia coli* has been one of the most studied bacteria since its metabolism has been well characterized as a model for metabolic engineering to synthesize food bioproducts.[5] *Bacillus subtilis*, a generally recognized as safe (GRAS) bacterial species, has also been intensively studied due to its biotechnological importance, versatility, and environmental adaptation. It can produce proteins, industrial enzymes, vitamins, terpenoids, and amino acids.[17] Numerous investigations have focused on optimizing *B. subtilis* as a bioproduct platform by SmF and solid-state fermentation using sustainable agro–industrial residues.[18]

LAB are some of the most used microorganisms to ferment food and beverages, especially dairy products.[7] However, in recent years, the fermentation of alternative plant-based beverages has been studied to search for LAB strains capable of fermenting cereal and nut-based beverages to obtain yogurt products. In this sense, Hozzein et al.[19] evaluated the fermentation of rice milk supplemented with skim milk (6%) with *Lactobacillus casei*, *Lactobacillus bulgaricus*, *Lactobacillus acidophilus*, and a culture consortium from a commercial yogurt. The results showed that *L. bulgaricus* was an efficient vegetal yogurt starter, reaching a viability of 8.5 Log CFU/mL. Other recent approaches for LAB applications have involved the biosynthesis and bioconversion of polyphenolic compounds because their enzyme complex (reductases and decarboxylases) allows the conversion of hydroxybenzoic and hydroxycinnamic acids.[7]

2.2.2 YEASTS

Yeasts are widely studied eukaryotic organisms because of their remarkable applications in the food industry.[8] *S. cerevisiae*, the most known and studied conventional yeast, has been extensively investigated and used in the food industry because of its GRAS status and versatility for bakery products and fermented beverage processing.[8, 9] This species has also been used as a cell factory to produce proteins, enzymes, amino acids, alcohol, and esters, among other essential products.[8] In addition, other metabolites have been obtained since genetic manipulation tools of this yeast have been investigated to conduct a metabolic pathway to increase a specific product.

Non-*Saccharomyces* yeasts have shown a unique metabolism promoting the release of aromatic compounds that can be used as flavoring ingredients by triggering a sequential hydrolysis pathway that involves several glycosidases.[20, 21] Native non-*Saccharomyces* yeasts have been isolated from the spontaneous fermentation of wine, beer, tequila, mescal, and fermented food, contributing to the synthesis of volatile aromatic compounds that impact the quality and final taste of beverages and food.[20–23]

Adame Soto et al.[21] isolated non-*Saccharomyces* yeasts from *Agave* fermentation, identifying species such as *Clavispora lusitaniae*, *Torulaspora delbrueckii*, *Kluyveromyces dobzhanskii*, *Kluyveromyces* sp., and *Kluyveromyces marxianus*. In addition, other yeasts have been reviewed to elucidate their importance and impact on wine, including *Pichia kluyveri*, *T. delbrueckii*, *Hanseniaspora vineae*, *Schizosaccharomyces pombe*, *Wickerhamomyces anomalus*, *Metschnikowia pulcherrima*, *Lachancea thermotolerans*, and *Candida stellata*.[20]

Nonconventional yeasts have been investigated due to their growth capacity in atypical environments, for example, oil, and industrial waste, among others, or for their propagation at extreme temperature, pressure, and pH conditions. This group also includes dimorphic yeasts such as *Pichia fermentans*, a biofilm-forming yeast, or *Yarrowia lipolytica*, which can shift its morphology from yeast-cell to pseudohyphal and/or hyphal shape, depending on the environmental and nutritional conditions.[9] For example, the yeast *Rhodosporidium kratochvilovae* can grow in oleaginous media to produce carotenoids.[4] *Y. lipolytica* has been widely studied because of its versatile metabolism to produce enzymes (lipases, proteases, and phosphatases), organic acids, and single-cell proteins.[3]

2.2.3 FILAMENTOUS FUNGI

Filamentous fungi are eukaryotic microorganisms that are abundant throughout the world. They are mainly saprophytes with branched hyphae and secrete a wide range of enzymes that can hydrolyze macromolecules and complex organic substrates.[24, 25]

Fungal growth is associated with increased amounts of cellular components leading to biomass production, forming extensive hyphae and filaments. As multicellular organisms, fungi evolve by the extension of individual hyphae located at the tip and biomass synthesis throughout the mycelium. In this sense, their growth can only be determined indirectly by impedimetry and conductimetry, immunological techniques, and the production of ergosterol, Adenosine triphosphate (ATP), pectin esterase, and volatile compounds.[24, 26–30]

Fungi are considered heterotrophs since they cannot produce chlorophyll and because they require macro- and micronutrients that contain sources of carbon and nitrogen as well as minerals and vitamins.[31]

Filamentous fungi can be used to obtain various bioproducts that play a role in human well-being, with advantages such as easy management, large-scale production, and versatile growth conditions.[32]

Fungi have been consumed since ancestral times and contain high amounts of proteins and essential amino acids. Other components of fungi are dietary fibers, polyunsaturated fatty acids, vitamins, L-carnitine, pigments, and enzymes. Therefore, they are of great importance in the food industry.[24, 33] *Aspergillus niger* has been used commercially to produce citric acid for over 100 years,[34] and *Aspergillus awamori* is applied in the production of amylase.[35] Other genera, such as *Penicillium*, *Fusarium*, and *Aspergillus*, can synthesize pigments with a broad spectrum of colors (melanins, flavins, quinines, and phenazines).[11]

2.2.4 MICROALGAE

Microalgae comprise a diverse group of rapidly growing unicellular microorganisms, including 72,500 cataloged species. Currently, there are 16 classes, according to the taxonomic standards, including diatoms (*Bacillariophyceae*), golden algae (*Chrysophyceae*), and green algae (*Crysophyceae*).[14, 15] These microorganisms can be found in saline or fresh water and extreme environments such as volcanic waters, deserts, and acidic and frozen soils.[15, 36] Microalgae have been considered the most efficient photosynthetic microorganisms, converting carbon dioxide into biomass. The photosynthetic efficiency of microalgae is 10–20%, whereas that of plants is 1–2%.[37]

Microalgae biomass is rich in proteins, lipids, and pigments (chlorophylls, carotenoids, and phycobiliproteins). In addition, other compounds have been found in the culture media, such as extracellular carbohydrates, lipids, and volatile compounds. Because of this, microalgae cultivation has gained interest regarding the formulation of microalgae-based food, the obtention of natural bioproducts, and the direct consumption of high-quality microbial protein.[14, 36]

Currently, *Arthrospira*, *Chlorella*, *Aphanizomenon*, and *Dunaliella salina* are the most cultivated microalgae for commercial purposes.[36] However, although microalgae consumption has been recognized as safe, there are only a few species authorized for human consumption, limiting their food application.[37]

2.3 SUSTAINABLE PRODUCTION OF HIGH-VALUE BIOPRODUCTS BY SMF

Microorganisms with relevance in the food industry have been cultivated by SmF according to their growth requirements. Solid-state fermentation has also been evaluated especially for fungi using agro–industrial by-products. However, SmF has shown advantages compared with solid-state fermentation, especially regarding the stability and control of growth parameters such as pH, temperature, and aeration.[6]

In this sense, efficient strategies have been conceptualized to produce bioproducts by SmF in flasks, followed by scaling up to a bioreactor, where the bioreactor design can improve yields and productivity by evaluating the configuration and controlling the culture parameters.[3]

Bacteria and yeast generally grow as cellular biomass in liquid systems. However, some bacteria can grow as biofilm, and yeasts (nonconventional) can present pellet formation, depending on the environmental factors, the carbon source, the viscosity of the medium, and the inoculum concentration.[38–40]

Industrially, multiple ways of SmF are employed to produce a wide variety of bioproducts. Batch culturing is one of the most used and the oldest system used for the fermentation process, followed by feed-batch culture (semi-open system) and continuous systems.[39, 40] Other approaches have involved partitioning bioreactors, where a second immiscible organic phase is added into the cell-containing aqueous phase as a sequester of the produced metabolite. The organic phase in these bioreactors has also been added as part of the bioconversion media.[41] Partitioning bioreactors might present oxygen mass transfer limitations, causing a slow uptake by the microorganism. For these reasons, multiphase partitioning bioreactors have been designed to overcome oxygen limitations and to study the pH, temperature, nature of the partitioning phase, agitation, and aeration for the optimal production of biomolecules and bioproducts.[42]

Microbial growth systems have recently been involved in using food by-products and waste to obtain sustainable bioprocessing strategies by the biochemical conversion of renewable sources to produce biomolecules and bioproducts and eventually generate novel proposals for the circular economy.[8] In this context, SmF has been used to treat industrial waste and agro–industrial food by-products as nonconventional media to obtain microbial biomass and to elucidate sustainable processes to obtain bioproducts (Fig. 2.1). Table 2.1 summarizes some important bioproducts obtained through microbial metabolisms by SmF.

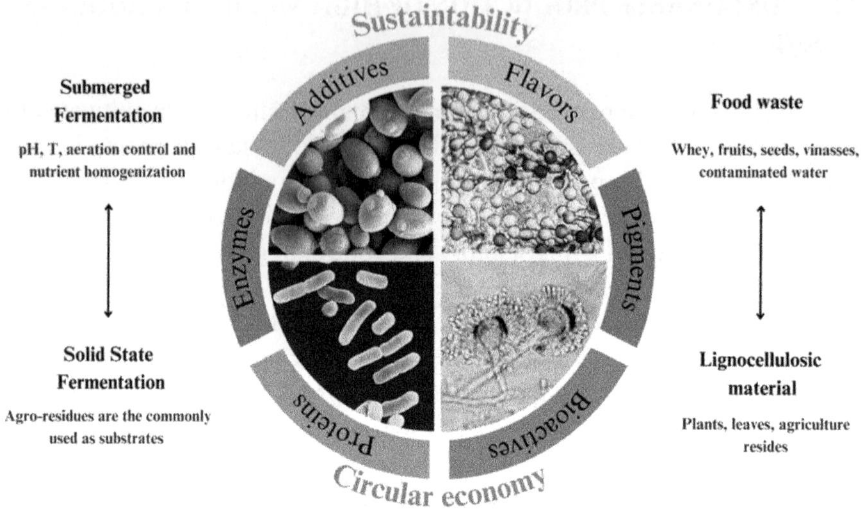

FIGURE 2.1 Microbial bioproducts as potential proposals for a sustainable circular economy.

TABLE 2.1 Sustainable Microbial Bioproducts Obtained via Submerged Fermentation.

Bioproducts	Application	Microorganism	Culture system	Reference
Flavors				
Vanillin	Food additive	*Enterobacter hormaechei*	Batch/250-mL Erlenmeyer flasks, 30°C, pH 5, 130 rpm, 32 h	[44]
		Bacillus safensis SMS1003	Batch/Erlenmeyer flasks, Eugenol bioconversion, 37°C; pH 7.0, 180 rpm	[92]
Isoamyl acetate	Banana flavor	*Pichia fermentans*	Batch/1-L Glass container with 600 mL of media culture, 28°C, 24/liquid–liquid extraction with decane	[45]
2-Phenyletanol	Rose-like flavor antimicrobial compound	*Kluyveromyces marxianus*	Batch/5-L Bioreactor with 3-L of media culture, 30°C pH 5/ membrane-based solvent extraction	[47]

TABLE 2.1 *(Continued)*

Bioproducts	Application	Microorganism	Culture system	Reference
2-Phenylethyl acetate, 2-phenyletanol	Floral and sweet flavors	*Wickerhamomyces anomalus, Candida glabrata, Candida utilis, and Candida parapsilosis*	Batch/Erlenmeyer flask, tequila vinasses, 30°C, 250 rpm	[93]
Isoamyl alcohol, 2-phenylethanol (2-PE), ethyl acetate, isoamyl acetate, 2-phenyl-ethyl acetate	Fruity/floral alcohols, acetate, and ethyl esters	*Kluyveromyces marxianus*	Batch/Erlenmeyer flask, tequila vinasses, 30°C, 250 rpm	[94]
EPS				
Levan	Prebiotic, soluble dietary fiber	*Saccharomyces cerevisiae*	Fed-batch/5-L scale jar fermenter, 30°C, pH 5.5, 42 h	[95]
Dextran	Prebiotics	*Weissella confusa*	Batch/static and dynamic conditions (100 rpm), 33°C, 48 h	[96]
Pullulan	Prebiotic	*Aureobasidium pullulans*	Batch/ 5-L stirred fermenter	[97]
Nondefined EPS	Rheological properties	*Anabaena* sp. *PCC 7120*	Two-stage cultivation	[98]
Pigments				
Carotenoids	Natural orange pigment	*Penicillium* sp. GBPI-P155	Batch/250-mL Erlenmeyer flask, containing 50-mL sterilized broth	[99]
Lutein, fucoxan-thin, β-carotene	Fat-soluble pigments	*Chlorella vulgaris* BIN	Photobioreactor	[100]
Astaxanthin, canthaxanthin, and adonixanthin	Natural pigments	*Chromochloris zofingiensis*	Batch/two-stage cultivation with acetate supplementation	[101]
β-carotene, lycopene	Food grade dye	*Blakeslea trispora* NRRL 2895 and 2896	Batch/250-mL Erlenmeyer flasks, 180 rpm, 4 days	[102]
Orange–red, green, maroon–red, orange–brown, and yellow	Natural pigments	*Penicillium multi-color, P. canescens, P. herquie, Talaromyces verruculosus, Fusarium solani*	Batch/250-mL Erlenmeyer flasks, for 14 days	[103]

TABLE 2.1 *(Continued)*

Bioproducts	Application	Microorganism	Culture system	Reference
Anthraquinonyl carboxylic acid	Food colorant and cosmetic colorant	*Penicillum oxyliums*	Not Defined (Patent)	[104]
Enzymes				
l-asparaginases	Reduce the acrylamide formation in potato chips or french fries, bakery products, and coffee by degradation of l-asparagine	*Aspergillus oryzae* *Rhizopus microsporus* IBBL, and *Trichosporon asahii* IBBLA1	Batch/Semi-pilot scale in bioreactor, working volume of 5.6 L	[75, 105]
Pectinase	Improve the quantity of the extracted orange juice, Efficiency in pectin degradation from grape juice	*Aspergillus niger*	Batch/Shake flasks for 120 h	[106]
Lipase	Food additives in flavor modification, synthesis of esters with significant antioxidant activity.	*Aspergillus niger* ultraviolet (UV)-10	Batch/shake flasks 250 mL	[107]
Xylanase	Improve yield and clarity of pineapple juice	*Aspergillus niger*	Batch/shake flasks for 144 h	[108]
Proteins				
Mycoprotein	Edible fungal biomass	*Aspergillus oryzae*	1200-L airlift bioreactor, working volume of 1200 L, 48 h	[32]
Unicellular protein	Functional ingredient	*Fusarium venenetum* rR372C	Batch/ shake flasks 200 mL and stirred-tank bioreactor (3 L) in the second phase (24–48 h).	[109]

TABLE 2.1 *(Continued)*

Bioproducts	Application	Microorganism	Culture system	Reference
Mycoprotein	Meat analogues	*Agaricus bisporus*	Batch/stirred-tank fermenter with 2.5 L and scale up to 15 L	[110]
Protein-rich biomass	Alternative source of food and feed given the amino acid and fatty acid	*Rhizopus delemar*	Glass bubble column bioreactors, with a working volume of 3, 26, and 1200 L for 48 h	[83]

2.3.1 FLAVORS

Flavors are related to the interaction between volatile and nonvolatile compounds and, thus, with the final quality of food and beverages. The interest in natural flavors and fragrances has increased in the last decade, and in the food industry, they account for 25% of the total food additives.[41]

One of the most representative flavors is vanillin (4-hydroxy-3-methoxy-benzaldehyde), which, for decades, was chemically synthesized. However, in recent years, it has been prohibited as a food additive because of the toxic compounds from the chemical reaction.[43] In this sense, natural and sustainable strategies have been developed, using microorganisms to develop "white" biotechnology processes.[43]

Biovanillin has recently been produced by extracting ferulic acid from banana peels, followed by bacterial bioconversion using *E. hormaechei* in SmF. The major concentration of 3.8 g/L was obtained from 7 g/L of ferulic acid after 24 h, controlling the pH and temperature (pH 7, 30°C), with agitation of 150 rpm.[44]

Esters derived from microbial fermentation are flavoring agents with great importance in the food industry because of their predominantly fruity and floral aroma. The most representative esters in food and beverages include ethyl hexanoate (apple flavor), ethyl acetate (fruit flavor), 2-phenylacetate (floral flavor), and isoamyl acetate (banana flavor).[41]

Shrivastava et al.[9] reviewed applications of *Pichia* as a cell factory, including short-chain flavoring ester production, such as isoamyl acetate. This banana flavor has been produced from L-leucine by *P. fermentans* ITD00165 by SmF of sugar cane molasses coupled to liquid–liquid *in situ* product removal (ISPR) using decane, reaching ≈2 mg/L of isoamyl acetate with 4 g/L of L-leucine addition.[45]

The compound 2-PE, an important flavoring higher alcohol related to the rose aroma, has been widely reported in yeast biosynthesis, using the species *Pichia anomala*, *P. fermentans*, *Pichia pastoris*, *Pichia kudriavzevii*, *Pichia membranaefaciens*, *C. utilis*, *Y. lipolytica*, *Zygosaccharomyces rouxii*, *Kluyveromyces lactis*, and, especially, *S. cerevisiae* and *K. marxianus*.[46] Of these, *K. marxianus* ITD0090 has been studied for the production of 2-PE and 2-phenylethyl acetate, both volatile aromatic compounds related to the rose threshold.[21, 47, 48] Further, 2-PE SmF was studied in conventional batch systems (flask and bioreactor) and coupling an ISPR process by membrane-based solvent extraction. The maximum concentration of 2-PE in the flask was 0.7 g/L, and the concentration could be increased by controlling aeration and agitation parameters (1.49 g/L 0.8 with 0.8 vvm and 350 rpm) and coupling with ISPR (3.0 g/L).[47] Yeast-derived flavors have high potential as natural alternatives to improve food and beverage quality.

2.3.2 EXOPOLYSACCHARIDES

EPSs are biopolymers produced by bacteria, fungi, yeasts, and microalgae.[49] Their biosynthesis and accumulation generally take place after the growth phase of the microorganism, and they are subsequently secreted into the extracellular medium as a capsule, constituting mucous layers that are attached to the cells or in a soluble form, where the polysaccharides are weakly attached to the cell surface.[50]

The EPS obtained has been evaluated in biotechnological processes by fermentation in discontinuous, fed-batch, and continuous-type systems.[51] The characteristics and amounts of EPS can be influenced by intrinsic factors such as the carbon and nitrogen source and by extrinsic factors such as incubation parameters, temperature, and pH.[51] Bacteria are the largest producers of EPS, of which the most representative is xanthan. Fungi and yeasts have also been investigated for the production of EPS such as scleroglucan and pullulan. Microalgae produce EPS with unique structural characteristics that include functional groups such as methyl and sulfate; however, their EPS yield is low compared to that of other microorganisms.[49]

The application of EPS has increased considerably in recent years in various industrial sectors such as the textile and cosmetic industries, agriculture, and the pharmaceutical, biomedical, and food industries. In the food industry, EPSs have been widely used in food formulations as stabilizers, emulsifiers, bio-thickeners, gelling agents, and prebiotics due to their rheological and techno-functional properties.[52] Other applications include biofilm formulations for food packaging, where quality standards need to be fulfilled, aiming to eliminate artificial additives to diminish the environmental impact (Table 2.2). The use of EPS-based materials presents a green technological solution by reducing dependence on fossil resources. Moreover, EPSs are nontoxic, biocompatible, and biodegradable.[53] However, one of the main challenges of EPS-based biopackaging is the mechanical properties and water solubility, characteristics that confer high degradability but limit up-scaling in industrial production.[53] Nevertheless, optimizing EPS production processes and evaluating their sustainability and economic viability are important for large-scale production. Overall, EPSs offer promising opportunities for sustainable solutions in the food industry.

TABLE 2.2 Exopolysaccharide-based Biofilms for Food Conservation.

EPS	Biofilm characteristics	References
Cellulose	Antimicrobial properties for meat packaging	[111]
Nanocellulose	Film for packaging with biosensor for detection of H_2S	[112]
Kefiran	Edible biodegradable film	[113]
Levan/ pullulan	Edible film for food packaging with bioactive properties	[114]
Pullulan	Functional composite films with antimicrobial and UV-barrier properties	[115]
Dextran	Biopackaging for the preservation of edible mushrooms *A. bisporus*	[116]

2.3.3 PIGMENTS

Color is the first and most pleasant characteristic of food products and is associated with the food's taste, safety, and nutritional value.[54] Nature has always been an abundant source of food dyes. However, due to the high demand for pigments, along with high costs, they have been obtained synthetically for years until the World Health Organization (WHO) limited their consumption in 2010 due to health risks and a high environmental impact.[55, 56]

For these reasons, consumers have become more selective regarding food, which has promoted the development of natural alternatives to produce attractive, pleasant, and nutritional food dyes.[57, 58]

Plants, algae, bacteria, filamentous fungi, and yeasts can produce a range of colors, including several chemical classes of pigments such as chlorophyll, curcumin, anthocyanins, betalains, azaphilones, quinones, flavins, and carotenoids.[11, 59, 60]

Carotenoids are natural lipid-soluble compounds widely used in the food industry as safe additives and natural colorants.[8, 61] Of these compounds, β-carotene is the most abundant and stable pigment produced naturally and of great importance because of its health benefits, which include vitamin A precursors in humans as well as anticancer, antioxidant, and anticardiovascular properties.[62] From an economic perspective, β-carotene is the most important carotenoid, followed by lutein and astaxanthin.[61] Some of the main microbial sources for carotenoids are microalgae (Table 2.1). However, nonconventional yeasts have recently been investigated to this end.

The red yeast strains *R. kratochvilovae* FMCC Y-42 and Y-43 have recently been investigated in terms of the production of carotenoids using synthetic media based on glucose and galactose at various carbon/nitrogen (C/N) ratios. Carotenoid production was related to the C/N ratio, reaching 2.59 mg/L of carotenoids on glucose-based media after 39 h of fermentation for the Y-42 strain. Strain Y-43 showed a total carotenoid production of 2 mg/L from glucose and glucose-galactose media.[4]

Rhodotorula glutinis DSM 70398 was evaluated in SmF with olive mill wastewater at 15°C, 150 rpm, and 10 L/min airflow to produce β-carotene, which was extracted using solid–liquid, ultrasound-assisted, and microwave-assisted separation technologies, with organic solvents. The highest β-carotene yield was 45 ± 3 μg/g, obtained using the ultrasound-assisted technique.[62] Yeast strains that can grow and synthesize lipids and carotenoids in agricultural waste and wastewater are promising microorganisms to obtain safe and sustainable products through circular economy proposals.[60]

Fungi produce various natural pigments. For example, *Monascus, Penicillium oxalicum* (natural red), *Aphis gossypii* (riboflavin), and *B. trispora* (β-carotene and antioxidant lycopene) are widely used.[63, 64] *Trypethelium eluteriae* produces a violet–red pigment (naphthoquinone),[65] and violet–red (betacyanin and phenolic compounds) is also produced by *Alternanthera brasiliana*. Moreover, strains of ascomycetous filamentous fungi that are considered potential pigment producers include *Talaromyces*, which produce red pigments, *Cordyceps unilateralis* (intense blood red pigment), *Herpotrichia rhodosticta* (orange), *Curvularia lunata*, and several species of *Drechslera* (diverse pigments); these fungi are promising because they are not mycotoxigenic or pathogenic to humans.[10]

Edible mushrooms traditionally consumed in several countries can synthesize and secrete pigments that are free of mycotoxins.[66] The success of pigments produced by these microorganisms depends on consumer acceptance, capital investment, and regulatory approval.[64] Several commercial pigments of fungal origin are available on the market, and Arpink red™ was the first commercial red dye produced using the fungus *P. oxalicum* var. *arrmeniaca* CCM 8242.[24, 67] The production of microbial pigments can be up-scaled by SmF, meeting the increasing demand.

2.3.4 ENZYMES

Enzymes can be derived from various sources, but microbial enzymes have advantages over those of animal and plant origin. Fungal enzymes are some of the most reported biocatalysts due to their ability to be produced in large quantities, their efficiency in the procurement process, and their high economic viability. The production of fungal enzymes has had an incredible influence due to their wide applications in the industry, environment, and biomedical processes.[68]

In the food industry, biotechnology offers different approaches to reduce production costs, improve gene discovery methods, and decrease enzyme production time by accelerating fungal growth.[69, 70]

Enzymes are present in many foods either as oxidant, inhibitory, or hydrolytic compounds to improve, eliminate, or restore the properties of a producer acting technologically in the food being manufactured. Among the market sectors of enzymes are lipases, pectins, xylanases, β-glucanases phytases, and cellulases.[71, 72]

The application of fungal enzymes in food production offers a promising approach to improve the preservation and shelf life of food without affecting the characteristics of the organoleptic properties and nutritional content of food.[73]

Fungal enzymes produced by SmF used in the food industry include l-asparaginase from *A. oryzae* and *A. niger* (GRAS microorganisms), which is used as a food additive (corn fritters, bakery) to prevent acrylamide formation, without altering the nutritional value.[74, 75]

The fungal genera *Trichoderma* and *Aspergillus* produce lignocellulolytic enzymes (cellulases, xylanases, and pectinases) widely used in the processing of fruit and vegetable juices, accelerating extraction, pressing, and sedimentation and improving color, appearance, flavor, texture, and the availability of bioactive compounds (phytochemical or phenolic) .[76] Protease enzymes produced by *A. Oryzae* DRDFS13MN726447 are used to produce cheese and improve the organoleptic and rheological characteristics of cheese;[77, 78] they can also reduce the allergenic properties of some fermented dairy products. To generate pectinase by *A. niger* AUMC 4156, *Paecilomyces* variety AUMC 4149 and *P. oxalicum* AUMC 4153 are applied, improving the viscosity of purees and for the extraction and clarification of fruit juice. *A. niger* is considered one of the most essential fungi in biotechnological applications.[12]

Commercial enzyme production needs robust investigation in the fermentation conditions to reach a maximum yield and specific enzyme activity. Moreover, further evaluation is necessary to optimize the production and extraction of enzymes when industrial wastes are evaluated.

2.4 MICROBIAL PROTEIN: INNOVATION OF ENVIRONMENTALLY VIABLE FOOD

At the beginning of the last century, microbial protein has been explored to complement the nutrient demands of animals and humans. Yeast obtained from beer production was studied and evaluated for possible applications. The first approaches include the use of yeast cells as protein for animal food supplements in Berlin. This application was that successful that Germany replaced a great number of imported proteins and implemented aerobic SmF, using a medium that included a nitrogen source based on ammonium salt. Later, a fed-batch process was implemented to increase biomass production for feed protein, and around 1936, cells were also used as human dietary supplements.[79]

Nowadays, people prefer the consumption of healthier and animal-friendly sources of protein. Yeast protein has shown several advantages and can be rapidly obtained from low-cost by-products or food industry waste.[80] Therefore, a non-conventional yeast, *Y. lipolytica*, was studied to obtain protein from food waste from an anaerobic digestion reactor of a company that treats food waste. The fermentation process was based on one and two stages, reaching 38.8 ± 0.2% w/w biomass dry weight in the two-stage fermentation. The biotechnological use of *Y. lipolytica* is promising since this yeast has shown a unique metabolism and remarkable ability to grow in various carbon sources and food waste.[81] Moreover, *Y. lipolytica* has been certified by the Food and Drug Administration as a GRAS yeast safe for food use.[82] Other sources of microbial protein are filamentous fungi, which commonly produce mycoprotein, a low-energy GRAS biomolecule.

Mushrooms are ideal for food as they contain high amounts of protein and all essential amino acids. Mycoprotein has been obtained from several fungal strains such as *R. delemar*,[83] *Neurospora crassa*,[84] *Rhizopus oryzae*, *A. oryzae*, *Neurospora intermedia*,[83] and *A. niger*[85] and a wide range of substrates (potato residues, whey, vegetable, and fruit residues) under different conditions (humidity, pH, temperature, size and age of the inoculum). Mycoprotein biomass was obtained via SmF with 60–90% of crude protein, showing a yield and nutritional value significantly higher than that of meat.[83, 86]

Other strains, such as *Fusarium venenatum*, have been recognized by the food industry for their use in large-scale industrial protein production.[33, 87] Such protein has been in high demand in recent years as a more sustainable and healthier alternative to red meat.[83] Recent studies showed that mycoprotein has excellent nutritional value, functional properties, and a meat-like texture, making it a replacement protein source for meat proteins of plant and animal origin. In addition, its consumption can improve the lipid profile since it is low in fat and cholesterol-free, reduces energy intake, and stimulates muscle protein synthesis.[70]

The main products made from mycoprotein already marketed include tofu, tempeh, miso, oncom, red rice, hamanatto, shoyu, sausages, hamburgers, and steaks, replacing chicken and beef. The fungal biomass obtained after fermentation can be used directly as food. It is easily digestible, and the chitinous walls of fungi are sources of dietary fiber.[24, 33, 70] Microbial protein production offers advantages such as scalability, cost-effectiveness, and a reduced environmental impact compared to traditional protein sources.

2.4.1 ROLE IN THE FOOD MARKET

Nowadays, filamentous fungi and yeasts are the most predominant microorganisms used for human consumption. These microorganisms have been used since ancestral times, especially to ferment food and beverages or as microbial protein.[88, 89] The increasing demand for new alternatives to health and nutritional food has favored the emergence of several companies and start-ups that develop microbial food for humans and animal feed, especially microbial protein, which can be obtained as single-cell protein or as a by-product from other bioprocesses by SmF.[90] One of the most significant companies is Marlow Foods Ltd (https://www.quorn.co.uk/), which produces Quorn TM mycoprotein. This industry has different products on the market, including alternative and sustainable meat analogs, for example, sausages, burgers, and steaks that replace chicken and beef.[24] Quorn is classified as a safe and well-tolerated food by the FDA and the UK's Food Standards Agency (FSA). Other companies, such as MycoRena, offer meat substitutes (https://mycorena.com/), Kernel MycoFood uses SmF with *F. venenatum* to obtain mycoprotein (https://www.kernel.bio/), and Innomia Laboratories also offers products such as meat alternatives (https://innomylabs.com/#!/-inicio/).

Lallemand. Inc. has specialized in the development, production, and marketing of yeasts (*Saccharomyces* and *Torula*) and bacteria as single-cell proteins for human consumption (LBI, Lake States®, Engevita™) (https://bio-lallemand.com/). The yeast *S. cerevisiae* yeast has also been cultivated for LeSaffre to obtain yeast-derived products, including proteins such as Lynside® Nutri, Lynside® ProteYn, and related products (Lesaffre Human Care products), as well as yeast-based flavor ingredients (Biospringer products) https://www.ulprospector.com/es/la/Food/Suppliers/4126/Lesaffre-Human-Care.

Microbial protein and its derivates have shown health benefits and nutrient bioavailability. In this sense, developing sustainable food products from microbiological proteins may increase consumption if some barriers are overcome.[90] Consumer acceptance is a crucial factor in expanding this market. However, some factors can affect the rentability of these products, such as lacking information on the microbial sources and their properties. The most controversial sources are fungi as they are commonly associated with molds and their ability to cause food deterioration, produce mycotoxins, or cause allergic reactions. However, another critical factor is the sensory part as food can be perceived as difficult to chew and bitter.[88, 91]

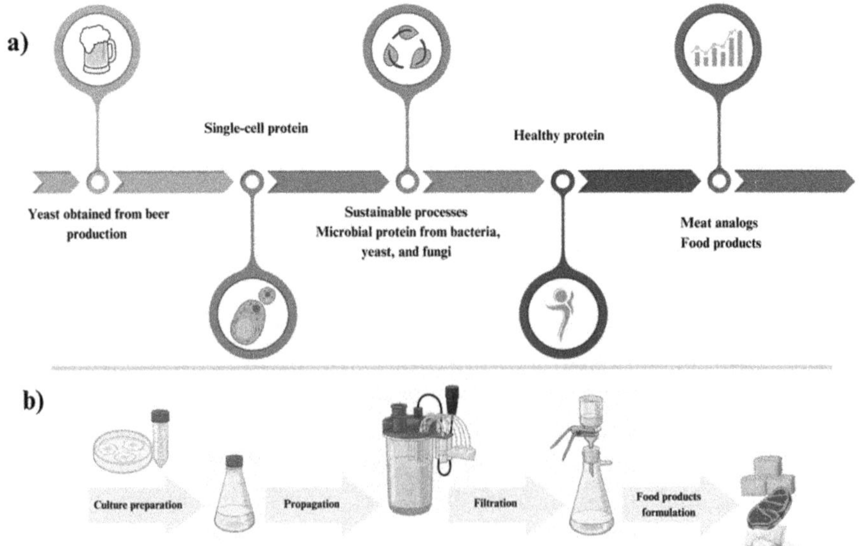

FIGURE 2.2 Microbial protein. (a) Line time of food applications. (b) SmF at laboratory scale to formulate microbial food products.

2.5 CONCLUSIONS AND PERSPECTIVES

The application of microorganisms and SmF in the production of food products can offer numerous advantages, including a diverse range of final products, improved sensory and nutritional properties of food, and the production of enzymes and metabolites useful in the food industry. These new technologies imply challenges and positive expectations that allow the constant and evolutionary development of novel food products, contributing to sustainability in industrial food production.

Among the main challenges to consider for food formulation from microbial sources is the control of food products since substrate composition, fermentation conditions, and contaminants can influence the final product. Another aspect to consider is the scaling process as adapting liquid fermentation processes to an industrial scale can be complex. When the production increases, it is necessary to ensure optimal fermentation conditions, guaranteeing the efficiency and quality of the final product. Finally, consumer acceptance needs to be explored according to the food formulation and bioproduct application.

The perspectives of using SmF for the development of novel bioproducts are related to the increase in the interest of consumers who look for sustainable, eco-friendly, and animal-free food to promote a market based on food product diversification. In addition, with these products, health and nutrition improvements are expected. Moreover, sustainability and waste reduction will develop a circular economy in which by-products and agricultural waste will be converted into food and value-added products.

KEYWORDS

- **cell factories**
- **bioproducts**
- **biomolecules**
- **liquid fermentation**
- **circular economy**

REFERENCES

1. Lin, L.; Xu, J. Production of Fungal Pigments: Molecular Processes and Their Applications. *J. Fungi.* **2022,** *9,* 44.
2. Lin, L.; Xu, J. Fungal Pigments and Their Roles Associated with Human Health. *J. Fungi.* **2020,** *6,* 1–37.
3. Rebello, S.; Abraham, A.; Madhavan, A.; Sindhu, R.; Binod, P.; Bahuleyan, K.; Aneesh, A.; Mathachan, E.; Ashok, P. Non-conventional Yeast Cell Factories for Sustainable Bioprocesses. *FEMS Microbiol. Lett.* **2018,** *365,* 1–10.
4. Sereti, F.; Papadaki, A.; Alexandri, M.; Kachrimanidou, V.; Kopsahelis, N. Exploring the Potential of Novel *R. kratochvilovae* Red Yeasts Towards the Sustainable Synthesis of Natural Carotenoids. *Sustain. Chem. Pharm.* **2023,** *31,* 1–8.
5. Nilam, T.; Tengku, B.; Bin Abas, A. *Escherichia coli* Cell Factory for Synthesis of Biomolecules. *Biomanufacturing Sustain. Prod. Biomol.* **2023,** 1–26.
6. Couto, S. R.; Sanromán, M. Á. Application of Solid-State Fermentation to Food Industry-A Review. *J. Food Eng.* **2006,** *76,* 291–302.
7. Gaur, G.; Gänzle, M. G. Conversion of (poly)phenolic Compounds in Food Fermentations by Lactic Acid Bacteria: Novel Insights Into Metabolic Pathways and Functional Metabolites. *Curr. Res. Food Sci.* **2022,** *6,* 100448.
8. Nandy, S. K.; Srivastava, R. K. A Review on Sustainable Yeast Biotechnological Processes and Applications. *Microbiol. Res.* **2017,** *207,* 83–90.
9. Shrivastava, A.; Pal, M.; Sharma, R. K. Pichia as Yeast Cell Factory for Production of Industrially Important Bio-products: Current Trends, Challenges, and Future Prospects. *J. Bioresour. Bioprod.* **2023,** *8,* 108–124.
10. Gmoser, R.; Sintca, C.; Taherzadeh, M. J.; Lennartsson, P. R. Combining Submerged and Solid State Fermentation to Convert Waste Bread into Protein and Pigment Using the Edible Filamentous Fungus *N. intermedia. Waste Manag.* **2019,** *97,* 63–70.
11. Kalra, R.; Conlan, X. A.; Goel, M. Fungi as a Potential Source of Pigments: Harnessing Filamentous Fungi. *Front. Chem.* **2020,** *8.*
12. Almowallad, S. A.; Aljobair, M. O.; Alkuraieef, A. N.; Aljahani, A. H.; Alsuhaibani, A. M.; Alsayadi, M. M. Utilization of Agro-Industrial Orange Peel and Sugar Beet Pulp Wastes for Fungal Endo- Polygalacturonase Production. *Saudi J. Biol. Sci.* **2022,** *29,* 963–969.
13. Awasthi, M. K.; Harirchi, S.; Sar, T.; Vs, V.; Rajendran, K.; Gómez-García, R.; Hellwig, C.; Binod, P.; Sindhu, R.; Madhavan, A.; Kumar, A. N.; Kumar, V.; Kumar, D.; Zhang, Z.; Taherzadeh, M. J. Myco-biorefinery Approaches for Food Waste Valorization: Present Status and Future Prospects. *Bioresour. Technol.* **2022,** *360,* 127592.
14. Jacob-Lopes, E.; Maroneze, M. M.; Deprá, M. C.; Sartori, R. B.; Dias, R. R.; Zepka, L. Q. Bioactive Food Compounds from Microalgae: An Innovative Framework on Industrial Biorefineries. *Curr. Opin. Food Sci.* **2019,** *25,* 1–7.
15. Tan, K. Y.; Low, S. S.; Manickam, S.; Ma, Z.; Banat, F.; Munawaroh, H. S. H.; Show, P. L. Prospects of Microalgae in Nutraceuticals Production with Nanotechnology Applications. *Food Res. Int.* **2023**. *169,* 112870.
16. Berlanga, M.; Guerrero, R. Living Together in Biofilms: The Microbial Cell Factory and its Biotechnological Implications. *Microb. Cell Fact.* **2016,** *15,* 1–11.
17. Su, Y.; Liu, C.; Fang, H.; Zhang, D. *Bacillus subtilis*: A Universal Cell Factory for Industry, Agriculture, Biomaterials and Medicine. *Microb. Cell Fact.* **2020,** *19,* 1–12.

18. Mishra, S. K.; Bhatt, I.; Paul, P. K. *Bacillus subtilis* Cell Factory. *Biomanufacturing Sustain. Prod. Biomol.* **2023,** 165–173.
19. Hozzein, W. N.; Hisham, S. M.; Alkhalifah, D. H. M. A Sustainable Method: Production of the Fermented Rice Milk Yogurt by Using Three Efficient Lactic Acid Bacteria. *Appl. Sci.* **2023,** *13.*
20. Maicas, S.; Mateo, J. J. The Life of *Saccharomyces* and Non-*Saccharomyces* Yeasts in Drinking Wine. **2023,** *11,* 1178.
21. Adame-Soto, P. J.; Aréchiga-Carvajal, E. T.; López, M. G.; González-Herrera, S. M.; Moreno-Jiménez, M. R.;Urtiz-Estrada, Norma; Rutiaga-Quiñones, O. M. Potential Production of 2-Phenylethanol and 2-phenylethylacetate by Non-*Saccharomyces* Yeasts from *Agave durangensis*. *Ann. Microbiol.* **2019,** *69,* 989–1000.
22. Miguel, G. A.; Carlsen, S.; Arneborg, N.; Saerens, S. M. G.; Laulund, S.; Knudsen, G. M. Non-*Saccharomyces* Yeasts for Beer Production: Insights into Safety Aspects and Considerations. *Int. J. Food Microbiol.* **2022,** *383,* 109951.
23. Segura-García, L. E.; Taillandier, P.; Brandam, C.; Gschaedler, A. Fermentative capacity of *Saccharomyces* and Non-*Saccharomyces* in Agave Juice and Semi-synthetic Medium. *LWT- Food Sci. Technol.* **2015,** *60,* 284–291.
24. Awasthi, M. K.; Kumar, V.; Hellwig, C.; Wikandari, R.; Harirchi, S.; Sar, T.; Wainaina, S.; Sindhu, R.; Binod, P.; Zhang, Z. Taherzadeh, M. J. Filamentous Fungi for Sustainable Vegan Food Production Systems within a Circular Economy: Present Status and Future Prospects. *Food Res. Int.* **2023,** *164,* 112318.
25. Wikandari, R.; Manikharda; Dewanti-Hariyadi, R.; Taherzadeh, M. J. Filamentous Fungi for Food. *Curr. Dev. Biotechnol. Bioeng.* **2023,** 343–397.
26. Meletiadis, J.; Meis, J. F. G. M.; Mouton, J. W.; Verweij, P. E.; Analysis of Growth Characteristics of Filamentous Fungi in Different Nutrient Media. *J. Clin. Microbiol.* **2001,** *39,* 478–484.
27. Osadolor, O. A.; Nair, R. B.; Lennartsson, P. R.; Taherzadeh, M. J. Empirical and Experimental Determination of the Kinetics of Pellet Growth in Filamentous Fungi: A Case Study Using *Neurospora intermedia*. *Biochem. Eng. J.* **2017,** *124,* 115–121.
28. Qin, S.; Wainaina, S.; Awasthi, S. K.; Mahboubi, A.; Liu, T.; Liu, H.; Zhou, Y.; Liu, H.; Zhang, Z.; Taherzadeh, M. J.; Awasthi, M. K. Fungal Dynamics During Anaerobic Digestion of Sewage Sludge Combined with Food Waste at High Organic Loading Rates in Immersed Membrane Bioreactors. *Bioresour. Technol.* **2021,** *335,* 125296.
29. Sawai, J.; Yoshikawa, T. Measurement of Fungi by an Indirect Conductimetric Assay. *Lett. Appl. Microbiol.* **2003,** *37,* 40–44.
30. Wormit, A.; Usadel, B. The Multifaceted Role of Pectin Methylesterase Inhibitors (PMEIs). *Int. J. Mol. Sci.* **2018,** *19,* 2878.
31. Troiano, D.; Orsat, V.; Dumont, M. J. Status of Filamentous Fungi in Integrated Biorefineries. *Renew. Sustain. Energy Rev.* **2020,** *117,* 109472.
32. Harirchi, S.; Rousta, N.; Varjani, S.; Taherzadeh, M. J. Sampling, Preservation, and Growth Monitoring of Filamentous Fungi. *Curr. Dev. Biotechnol. Bioeng.* **2023,** 149–180.
33. Wikandari, R.; Hasniah, N.; Taherzadeh, M. J. The Role of Filamentous Fungi in Advancing the Development of a Sustainable Circular Bioeconomy. *Bioresour. Technol.* **2022,** *345,* 126531.
34. Papagianni, M. Advances in citric acid fermentation by *Aspergillus niger*: Biochemical Aspects, Membrane Transport and Modeling. *Biotechnol. Adv.* **2007,** *25,* 244–263.

35. Saleh, A. A.; Ohtsuka, A.; Yamamoto, M.; Hayashi, K. *Aspergillus awamori* Feeding Modifies Lipid Metabolism in Rats. *Biomed Res. Int.* **2013,** *2013.*
36. Levasseur, W.; Perré, P.; Pozzobon, V. A Review of High Value-added Molecules Production by Microalgae in Light of the Classification. *Biotechnol. Adv.* **2020,** *41,* 107545.
37. Chen, C.; Tang, T.; Shi, Q.; Zhou, Z.; Fan, J. The Potential and Challenge of Microalgae As Promising Future Food Sources. *Trends Food Sci. Technol.* **2022,** *126,* 99–112.
38. Jach, M. E.; Serefko, A.; Ziaja, M.; Kieliszek, M. Yeast Protein as an Easily Accessible Food Source. *Metabolites* **2022,** *12.*
39. Saran, S.; Malaviya, A.; Chaubey, A. Introduction, Scope and Significance of Fermentation Technology. In *High Value Fermentation Products,*John Wiley & Sons, Inc.: Hoboken, NJ, USA, **2019;** 1–25.
40. Singhal, G.; Verma, V.; Bhagyawant, S.; Srivastava, N. Fermentation Technology Prospecting on Bioreactors/Fermenters: Design and Types. In *Principles and Applications of Fermentation Technology,*John Wiley & Sons, Inc.: Hoboken, NJ, USA, **2018**; pp 65–83.
41. Palmerín-Carreño, D. M.; Cordero-Soto, I. N.; Valle-Cervantes, S.; Ríos-Nuñez, L. G.; Prado-Barragán, L. A.; Rutiaga-Quiñones, O. M. Multiphase Bioreactors in the Food Industry: Aroma production, *Adv. Chem. Eng.* **2019,** *54,* 151–193.
42. Raguraman, S.; Chia, W. Y.; Chew, K. W.; Show, P. L. Considerations in Designing A Multiphase Separation System. *Princ. Mult. Sep. Syst.* **2023,** 331–344.
43. Jiang, W.; Chen, X.; Feng, Y.; Sun, J.; Jiang, Y.; Zhang, W.; Xin, F.; Jiang, M. Current Status, Challenges, and Prospects for the Biological Production of Vanillin. *Fermentation* **2023,** *9,* 1–15.
44. Saeed, S.; Ur Rehman Baig, U.; Tayyab, M.; Altaf, I.; Irfan, M.; Raza, S. Q.; Nadeem, F.; Mehmood, T. Valorization of Banana Peels Waste into Biovanillin and Optimization of Process Parameters Using Submerged Fermentation. *Biocatal. Agric. Biotechnol.* **2021,** *36,* 102154.
45. Sánchez-Castañeda, A. K.; Athès, V.; Moussa, M.; López-Miranda, J.; Páez-Lerma, J. B.; Soto-Cruz, N. O.; Trelea, I. C. Modeling of Isoamyl Acetate Production by Fermentation with *Pichia fermentans* in an Aerated System Coupled to In Situ Extraction. *Process Biochem.* **2017,** *65,* 11–20.
46. Mitri, S.; Koubaa, M.; Maroun, R. G.; Rossignol, T.; Nicaud, J. M.; Louka, N. Bioproduction of 2-Phenylethanol through Yeast Fermentation on Synthetic Media and on Agro-Industrial Waste and By-Products: A Review. *Foods* **2022,** *11.*
47. Cordero-Soto, I. N.; Castillo-Araiza, C. O. Rutiaga-Quiñones, O. M.; Moussa, M.; Béal, C.; Gallegos-Infante, A.; Soto-Cruz, O. N.; Ochoa-Martínez, L. A.; Huerta-Ochoa, S. Intensification of 2-phenylethanol Production using an Aerated System Assisted by A Membrane-Based Solvent Extraction Technique. *Rev. Mex. Ing. Química.* **2021,** *20,* 739–750.
48. Cordero-Soto, I. N.; Huerta-Ochoa, S.; Moussa, M.; Ochoa-Martínez, L. A.; Soto-Cruz, N. O.; Rutiaga-Quiñones, O. M. Recent Advances in the Bioconversion of 2-Phenylethanol Through Biotechnological Processes for Using as a Natural Food Additive. In *Food Microbiology and Biotechnology, Safe and Sustainable Food Production*; Nevárez-Moorillón, G. V., Prado-Barragán. A., Martínez-Hernández, J. L., Noé, C. A., Eds.; Apple Academic Press, Inc., **2020**; pp 231–247.

49. Medina-Cabrera, E. V.; Rühmann, B.; Schmid, J.; Sieber, V. Optimization of Growth and EPS Production in two Porphyridum Strains. *Bioresour. Technol. Rep.* **2020,** *11,* 100486.
50. Nwodo, U. U.; Green, E.; Okoh, A. I. Bacterial Exopolysaccharides: Functionality and Prospects. *Int. J. Mol. Sci.* **2012,** *13,* 14002–14015.
51. Freitas, F.; Torres, C. A. V.; Reis, M. A. M. Engineering Aspects of Microbial Exopolysaccharide Production. *Bioresour. Technol.* **2017,** *245,* 1674–1683.
52. Korcz, E.; Varga, L. Exopolysaccharides from Lactic Acid Bacteria: Techno-functional Application in the Food Industry. *Trends Food Sci. Technol.* **2021,** *110,* 375–384.
53. Nešić, A.; Cabrera-Barjas, G.; Dimitrijević-Branković, S.; Davidović, S.; Radovanović, N.; Delattre, C. Prospect of Polysaccharide-based Materials as Advanced Food Packaging. *Molecules.* **2020,** *25.*
54. Spence, C. On the Psychological Impact of Food Colour. *Flavour* **2015,** *4,* 21.
55. Brewster, E. Flavors of Change: Taste Trends Evolve in Pandemic's Wake. *Food Technol.* **2020,** *74,* 53–64.
56. Tuli, H. S.; Chaudhary, P.; Beniwal, V.; Sharma, A. K. Microbial Pigments as Natural Color Sources: Current Trends and Future Perspectives. *J. Food Sci. Technol.* **2015,** *52,* 4669–4678.
57. Gebhardt, B.; Sperl, R.; Carle, R.; Müller-Maatsch, J. Assessing the Sustainability of Natural and Artificial Food Colorants. *J. Clean. Prod.* **2020,** *260,* 120884.
58. Bora, P.; Das, P.; Bhattacharyya, R.; Barooah, M. S.; Saikia Barooah, M. Biocolour: The Natural Way of Colouring Food. *J. Pharmacogn. Phytochem.* **2019,** *8,* 3663–3668.
59. Corrêa, R. C. G.; Garcia, J. A. A.; Correa, V. G.; Vieira, T. F.; Bracht, A; Peralta, R. M. Pigments and Vitamins From Plants as Functional Ingredients: Current Trends and Perspectives. *Adv. Food Nutr. Res.* **2019,** *90,* 259–303.
60. Paul, D.; Kumari, P. K.; Siddiqui, N. Yeast Carotenoids: Cost-Effective Fermentation Strategies for Health Care Applications. *Fermentation* **2023,** *9,* 147.
61. Rapoport, A.; Guzhova, I.; Bernetti, L.; Buzzini, P.; Kieliszek, M.; Kot, A. M. Carotenoids and Some Other Pigments from Fungi and Yeasts. *Metabolites* **2021,** *11,* 1–17.
62. Hladnik, L.; Vicente, F. A.; Košir, A.; Grilc, M.; Likozar, B. Stirred, Ultrasound-assisted and Microwave-assisted Extraction Process of β-carotene from *Rhodotorula glutinis* in Biorefinery Downstream. *Sep. Purif. Technol.* **2022,** 311.
63. Dufossé, L. Red Colourants from Filamentous Fungi: Are They Ready for the Food Industry? *J. Food Compos. Anal.* **2018,** *69,* 156–161, doi: 10.1016/J.JFCA.2017.11.002.
64. Dufossé, L. Production of Bioactive Pigmented Compounds by Filamentous Fungi. *Curr. Dev. Biotechnol. Bioeng.* **2023,** 325–341.
65. Basnet, B. B.; Liu, L.; Zhao, W.; Liu, R.; Ma, K.; Bao, L.; Ren, J.; Wei, X.; Yu, H.; Wei, J.; Liu, H. New 1, 2-naphthoquinone-derived Pigments from the Mycobiont of Lichen *Trypethelium eluteriae* Sprengel. *Nat. Prod. Res.* **2019,** *33,* 2044–2050.
66. Blumenthal, C. Z. Production of toxic metabolites in *Aspergillus niger*, *Aspergillus oryzae*, and *Trichoderma reesei*: Justification of Mycotoxin Testing in Food Grade Enzyme Preparations Derived from the Three Fungi. *Regul. Toxicol. Pharmacol.* **2004,** *39,* 214–228.
67. Caro, Y.; Venkatachalam, M.; Lebeau, J.; Fouillaud, M.; Dufossé, L. Pigments and Colorants from Filamentous Fungi. In *Fungal Metabolites*, Springer International Publishing: Cham, **2017**; pp 499–568.

68. El-Gendi, H.; Saleh, A. K.; Badierah, R.; Redwan, E. M.; El-Maradny, Y. A.; El-Fakharany, E. M. A Comprehensive Insight into Fungal Enzymes: Structure, Classification, and Their Role in Mankind's Challenges. *J. Fungi.* **2021,** *8,* 23.
69. Ouedraogo, J. P.; Tsang, A. Production of Native and Recombinant Enzymes by Fungi for IndustrialApplications. *Encycl. Mycol.* **2021,** 222–232.
70. Karimi, S.; Ferreira, J. A.; Taherzadeh, M. J. The Application of Fungal Biomass as Feed. *Encycl. Mycol.* **2021,** 601–612.
71. Raveendran, S.; Parameswaran, B.; Ummalyma, S. B.; Abraham, A.; Mathew, A. K.; Madhavan, A.; Rebello, S.; Pandey, A. Applications of Microbial Enzymes in Food Industry. *Food Technol. Biotechnol.* **2018,** *56,* 16–30.
72. Toushik, S. H.; Lee, K. T.; Lee, J. S.; Kim, K. S. Functional Applications of Lignocellulolytic Enzymes in the Fruit and Vegetable Processing Industries. *J. Food Sci.* **2017,** *82,* 585–593.
73. Okpara, M. O. Microbial Enzymes and Their Applications in Food Industry: A Mini-Review. *Adv. Enzym. Res.* **2022,** *10,* 23–47.
74. da Cunha, M. C.; dos Santos Aguilar, J. G.; de Melo, R. R.; Nagamatsu, S. T.; Ali, F.; de Castro, R. J. S.; Sato, H. H. Fungal L-asparaginase: Strategies for Production and Food Applications. *Food Res. Int.* **2019,** *126,* 108658.
75. Doriya, K.; Jose, N.; Gowda, M.; Kumar, D. S. Solid-State Fermentation vs Submerged Fermentation for the Production of l-Asparaginase. *Adv. Food Nutr. Res.* **2016,** *78,* 115–135.
76. Florencio, C.; Cunha, F. M.; Badino, A. C.; Farinas, C. S.; Ximenes, E.; Ladisch, M. R. Secretome Analysis of *Trichoderma reesei* and *Aspergillus niger* cultivated by Submerged and Sequential Fermentation Processes: Enzyme Production for Sugarcane Bagasse Hydrolysis. *Enzyme Microb. Technol.* **2016,** *90,* 53–60.
77. Bhat, M. K. Cellulases and Related Enzymes in Biotechnology. *Biotechnol. Adv.* **2000,** *18,* 355–383.
78. Mamo, J.; Getachew, P.; Samuel Kuria, M.; Assefa, F. Application of Milk-Clotting Protease from A*spergillus oryzae* DRDFS13 MN726447 and *Bacillus subtilis* SMDFS 2B MN715837 for Danbo Cheese Production. *J. Food Qual.* **2020,** 2020.
79. Ugalde, U. O.; Castrillo, J. I. Single Cell Proteins from Fungi and Yeasts. *Appl. Mycol. Biotechnol.* **2002,** *2,* 123–149.
80. Ma, J.; Sun, Y.; Meng, D.; Zhou, Z.; Zhang, Y.; Yang, R. Yeast Proteins: The Novel and Sustainable Alternative Protein in Food Applications. *Trends Food Sci. Technol.* **2023,** *135,* 190–201.
81. Yang, R.; Chen, Z.; Hu, P.; Zhang, S.; Luo, G. Two-stage Fermentation Enhanced Single-cell Protein Production by *Yarrowia lipolytica* from food waste. *Bioresour. Technol.* **2022,** *361,* 1–8.
82. Carvalho, A. S. S.; Sales, J. C. S.; Nascimento, F. V. R; Bernardo Dias Souza, C. E.; Conceição de Lemes, A. C.; Coelho, M. A. Z. Lipase Production by *Yarrowia lipolytica* in Solid-State Fermentation Using Amazon Fruit By-Products and Soybean Meal as Substrate. *Catalysts*. **2023,** *13,* 1–15.
83. Sar, T.; Larsson, K.; Fristedt, R.; Undeland, I.; Taherzadeh, M. J. Demo-scale Production of Protein-rich Fungal Biomass from Potato Protein Liquor for use as Innovative Food and Feed Products. *Food Biosci.* **2022,** *47,* 101637.
84. Bartholomai, B. M.; Ruwe, K. M.; Thurston, J.; Jha, P.; Scaife, K.; Simon, R.; Abdelmoteleb, M.; Goodman, R. E.; Farhi, M. Safety Evaluation of *Neurospora crassa*

Mycoprotein for use as a Novel Meat Alternative and Enhancer. *Food Chem. Toxicol.* **2022,** *168,* 113342.

85. Oshoma, C.; Eguakun-Owie, S. Conversion of Food Waste to Single Cell Protein using *Aspergillus Niger*. *J. Appl. Sci. Environ. Manag.* **2018,** *22,* 350.
86. Molfetta, M.; Morais, E. G.; Barreira, L.; Bruno, G. L.; Porcelli, F.; Dugat-Bony, E.; Bonnarme, P.; Minervini, F. Protein Sources Alternative to Meat: State of the Art and Involvement of Fermentation. *Foods*. **2022,** 11.
87. Gilani, G. S.; Lee, N. Sources of Food-grade Protein. *Encycl. Food Sci. Nutr.* **2003,** 4873–4879.
88. Derbyshire, E. J.; Delange, J. Fungal Protein – What Is It and What Is the Health Evidence? A Systematic Review Focusing on Mycoprotein. *Front. Sustain. Food Syst.* **2021,** 5.
89. Ritala, A.; Häkkinen, S. T.; Toivari, M.; Wiebe, M. G. Single Cell Protein-state-of-the-art, Industrial Landscape and Patents 2001–2016. *Front. Microbiol.* **2017,** 8.
90. Graham, A. E.; Ledesma-Amaro, R. The Microbial Food Revolution. *Nat. Commun.* **2023,** *14,* 2231.
91. Ahmad, M. I.; Farooq, S.; Alhamoud, Y.; Li, C.; Zhang, H. A Review on Mycoprotein: History, Nutritional Composition, Production Methods, and Health Benefits. *Trends Food Sci. Technol.* **2022,** *121,* 14–29.
92. Singh, A.; Mukhopadhyay, K.; Ghosh Sachan, S. Biotransformation of Eugenol to Vanillin by a Novel Strain *Bacillus safensis* SMS1003. *Biocatal. Biotransformation.* **2019,** *37,* 291–303.
93. Rodríguez-Romero, J. J.; Aceves-Lara, C. A.; Silva, C. F.; Gschaedler, A.; Amaya-Delgado, L.; Arrizon, J. 2-Phenylethanol and 2-phenylethylacetate Production by Nonconventional Yeasts Using Tequila Vinasses as a Substrate. *Biotechnol. Rep.* **2020,** *25.*
94. Adame-Soto, P. J.; Aréchiga-Carvajal, E. T.; González-Herrera, S. M.; Moreno-Jiménez, M. R.; Rutiaga-Quiñones, O. M. Characterization of Mating Type on Aroma Production and Metabolic Properties Wild *Kluyveromyces marxianus* Yeasts. *World J. Microbiol. Biotechnol.* **2023,** *39,* 1–17.
95. Ko, H.; Bae, J. H.; Sung, B. H.; Kim, M. J.; Kim, C. H.; Oh, B. R.; Sohn, J. H. Efficient Production of Levan Using a Recombinant Yeast *Saccharomyces cerevisiae* Hypersecreting A Bacterial Levansucrase. *J. Ind. Microbiol. Biotechnol.* **2019,** *46,* 1611–1620.
96. Rosca, I.; Petrovici, A. R.; Peptanariu, D.; Nicolescu, A.; Dodi, G.; Avadanei, M.; Ivanov, I. C.; Bostanaru, A. C.; Mares, M.; Ciolacu, D. Biosynthesis of Dextran by *Weissella confusa* and its In Vitro Functional Characteristics. *Int. J. Biol. Macromol.* **2018,** *107,* 1765–1772.
97. Wang, D.; Ju, X.; Zhou, D.; Wei, G. Efficient Production of Pullulan Using Rice Hull Hydrolysate by Adaptive Laboratory Evolution of *Aureobasidium pullulans*. *Bioresour. Technol.* **2014,** *164.*
98. Roy, K. S.; Bannerjee, S.; Hazra, T.; Das, D.; Pandit, S.; Lahiri, D.; Nag, M.; Ray, R. R; Sarkar, T.; Moovendhan, M.; Kavisri, M. Exopolysaccharide Production by *Anabaena sp.* PCC 7120: Physicochemical Parameter Optimization and Two-Stage Cultivation Strategy to Maximize the Product Yield. *Biomass Convers. Biorefinery*. **2023,** 1–13.

99. Pandey, N.; Jain, R.; Pandey, A.; Tamta, S. Optimisation and Characterisation of the Orange Pigment Produced by a Cold Adapted Strain of *Penicillium sp.* (GBPI_P155) Isolated from Mountain Ecosystem. *Mycology*. **2018,** *9,* 81–92.
100. Bazarnova, J.; Smyatskaya, Y.; Shlykova, A.; Balabaev, A.; Đurović, S. Obtaining Fat-Soluble Pigments-Carotenoids from the Biomass of Chlorella Microalgae. *Appl. Sci.* **2022,** *12.*
101. Minyuk, G.; Sidorov, R.; Solovchenko, A. Effect of Nitrogen Source on the Growth, Lipid, and Valuable Carotenoid Production in the Green Microalga *Chromochloris zofingiensis*. *J. Appl. Phycol.* **2020,** *32,* 923–935.
102. Choudhari, S. M.; Ananthanarayan, L.; Singhal, R. S. Use of Metabolic Stimulators and Inhibitors for Enhanced Production of β-carotene and Lycopene by *Blakeslea trispora* NRRL 2895 and 2896. *Bioresour. Technol.* **2008,** *99,* 3166–3173.
103. Molelekoa, T. B. J.; Regnier, T.; da Silva, L. S.; Augustyn, W. Production of Pigments by Filamentous Fungi Cultured on Agro-industrial by-products Using Submerged and Solid-state Fermentation Methods. *Fermentation* **2021,** *7.*
104. Sardaryan, E. Strain of the Microorganism *Penicillium oxalicum* var. *Armeniaca* and its Application. **2018**.
105. Ashok, A.; Kumar, D. S. Laboratory Scale Bioreactor Studies on the Production of l-Asparaginase Using *Rhizopus microsporus* IBBL-2 and *Trichosporon asahii* IBBLA1. *Biocatal. Agric. Biotechnol.* **2021,** *34,* 102041.
106. Ahmed, I.; Zia, M. A.; Hussain, M. A.; Akram, Z.; Naveed, M. T.; Nowrouzi, A. Bioprocessing of Citrus Waste Peel for Induced Pectinase Production by *Aspergillus niger*; Its Purification and Characterization. *J. Radiat. Res. Appl. Sci.* **2016,** *9,* 148–154.
107. Mahadik, N. D.; Bastawde, K. B.; Puntambekar, U. S.; Khire, J. M.; Gokhale, D. V. Production of Acidic Lipase by a Mutant of *Aspergillus niger* NCIM 1207 in Submerged Fermentation. *Process Biochem.* **2004,** *39,* 2031–2034.
108. Hmida-Sayari, A.; Taktek, S.; Elgharbi, F.; Bejar, S. Biochemical Characterization, Cloning and Molecular Modeling of a Detergent and Organic Solvent-Stable Family 11 xylanase from the Newly Isolated *Aspergillus niger* US368 Strain. *Process Biochem.* **2012,** 1839–1847.
109. Fatemeh, S.; Reihani, S.; Khosravi-Darani, K. Mycoprotein Production from Date Waste Using *Fusarium venenatum* in a Submerged Culture. *Appl. Food Biotechnol.* **2018,** *5,* 243–252.
110. Kim, K.; Choi, B.; Lee, I.; Lee, H.; Kwon, S.; Oh, K.; Kim, A. Y. Bioproduction of Mushroom Mycelium of *Agaricus bisporus* by Commercial Submerged Fermentation for the Production of Meat Analogue. *J. Sci. Food Agric.* **2011,** *91,* 1561–1568.
111. Rasouli, Y.; Moradi, M.; Tajik, H.; Molaei, R. Fabrication of anti-Listeria Film Based on Bacterial Cellulose and *Lactobacillus sakei*-Derived Bioactive Metabolites; Application in Meat Packaging. *Food Biosci.* **2021,** *42,* 1–9.
112. Sukhavattanakul, P.; Manuspiya, H. Fabrication of Hybrid Thin Film Based on Bacterial Cellulose Nanocrystals and Metal Nanoparticles with Hydrogen Sulfide Gas Sensor Ability. *Carbohydr. Polym.* **2020,** *230,* 115566.
113. Ghasemlou, M.; Khodaiyan, F.; Oromiehie, A.; Yarmand, M. S. Development and Characterisation of a New Biodegradable Edible Film Made from Kefiran, An Exopolysaccharide Obtained From Kefir Grains. *Food Chem.* **2011,** *127,* 1496–1502.

114. Gan, L.; Jiang, G.; Yang, Y.; Zheng, B.; Zhang, S.; Li, X.; Tian, Y.; Peng, B. Development and Characterization of Levan/pullulan/chitosan Edible Films Enriched with ε-polylysine for Active Food Packaging. *Food Chem.* **2022,** *388,* 1–8.
115. Roy, S.; Rhim, J. W. Fabrication of Copper Sulfide Nanoparticles and Limonene Incorporated Pullulan/carrageenan-based Film with Improved Mechanical and Antibacterial Properties. *Polymers* **2020,** *12,* 1–14.
116. Díaz-Montes, E.; Yáñez-Fernández, J.; Castro-Muñoz, R. Dextran/chitosan Blend Film Fabrication for Bio-packaging of Mushrooms (*Agaricus bisporus*). *J. Food Process. Preserv.* **2021,** *45,* 1–12.

CHAPTER 3

Pomace from Pome Fruits as a Source of Functional Foods and Extracts with Nutraceutical Potential

SALAS-RAMÍREZ, CARLOS ALONSO[1], GONZÁLEZ-LAREDO[1], RUBÉN FRANCISCO[1], ROCHA-GUZMÁN[1], NURIA ELIZABETH[1], GALLEGOS-INFANTE[1], JOSÉ ALBERTO[1], HERRERA-ROCHA[1], KAREN MARLENNE[1], GONZÁLEZ-MERCADO[1], MANUEL EFRAÍN[2], and MORENO-JIMÉNEZ[1], MARTHA ROCÍO[1]

[1]*Grupo de Investigación en Laboratorio Nacional CONAHCYT de Apoyo a la Evaluación de Productos Bióticos (LaNAEPBi), Unidad de Servicio, Tecnológico Nacional de México/ I.T. de Durango (TecNM/ITD), Dgo., México*

[2]*Grupo de Investigación en Ciencias del Movimiento Humano, Universidad Juárez del Estado de Durango Durango, Dgo., México*

ABSTRACT

The trend toward increasing the consumption of fruits is associated with the promising health effects derived from their consumption. Fruits are consumed fresh and most of the population consumes them as juices, juice concentrates, and other products obtained from them, due to their accessibility and easy preparation. During this processing, subproducts are generated from the harvest to the processing of the products in the juice industries, promoting the disposal of residues with nutritional and functional potential. This chapter addresses the functional potential of the compounds present in pomaces of apple (*Malus domestica*), pear (***Pyrus vulgaris*)**, and quince (*Cydonia oblonga* Mill.). The conventional and non-conventional methods

The Food-Energy-Water Triangle for Sustainable Development: Integrative and Synergic Solutions.
Pedro Aguilar-Zárate, Mónica Chávez-González, Ann Rose Abraham, and A. K. Haghi (Eds.)

to extract bioactive compounds of interest for prevention or treatment of different pathologies are also commented.

3.1 INTRODUCTION

Food waste is an issue of main interest to the world, currently, it is derived mainly for reasons of inequality in the production, distribution, and waste of food between developed and developing countries. It is estimated that according to the Food and Agriculture Organization of the United Nations,[1] that developed countries waste is equivalent to 95–115 kg/year per inhabitant, while in countries with a lower degree of development, it amounts to 6–11 kg/year.

Food is lost or wasted along the entire food chain, from initial agricultural production to final consumption in households. Food loss and waste also account for 8–10% of global greenhouse gas emissions, contributing to an unstable climate and extreme weather events such as droughts and floods. These changes negatively impact crop yields, potentially reduce the nutritional quality of crops and cause disruptions in the supply chain.[2]

Fruits represent 15–30% of the total food waste, depending on the type of consumption. In countries with a lower degree of development, the consumption of unprocessed (fresh) fruits is approximately 95%. Meanwhile, in Latin America, it is around 50%, and in countries with higher incomes (North America and Europe), the estimates amount to less than 40%,[1] which indicates that processing is key to the production of waste and its potential use in the West. It has been described that in the world 50% of the harvested fruit is destined for the production of juices and concentrates to prepare instant juices.[3] Within the juices highlights the growing demand for the juice of pome fruits such as apples, pears, and quinces, which are members of the Rosaceae family, subfamily *Pomoideae*, and whose fruits have a core with several small seeds, surrounded by a hard membrane, with an edible layer of pulp.[4] From these fruits, the juice extracted from the apple fruit stands out, since it is currently the second most popular only behind the orange juice.[5]

Subproducts from the juice production of pome fruits include peels, seeds, and pulp. The apple bagasse or pomace is the main residue of the apple extraction process, representing up to 30% of the original fruit.[6] In the case of quince, the extraction of juice is marginal, focusing its use to produce fermented foods and liqueurs, while the whole fruit's main use is for the production of jellies and candies.[7]

Residues of fruit, such as skin, bagasse, leaves, or seeds, are rich in diverse phytochemical compounds, both nutrients and non-nutrients, that potentially may be used by the food and health industry. The list includes fibers (e.g., cellulose, hemicellulose, pectins, etc.), as well as bioactive compounds of mainly phenolic nature (e.g., phenolic acids, flavonoids, condensed tannins, and hydrolyzable tannins) (Fig. 3.1). Particularly, polyphenols are secondary products of plants metabolism. Many of these metabolites have proven human health benefits, highlighting their antioxidant, anti-inflammatory, antimicrobial, immunomodulatory and even anticancer activities. Therefore, it is necessary to look for strategies that allow the optimal extraction of these compounds of interest with the minimum use of resources.

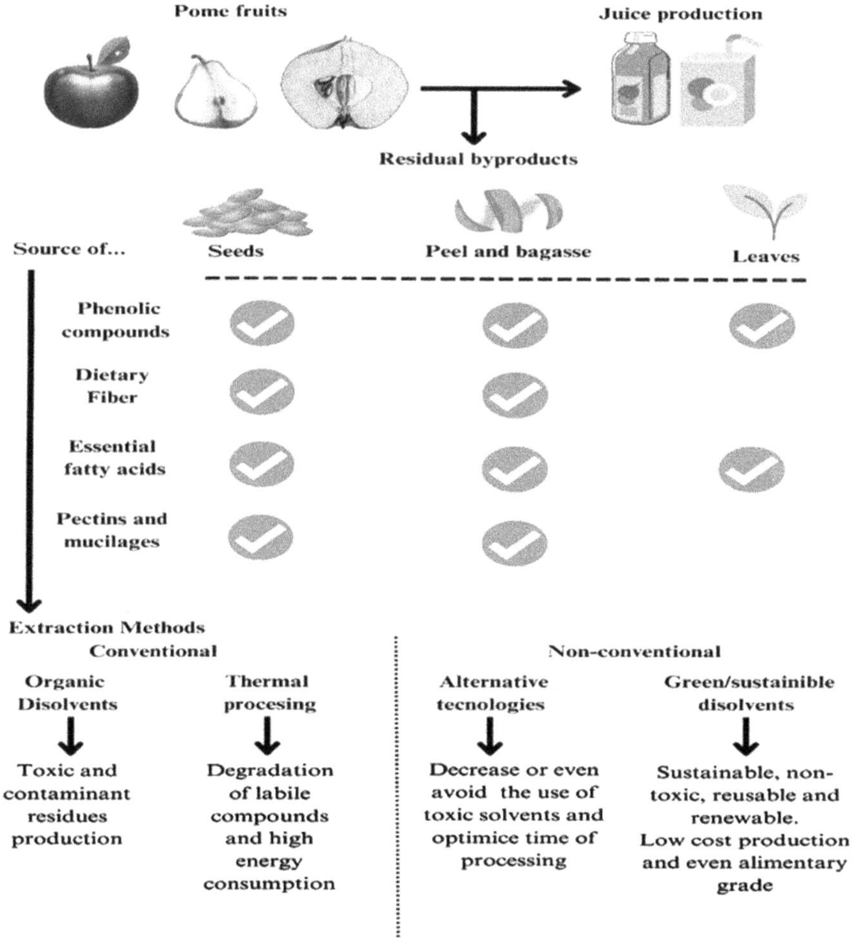

FIGURE 3.1 Potential use of agro-industrial residues from the production of pomaces of fruit juices as a source of bioactive compounds from a sustainable perspective.

3.2 JUICE INDUSTRY AND TRENDS IN THE CONSUMPTION OF POME FRUIT JUICES

Fruit juices are non-fermented beverages obtained from mechanical squeezing or macerating fruits. In 2022, the global juice market reached an economic value of USD 141 billion, with a production volume of 67.6 billion liters,[8] while the market value of the global fruit industry will reach USD 189.9 billion by 2028.[9] Among the main expectations of consumers are associated with gaining benefits from such fruit juices, principally with products free of added sugars and preservatives.[8]

Nowadays, the population considers food as a key process to promote a healthy condition. This is why consumers are demanding the generation of healthy beverages. During the COVID-19 pandemic, the consumption of fruit and vegetable juices increased worldwide, generating a growing demand for innovative products based on juices with functional ingredients such as Omega-3, fiber, bioactive compounds, vitamins, and probiotic content to support the immune system and prevent diseases. This has created opportunity for the juice industry to generate different types of products to satisfy these needs.

The fruit juice of greater consumption and production in the world is orange juice.[10] During recent years, the consumption of pome fruit juice, especially from apples has been relevant. In the global apple juice market, revenues in the segment are expected to amount 17,910 billion dollars in 2023, as well as an annual growth rate of around 3.80% .[11]

It has been reported that 25–30% of the harvested apple is used for the production of processed products, with apple juice concentrate (65%) as the main product.[12] Apple juice processing produces significant amounts of waste, representing 25 and 40% of the processed mass of apples, depending on the juice extraction method.[13]

3.3 ENVIRONMENTAL IMPACT OF FRUIT JUICE PRODUCTION: WATER FOOTPRINT, CARBON FOOTPRINT, AND BY-PRODUCT PRODUCTION

The production of fruit juice implies a significant consumption of water and energy. In the fruit juice sector, the water consumption derived from agricultural practices represents 70%, followed by the industrial sector with 20% of water consumption.[14, 15] Importantly, the demand for water is different if the juice is produced in the facility or imported as a juice concentrate. In this sense, a much lower demand for water and energy is established for the latter case. However, the water footprint has been reported to vary from 0.6 to 1.48

L for each liter of juice produced.[16] As for power consumption, it depends on the processing times and temperatures. In juice bottling facilities, energy consumption comes mainly from juice pasteurization, bottle cleaning, and refrigeration. As an average value, 0.71 MJ per liter of juice is required.[16]

Another important aspect to consider is the generation of food waste by the fruit juice industry during the processing phase. Fruit and vegetable processing operations produce waste or by-products, the main ones being leaves, seeds, bagasse, and peels.

3.4 SUBPRODUCTS OF FRUITS WITH FUNCTIONAL AND NUTRACEUTICAL POTENTIAL

The production of agro-industrial waste, in addition to the environmental impact associated with its decomposition and production of greenhouse gases, represents the loss of valuable constituents in the biological material. This includes nutrients and compounds with the potential to become valuable resources for the food and pharmaceutical industry. Among the compounds of biological importance are those with the potential to be used as ingredients for the preparation of functional foods, isolation of compounds with biological activity for the formulation of nutraceuticals, as well as antibiotics, enzymes, or biopolymers with multiple purposes (Table 3.1).

TABLE 3.1 Nutraceutical Potential of By-products Obtained from Pome Fruits.

Nutraceutical compound	Biological effect	References
Quinic acid	Decrease inflammation process by inhibiting the transcription of kappa nuclear factor β (NF-κβ)	[17]
	Stimulation of tryptophan and nicotidamide synthesis with effects on intestinal inflammation through inhibition of NF-κβ	[18]
	Modulation of intracellular Ca2+ homeostasis, improving glucose-stimulated insulin secretion.	[19]
Chlorogenic acid	Antimicrobial activity against pathogenic bacterial strains (*Shigella dysenteria* and *S. pneumoniae*)	[20]
	Decreased concentration of triglycerides and circulating cholesterol in plasma, as well as a decrease in triglycerides in liver tissue in rats	[21]
	Decrease body weight and total body fat by decreasing the accumulation of lipids at the hepatic level	[22]
	Potential cardiovascular health benefits associated with improved endothelial function, bioavailability of nitric oxide (NO) with decreased blood pressure	[23]

TABLE 3.1 *(Continued)*

Nutraceutical compound	Biological effect	References
Caffeic acid	Increased apoptotic morphological changes at cancer cell lines HeLa and ME-180	[24]
	Caffeic acid mimics the actions of antiestrogens and modifies major growth regulatory pathways ER/cyclin D1 and IGF-IR/pAkt resulting in impaired cell cycle progression and reduced cellular proliferation	[25]
	Doses of 50 and 100 μm suspended the migration rate of breast cancer MDA-MB-231 cells	[26]
Shikimic acid	Antiplatelet effect on PAC-1 and CD62P expression in addition to lowering monocyte-platelet aggregate formation	[27]
	Decreased oxidative stress and protect against kidney injury in cisplatin-treated mice by reducing serum creatinine levels.	[28]
Gallic acid	Tumor-suppressive effect in non-small-cell lung cancer by triggering the activation of p53	[29]
	Up-regulated pro-apoptotic genes such as TNF-α, TP53BP2, and GADD45A resulted in cell death in human oral squamous carcinoma cell lines (HOSCC)	[30]
	Ameliorate high glucose levels, insulin, inflammation	[31]
Protocatechuic acid	Influence on the activity of inducible isoenzyme of cyclooxygenase and NO synthase, cell cycle regulating and inflammatory cytokines involved in oncogenesis	[32]
	Reduced the incidence of liver lesions, including hepatocyte swelling, leukocyte infiltration, and necrosis	[33]
Benzoic acid derivatives	Oral administration of *p*-Hydroxybenzaldehyde (40 mg/kg) significantly ameliorated DSS-induced colitis. Oral administration of hydroxybenzoic acid (10–40 mg/kg) attenuated various colitis phenotypes in DSS-treated mice	[34]
	Hydroxybenzoic acid acts as mitochondriotropic antioxidants and cholinesterase (ChE) inhibitors, having potential benefits for the treatment and prevention of Alzheimer's disease	[35]
	4-hydroxybenzoic acid (4-HBA) exhibited an antihyperglycemic effect, having the potential as a dual agonist on PPARc and GPR40 receptors	[36]
Vanillic acid	Attenuated streptozotocin-induced body weight loss and hyperglycemia, along with improved lipid profile and HbA1c, without significant alteration of serum insulin levels	[37]
	Decreased the β-amyloid protein (Aβ) effects on learning and memory by suppressing oxidative stress and can be regarded as a neuroprotective substance in Alzheimer's disease	[38]
Quercetin-3-O-glucuronide	Modulates insulin resistance through the gut-brain axis in Alzheimer's disease	[39]
	Neurogenesis promotion	[40]

TABLE 3.1 *(Continued)*

Nutraceutical compound	Biological effect	References
	Sleep quantity and quality enhancing activity via the GABAergic pathway	[41]
Quercetin	Toxicity in kidney and tumor tissues	[42]
	Cell cycle arrest and mitochondria-mediated apoptosis on human cervical cancer	[43]
	Anticancer effects in acute myeloid leukemia (AML) cells	[44]
	Induced caspase-8, caspase-9, and caspase-3 activation, PARP cleavage, mitochondrial membrane depolarization, induced intratumoral oxidative stress	
	Anticancer effects prostate cancer through ER Stress Induction and ROS Production	[45]
Kaempferol-3-O-glucuronide	Anti-inflammatory effect *in vitro* suppresses the release of IL-6, IL-1β, IL-18 and TNF-α	[46]
	Decreased the levels of inflammatory biomarkers (C-reactive protein (CRP), IL-6, and TNF-α)	[47]
	Reduced the TNF-α and IL-6 levels	[48]
Rutin	Neuroprotective effect	[49, 50]
	Antimicrobial activity against pathogens (*Staphylococcus aureus*, *Pseudomonas aeruginosa*, and *Bacillus cereus*)	
Catechin	Promotes cell cycle arrest and apoptosis	[51]
	Inhibition of nuclear factor-κB (NF-κB)	[52]
	Inhibit the proliferation of cervical cancer cells	[53]
	Induced apoptosis by downregulating in cutaneous squamous cell carcinoma	[54]
Epicatechin	Ameliorating neuroinflammation, regulating mitochondria function, enhancing synaptic plasticity, and reducing neuronal by AMPK	[55]
	Antimicrobial activity	[56]
Luteolin	Regulation of glycolipid metabolism and antioxidation processes	[57]
	Neuroprotective, anti-inflammatory, and cardioprotective effects	[58]
Apigenina	Antimutation, anti-inflammatory, antioxidation, and anticancer effects	[59]
	Antitumor effects by inhibiting EGFR signaling	[60]
Naringenin	Lipid lowering and insulin	[61]
	Attenuating inflammation downregulating the NLRP3/NF-κB pathway	[62]
	Inhibits cancer progression by apoptosis induction, cell cycle arrest, Wnt/β-catenin, PI3K/Akt, NF-κβ, and TGF-β pathways.	[63]
Acaetin	Reduces cardiac hypertrophy by reducing oxidation, inflammation, and apoptosis.	[64]
	Antitumor effects on gastric cancer by targeting EGFR	[65]

TABLE 3.1 *(Continued)*

Nutraceutical compound	Biological effect	References
Phloridzin	Modulating intestinal inflammation decreased the synthesis of PGE2 and IL-8	[66]
	Antimicrobial and anti-inflammatory activity	[67]
	Antiobesity effect, attenuates obesity, inflammation, and hyperglycemia	[68]
	Hepatoprotective reduction of mitigation of oxidative stress, inflammation, and apoptosis	[69]
Mangiferin	Antineuroinflammatory effect, suppressing NF-κB and NLRP3 signaling pathway	[70]
	Prostate protection reduces pro-tumor cytokines	[71]
	Anti-diabetes anti-hyperlipidemia reduces blood glucose levels and total plasma cholesterol	[72]
Quercitrin	Antioxidant *in vitro* activity (ABTS, DPPH)	[73]
Phloretin	Anticarcinogenic activity, modulation of metastasis and inflammation in cancer cells	[74, 75, 76]
Ursolic acid	Anti-inflammatory activity and decrease in xylene-induced edema generation	[77]
Betulinic acid	*In vitro* antioxidant activity (DPPH y FRAP)	[78]
Oleic and linoleic acids	Decreased levels of triglycerides and cholesterol in circulation	[79, 80]
Polyunsaturated fatty acids (PUFAs)	Antioxidant activity and regulation of membrane stability in biological systems	[81]

Some authors indicate that parts of fruits and vegetables such as stems, leaves, peels, pulps, seeds, shells, bran, and roots, as well as the residues obtained after the juice extraction, are potential sources of compounds with biological activity in high quantities.[82] These results are related to the adequate extraction process of the products of interest. Therefore, it is important to determine the appropriate technologies that guarantee higher yields without causing significant environmental impacts.[83]

The residues are mainly composed of seeds, skin, bark, and bagasse. These contain potentially valuable bioactive compounds such as carotenoids, polyphenols, dietary fibers, vitamins, enzymes and oils, among others.[84] These phytochemicals can be used by the food industry for the development of functional or enriched foods and the health industry by generating nutraceutical products, among others.[85] For example, subproducts such as bagasse contain significant amounts of dietary fiber (e.g., pectin, hemicelluloses, and cellulose) that can be recycled into healthy food or feed.[86, 87] This is especially relevant considering the high content of bagasse from pome fruits. For example, in apple juice production, the waste can be 10 to 30% of the weight of the fruit and these wastes are composed mainly of peel, stems, seeds, calyx, and core.[88]

3.5 POME FRUITS OF NUTRITIONAL IMPORTANCE AND SOURCE IN FOOD INDUSTRY

Apple, pear, and quince are fruits of the Rosaceae family. They are consumed in various parts of the world for their high nutritional value, highlighting the content of fiber, vitamins, and polyphenolic content (Fig. 3.2). Polyphenolic compounds are of great interest because they have been linked to various health benefits. These can generate effects in different pathologies by modulating physiological processes related to oxidative stress, cell proliferation, enzymatic activities, ways.

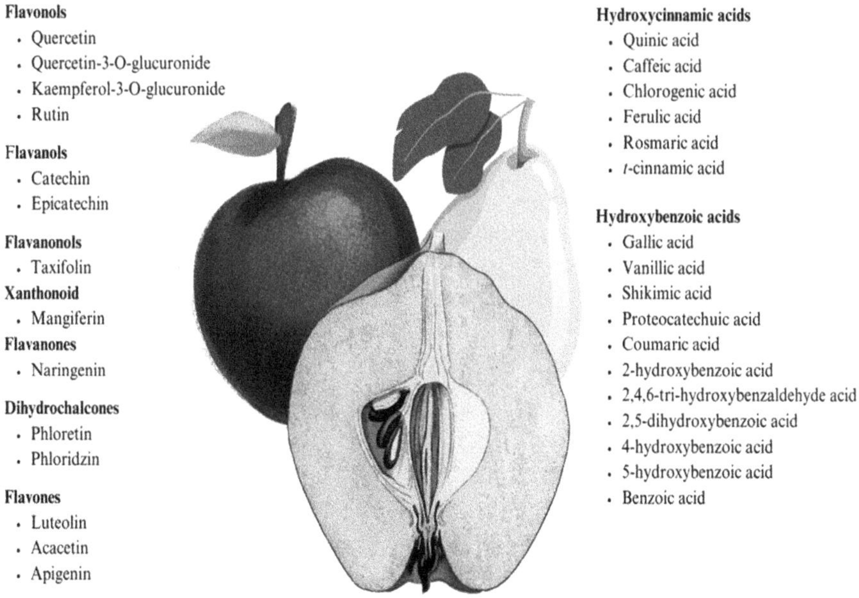

FIGURE 3.2 Polyphenolic compounds found in pome fruits (apple, quince, and pear).

3.5.1 *APPLE (MALUS DOMESTICA)*

Fruit of the apple tree (*Malus domestica* Borkh.) is a food source that is considered nutritive and beneficial to human health. Chemical profiling of phytochemical compounds consists of micronutrients, as well as physiologically active compounds, including vitamins, minerals, organic acids, sugars, and polyphenolic compounds.[89, 90]

It is reported that the content of polyphenolic compounds in apple is very similar between the species, only their concentration differs.[91] Apples contain different phenolic predominating mainly for their high content of acids (caffeic and coumaric acid), flavan-3-ols (epicatechin and procyanidins), and flavonols (quercetin, kaempferol, and rutin), and dihydrochalcones (phloretin and phloridzin). Specifically in the leaves are 3-hydroxyfloridzine, chloridzine, quercetin-3-O-arabinoside, and rutin.[92] While in the fruits are epicatechin, procyanidin B2, chlorogenic acid, phloridzin, caffeic acid, and quercetin. The dihydrochalcones have also been found in root bark and shoots.[93] On the other hand apple bagasse, a product of the apple juice, cider, jam, and vinegar processing industry, has compounds such as chlorogenic acid and gallic acid, and quercetin (and its derivatives), catechins and procyanidins.[94–96]

Vegetable oils from oilseeds and fruit seeds can be obtained using Soxhlet extraction using extraction solvents such as n-hexane or petroleum ether. Arain et al.[97] and Adebayo et al.[98] used n-hexane, obtaining profiles rich in linoleic acid (C18:2), oleic acid (C18:1), palmitic acid (C16:0), and stearic acid (C18:0) in apple seeds. While Matthäus and Özcan[97] used petroleum ether for the extraction of apple seed oil. Apple seed oil contains mainly oleic acid (about 51% of total fatty acids), followed by linoleic acid (about 29% of total fatty acids) and palmitic acid (about 10% of total fatty acids), also finding other compounds with biological activity such as tocopherols and tocotrienols.

3.5.2 PEAR (PYRUS VULGARIS)

The pear (*Pyrus vulgaris*) is a fruit native to Asia and has been cultivated in Europe and America for centuries.[99] It is recognized for its sweet and juicy taste, and its nutritional value, being rich in essential nutrients such as vitamin C, fiber, potassium, calcium, and iron, as well as for its fiber content and phenolic compounds. Thus, it has been associated with potential beneficial effects on health, including cardiovascular benefits.[100]

The content of polyphenolic compounds is distributed in all parts of the plant. Establishing the highest concentration in the skin, mainly chlorogenic acid, quercetin, and catechin. The pulp contains ferulic acid and rutin, among others, while the seeds contain procyanidins, flavonoid oligomers and ellagic acid.[78]

Regarding the extractions of fat-soluble or lipid compounds, Yukui et al.[101] indicated that seeds of *Pyrus communis* contain 179 g of fat-soluble compounds/kg of seeds (17.9%), using light petroleum as the extraction solvent, and extraction of saturated, monounsaturated, and PUFAs was mainly determined. Matthäus and Özcan[79] produced 31.7 g of pear seed oil when using petroleum ether as a solvent, in these extracts the fatty acid profile was established, determining that its composition is mainly oleic acid and linoleic acid.

3.5.3 QUINCE (CYDONIA OBLONGA)

Persian quince is a plant also known as *Cydonia vulgaris* Persoon or *Cydonia oblonga* Miller. This is a monotypic genus comprising the family Rosaceae, subfamily Maloideae and genus *Cydonia*. It grows as a multi-stemmed shrub or small tree and has pubescent to tomentose shoots, petioles, leaves, and fruits.[102] The edible portion of the fruit exceeds 90.6% and has a humidity of around 84.6%, an average pectin content of 1.8% and fiber content of 1.6%, minerals such as potassium (248.0 ± 0.02 mg/100 g), phosphorus (26.0 ± 0.04 mg/100 g), sodium (8.0 ± 0.03 mg/100 g), and calcium (18.0 ± 0.02 mg/100 g).[103] Due to sensory characteristics, such as hardness, the acidity of its pulp, and the sensation of astringency in the mouth after chewing, it is considered unattractive and feasible to consume as a fresh fruit.[104]

Traditional consumption involves different types of processing, leading to thermally processed presentations such as jams[81, 105] and jellies,[104] as well as in beverages products of mechanical extraction such as juices[106] and nectars.[107] Although thermal processing favors its acceptability and increases the concentration per gram of phenolic compounds, changes such as the increase in the content of total soluble solids derived from the addition of sugars, as well as the loss of thermolabile vitamins such as vitamin C, make these forms of consumption not entirely healthy.[108]

According to Fattouch et al. (2008), chlorogenic acid (5-O-caffeoylquinic acid) is the most abundant phenolic compound in the pulp (37%), while rutin (quercetin 3-O-rutinoside) is the most abundant compound in the peel (36%). Herrera-Rocha et al.[109] reported a special abundance of hydroxycinnamic acids (quinic acid and chlorogenic acid), flavonols (procyanidin B2), and flavonols (rutin) in the whole fruit.

Quince leaves are an important source of polyphenol compounds. From Portuguese, quince leaves have been isolated with organic acids such as oxalic, citric, malic, quinic, shikimic, and fumaric acids. Likewise, different polyphenols as caffeoylquinic acids (3,4,5-O-caffeoylquinic acids, 3,5-O-dicaffeoylquinic acid) and flavonoids (quercetin-3-O-galactoside, quercetin-3-O-rutinoside, kaempferol-3-O-glucoside, kaempferol-3-O-glucoside and kaempferol-3-orutinoside).[35,101] Rutin (36%) was the most abundant flavonoid found in its leaves.[111]

3.6 TECHNIQUES FOR OBTAINING BIOACTIVE COMPOUNDS

Extraction is a key process in the processing of food matrices, including fruit and vegetable waste to obtain compounds of interest for their biological activity. The objective of these methods is to maintain a balance between the quality of the target compound, the efficiency of the extraction process, minimum costs, and respect for the environment. There are several methods for the extraction of bioactive compounds, conventional and unconventional. They have different characteristics such as an extraction efficiency, technological feasibility, reproducibility, and conservation of bioactivity of the extracted compounds.[112]

3.6.1 CONVENTIONAL TECHNIQUES EXTRACTION

The methods for obtaining bioactive compounds with biological activity are those based on conventional extraction with organic solvents such as methanol, acetone, ethyl acetate, and their corresponding aqueous mixtures.[113] Conventional techniques for the extraction of bioactive compounds are maceration in conjunction with Soxhlet extraction and hydrodistillation. The extraction efficiency depends on operating parameters including agitation, temperature, nature of the matrix, and chemical structure of the bioactive compounds of interest.[114]

Soxhlet extraction is suitable when the matrix has limited solubility in the solvent.[115] Therefore, solvent selection is a critical aspect of extraction efficiency, and aspects such as analyte solubility and polarity must be considered.[116] In this sense, the use of traditional solvents in this process has several disadvantages, among them, are high amounts of organic solvents, selectivity problems, and the co-extraction of unwanted interferences.[117] In addition, the use of some solvents has restrictions to extract food and pharmaceutical products among these are methanol and chloroform.[118]

One form of maceration is Solid-Liquid Extraction (SLE) a simple and low-cost technique to extract compounds from a solid or semisolid matrix.[119] At this technique, the solvent is mixed during a specific time, waiting for the migration of the solute from the matrix to the solvent.[120] The disadvantage is that it is carried out over long periods of time and a great amount of organic solvents, mass transfer problems, and temperature conditions can alter the molecules of interest.[121] Therefore, different factors must be considered, among them polarity and solubility of bioactive compounds and solvents, the relationship between them, the sample size, temperature, pH, and pressure.[116]

Phenolic compounds are easily extracted by polar solvents derived from the presence of hydroxyl groups. Ethanol-water mixtures are economical, widely used, and allowed in the food industry.[122] In the extraction of other compounds with hydrophobic nature and limited solubility in water (e.g., carotenoids), the extraction process requires the use of organic solvents (e.g., methanol, acetone, hexane, petroleum ether, ethyl acetate). Therefore, it is required a careful separation and recovery of these solvents due to their toxicity and damage to the environment.[123]

Sowa et al.[124] performed extractions of *Malus domestica* leaves with methanol, ethyl acetate, and water extracts, determining that the highest polyphenolic content was obtained with ethyl acetate and methanol. They extracted mainly phloridzin, quercetin glycosides, and phenolic acids such as p-hydroxybenzoic and chlorogenic acids, which were present in traces.

Phytochemical analyses of quince seeds show a differential extraction profile of polyphenolic compounds. The ethanolic extracts show the presence of tannins, glycosides, and phenolic compounds.[125] While the use of methanol promoted the extraction of luteolin and apigenin. On the other hand, the use of methanol extracted a higher content of polyphenols in leaves and seeds, unlike the fruit.[126]

In summary, conventional extraction techniques are widely used due to their low costs, simple methodology, and ease of extraction from large solid samples. However, the long extraction time, the amount of solvents, and the possible affectation of the compounds by temperature make these techniques not environmentally friendly and are only used as reference techniques.

3.6.2 GREEN SOLVENTS FOR ISOLATION AND PURIFICATION OF COMPOUNDS

Green solvents are claimed as alternatives to the use of traditional solvents by characteristics such as (1) low toxicity; (2) high biodegradability; and

(3) sustainable in environmental terms. An example of a green solvent is ethyl lactate, which is completely biodegradable and naturally present in various foods. It is miscible with both water and hydrophobic solvents and is capable of extracting compounds in a wide range of polarities, that makes it a promising candidate for obtaining extracts rich in compounds with potential biological activity.[127] In addition, it is one of the most important bio-based solvents, it can be obtained from the fermentation of carbohydrate raw materials such as corn and soybean waste. Recent studies have demonstrated the effectiveness of ethyl lactate in the extraction of alkaloids,[128] curcuminoids[129], and especially carotenoids.[130] It is considered GRAS (Generally Recognized as Safe) and due to its low toxicity, it was approved by the United States Food and Drug Administration (FDA) and the European Food Safety Authority (EFSA) as a pharmaceutical and food additive.[131] Renewable, environmentally friendly, and biocompatible choline chloride-based natural deep eutectic solvents (NADES) are considered potential alternatives to conventional solvents in the extraction of phenolic compounds and other important natural products.[132]

Studies by Patil et al.[133] who tested nine NADES based on choline chloride, organic acids, alcohols, and sugars developed to study the extraction of curcuminoids from *Curcuma longa* L. In that study, they found that deep eutectic solvents based on choline chloride and lactic acid in a ratio of 1:1 provided the maximum extraction yield. In addition, the extraction based on deep eutectic solvents was intensified by ultrasound assistance obtaining a curcuminoid yield of 77.13 mg/g.

3.6.3 UNCONVENTIONAL TECHNIQUES EXTRACTION

In the use of alternative technologies for extraction, there are benefits, including shorter processing times, lower solvent consumption, lower energy demands, lower processing temperatures, and even the reduction of risks at work.[134] Several emerging technologies have been tested successfully and are promising options for the valorization of pome fruit baggases. Among them are ultrasound-assisted extraction (UAE), the supercritical fluid extraction (SFE), and the enzyme-assisted extraction (EAE).

3.6.3.1 UAE

Ultrasound (US) is an electromagnetic radiation, whose frequency range is between 18 and 100 kHz. It is considered as high power ultrasound when the intensity used (greater than 1 $W \cdot cm^{-2}$) usually ranges between 10 and 1000 $W \cdot cm^{-2}$. The use of high-power and low-frequency ultrasound (f = 20 to 100 kHz) is considered power ultrasound, according to Paniwnyk (2017). The application of ultrasound in such ranges causes cavitation, which makes it an effective application in the food industry.

The benefits of US extraction are attributed to acoustic cavitations. These are local microbubbles created in a liquid phase by subjecting a sample to ultrasound, which grow and oscillate rapidly before collapsing due to internal pressure changes. The collapse of the membranes of biological materials causes violent implosions, which allow fragmentation or interruption of the surface of the solid matrix, improving mass transfer and accelerating the diffusion of its contents to the solvent.[135, 136] Although UAEs show several advantages compared to conventional extractions, there are some aspects to consider to optimize sustainable techniques for obtaining compounds of interest. The cavitation process associated with the use of ultrasound is affected by the physical properties of the solvent used, among them viscosity, surface tension, and saturation vapor pressure.[137] Also, the temperature generated during ultrasonically assisted extractions can damage thermal labile compounds, including phenolic compounds.[135]

Generally, UAE requires less use of solvents, and the use of toxic solvents such as hexane is still quite common today, particularly for the extraction of nonpolar compounds (e.g., curcuminoids, terpenoids). Therefore, a possible application to solve this complication is the use of green solvents (e.g., deep eutectic solvents).[138]

Qui and Gia[139] reported an extraction of oil from apple seeds performing its extraction of fatty acids compared to the conventional Soxhlet method, finding a total of eight different fatty acids, of which 89.39% were unsaturated, making this extraction method efficient and profitable compared to the reference method. In another study by Belwal et al.,[140] to optimize the extraction of anthocyanins from pear shell UAE was used. They used ethanol 57% and trifluoroacetic acid 3% as solvents, determining improvements in the extraction and yields of anthocyanins compared to conventional methods, especially when the processing was carried out using batch runs and not continuously.

3.6.3.2 SFES

Supercritical fluids (SCFs) represent an alternative to reduce the use of organic solvents.[141] By definition, a fluid is considered to be in its critical state when heated above its critical temperature (Tc) and pressurized above its critical pressure (Pc).[142] SFE of solids is divided into two main steps: extraction, and separation of the solute from the solvent. To perform SFE, the fluid has to be carried in its supercritical state. Once it has reached the desired pressure and temperature, the fluid is filtered into the extractor, with an upward or downward flow. Then, the fluid extracts the solute contained in the matrix. Finally, the separation of the solute from the SCF will be achieved in the separator, where the SCF will turn into gaseous state, and the solute, once no longer solubilized in the SCF will be separated by gravity.[134]

The relatively low viscosity and high diffusivity of SCFs improve diffusion and mass transfer, thereby reducing extraction time.[143] SCFs have similar density to that of a liquid, which gives them a solvation capacity similar to traditional solvents.[144] Low temperature processing ensures the preservation of thermally labile compounds, superior quality, and better functionality of recovered compounds.[145] Previous reports by Uwineza and Waśkiewicz[143] suggest a temperature ranging between 35 and 60°C, as well as a pressure of 400 bar, to allow the extraction of thermolabile compounds at good yields and preserving functionality of the compounds of interest.

To perform extraction with SCFs, supercritical CO_2 (SC–CO_2) is the most used fluid for these purposes as part of the process.[146] The implementation of the process is facilitated due to its low critical conditions (Tc: 31°C, Pc:7.38 MPa). In addition, some advantages include non-flammability, low cost, abundance, and volatility at atmospheric pressure, which implies that after depressurization the extracts are solvent-free. SC–CO_2 is a non-polar solvent whose solvent power is between pentane and toluene.[147]

While this makes CO_2 extraction a SCF very good for the recovery of non-polar or low-polarity compounds, it is necessary to mention that the compatibility with other organic solvents is low, which makes the extraction of medium to high polarity compounds complicated. The addition of cosolvents, such as water, ethanol, and methanol, improves the solubilizing power of carbon dioxide and allows the extraction of more polar compounds.[148] According to Picot-Allain et al.[138] the use of ethanol as well as water as cosolvents for extraction by SCFs are more attractive for use from a sustainability perspective and care for the environment and health as they are not as toxic as methanol, hexane, and chloroform.

To optimize the extraction of fatty acids from quince seeds, Daneshvand et al.[149] made a comparison of the extraction performed with SCFs, using CO_2 as a fluid, compared to an UAE, finding higher yields without significant changes in the fatty acid profile, which consisted mainly of linoleic, palmitic, oleic, stearic, and eicosanoic acids. In apple, Ferrentino et al.[150] compared the extraction of apple seed oils using as a control the traditional method of Soxhlet with hexane versus an extraction with SCFs using CO_2 as a solvent. They achieved similar yields to those obtained by Soxhlet, and higher content of phenolic compounds and monounsaturated fatty acids, conferring better functional characteristics explored through *in vitro* antioxidant tests compared to the extract obtained by the traditional method.

Comparative studies were also carried out to extract and improve hydroxytyrosol yields from olive leaves with prepared eight NADES based on sugars and four based on organic acids. Within their experimental results, they showed that the eutectic mixture with citric acid, glycine and water (2:1:1,v:v) reached values of 87 ppm of hydroxytyrosol. It was more than four times the values obtained with water, the conventional solvent used for the extractions of such polyphenols. Thus, it could be an interesting option for its application in obtaining compounds of interest in pome fruit residues.[151]

3.6.3.3 EAE

The principle of enzymatic technology is the use of enzymes as catalysts to hydrolyze the plant cell wall components through the binding of the desired substrate to the active site of the enzyme. This interaction occurs when the enzyme changes its shape to fit the substrate on its active site, thus increasing the enzyme–substrate interaction, which leads to the disruption of cell wall bonds.[152] This technique can be combined with other unconventional methods, and depending on the enzyme selection to hydrolyze the cell wall components of the vegetal matrix, will allow the release of bioactive compounds. This technique involves relevant parameters such as reaction temperature, pH, substrate particle size, enzyme concentration, extraction time, and solvent type.[153]

Among the environmental advantages associated with enzymatic extractions, these methods are usually aqueous based, avoiding the use of toxic organic solvents, low energy demand, reduced reaction times, and use of milder temperatures. Additionally, enzymes are able to accurately extract specific biomolecules of interest. Depending on the operational conditions, purer extracts with minimal contaminants and even allowing for the separation of bioactive compounds without changing their biochemical structures and biological properties can be achieved.[154]

An example of this is the findings of Meini et al.,[155] who optimized the extraction of phenolic compounds from Syrah grape pomace. The enzymatic treatment enhanced the extraction yield of phenolics by up to 66% and its antioxidant capacity by up to 80%. In the same way, Wu et al.[156] performed ultrasound extraction using pectinases, cellulases, and papain to obtain phenolic compounds in broccoli, getting higher yields compared to conventional solvent extraction and improving their antioxidant capacities. Likewise, Li et al.[157] performed the extraction optimization of phenolic compounds of *Trapa quadrispinosa* using ultrasound-assisted cellulases using ethanol as solvent (30%), obtaining the best extraction yields at low temperatures (49°C) for short periods of time (25.5 min), and preserving the beneficial effects such as antioxidant and antitumor activity against Hela, HepG-2 and U251 cells.

3.7 CONCLUDING REMARKS

Pome fruits such as apple, pear, and quince are commonly used mainly to produce juices. Wastes or by-products generated through this manufacturing process are rich sources of bioactive compounds, especially polyphenols, and fiber, with potential benefits for human health. Conventional methods for obtaining these compounds are usually unfriendly to the environment, either by the use of polluting solvents or degradation of bioactive by high temperatures and longer processing times. Therefore, the use of alternative methods such as green solvents and extraction assistance technologies as well as their combined use, are promising procedures for obtaining these compounds of interest with a lower environmental impact.

KEYWORDS

- **pome fruits**
- **by-products**
- **green extraction**
- **bioactive compounds**
- **polyphenols**

REFERENCES

1. FAO. Pérdidas y Desperdicio De Alimentos en el Mundo – Alcance. Causas y Prevención, **2012**.
2. FAO. Hacer Frente a la Pérdida y el Desperdicio De Alimentos: Una Oportunidad de Ganar Por Partida Triple. In *Organización de las Naciones Unidas para la Agricultura y la Alimentación,* **2022**; pp 1–6.
3. Pacheco, M. T.; Moreno, F. J.; Villamiel, M. Chemical and Physicochemical Characterization of Orange By-Products Derived from Industry. *J. Sci. Food Agric.* **2019,** *99* (2), 868–876. https://doi.org/10.1002/jsfa.925
4. Department of Primary Industries Australia. Apples, Pears and Other Pome Fruit. **2019,** p 57. https://www.dpi.nsw.gov.au/agriculture/horticulture/pomes
5. USDA. Fresh Deciduous Fruit (Apples, Grapes, & Pears): World Markets and Trade. *USDA* **2014,** 9.
6. Coman, V.; Teleky, B. E.; Mitrea, L.; Martău, G. A.; Szabo, K.; Călinoiu, L. F.; Vodnar, D. C. Bioactive Potential of Fruit and Vegetable Wastes. *In Adv. Food Nutrit. Res.* **2020,** *91*, 157–225. https://doi.org/10.1016/bs.afnr.2019.07.001
7. Terrazas-Gómez, M. I.; Uranga-Valencia, L. P.; Armendáriz-Mendoza, M.; Palacios-Monarrez, A.; Villarreal-Ramírez, V. H. Desarrollo De Mercado De Productos Derivados Del Membrillo (Cydonia oblonga) en la Ciudad De Delicias, Chihuahua. *Revista Biológico Agropecuaria Tuxpan* **2022,** *10* (1), 48–55.
8. Goldstein Market Intelligence. Global Fruit Juice Market Outlook 2024. **2020** https://www.goldsteinresearch.com
9. IMARC, G. Fruit Juice Market: Global Industry Trends, Share, Size, Growth, Opportunity and Forecast 2020–2025. **2019** https://www.researchandmarkets.com/reports/5546890/fruit-juice market-global-industry-trends%0Ahttps://www.imarcgroup.com/fruit-juice-manufacturing-plant
10. Euromonitor. The Global Juice Market Report; Euromonitor International: London, UK, 2022; **2022**.
11. Statista. *Apple Juice-Worldwide*. Consumer Market Insights, 2023. https://www.statista.com/outlook/cmo/non-alcoholic-drinks/juices/apple-juice/worldwide
12. Bhushan, S.; Kalia, K.; Sharma, M.; Singh, B.; Ahuja, P. S. Processing of Apple Pomace for Bioactive Molecules. *Crit. Rev. Biotechnol.* **2008,** *28* (4), 285–296. https://doi.org/10.1080/07388550802368895
13. Catană, M.; Catană, L.; Iorga, E.; Lazăr, M.-A.; Lazăr, A.-G.; Teodorescu, R. I.; Asănică, A. C.; Belc, N. Achieving of Functional Ingredient From Apple Wastes Resulting From the Apple Juice Industry. *AgroLife Sci. J.* **2018,** *7* (1), 9–17.
14. Leal-Filho,W. Climate Change and the Sustainable Use of Water Resources; Climate Change Management; Springer: Berlin/Heidelberg, Germany, **2012**. ISBN: 978-3642222658
15. Martins, V. W. B.; Rampasso, I. S.; Siltori, P. F. S.; Cazeri, G. T.; Anholon, V.; Quelhas, O. L. G.; Leal Filho,W. Contributions from the Brazilian Industrial Sector to Sustainable Development. *J. Clean. Prod.* **2020,** *272,* 122762. https://doi.org/10.1016/j.jclepro.2020.122762
16. Walker, C.; Beretta, C.; Sanjuán, N.; Hellweg, S. Calculating the Energy and Water use in Food Processing and Assessing The Resulting Impacts. *Int. J. Life Cycle Assess* **2018,** *23,* 824–839. https://doi.org/10.1007/s11367-017-1327-6

17. Zeng, K.; Thompson, K. E.; Yates, C. R.; Miller, D. D. Synthesis and Biological Evaluation of Quinic Acid Derivatives as Anti-inflammatory Agents. *Bioorg. Med. Chem. Lett.* **2009,** *19* (18), 5458–5460. https://doi.org/10.1016/j.bmcl.2009.07.096
18. Pero, R. W.; Lund, H.; Leanderson, T. Antioxidant Metabolism Induced by Quinic Acid. Increased Urinary Excretion of Tryptophan and Nicotinamide. *Phytother. Res.* **2009,** *23* (1), 335–346. https://doi.org/10.1002/ptr.2628
19. Heikkilä, E.; Hermant, A.; Thevenet, J.; Bermont, F.; Kulkarni, S. S.; Ratajczak, J.; Santo-Domingo, J.; Dioum, E. H.; Canto, C.; Barron, D.; Wiederkehr, A.; De Marchi, U. The Plant Product Quinic Acid Activates Ca2+-dependent Mitochondrial Function and Promotes Insulin Secretion from Pancreatic Beta Cells. *Br. J. Pharmacol.* **2019,** *176* (17), 3250–3263. https://doi.org/10.1111/bph.14757
20. Lou, Z.; Wang, H.; Zhu, S.; Ma, C.; Wang, Z. Antibacterial Activity and Mechanism of Action of Chlorogenic Acid. *J. Food Sci.* **2011,** *76* (6). https://doi.org/10.1111/j.1750-3841.2011.02213.x
21. Rodriguez de Sotillo, D. V.; Hadley, M. Chlorogenic Acid Modifies Plasma and Liver Concentrations of: Cholesterol, Triacylglycerol, and Minerals in (fa/fa) Zucker Rats. *J. Nutrit. Biochem.* **2002,** *13* (12), 717–726. https://doi.org/10.1016/S0955-2863(02)00231-0
22. Shimoda, H.; Seki, E.; Aitani, M. Inhibitory Effect of Green Coffee Bean Extract on Fat Accumulation and Body Weight Gain in Mice. *BMC Complement. Alternat. Med.* **2006,** *6,* 1–9. https://doi.org/10.1186/1472-6882-6-9
23. Bondonno, C. P.; Mubarak, A.; Liu, A. H.; Considine, M. J.; Mas, E.; Croft, K. D. Acute Effects of Chlorogenic acid on Nitric Oxide Status, Endothelial Function and Blood Pressure in Healthy Volunteers: A Randomised Trial. *Free Radical Biol. Med.* **2012,** *53,* S191–S192. https://doi.org/10.1016/j.freeradbiomed.2012.08.402
24. Kanimozhi, G.; Prasad, N. R. Anticancer Effect of Caffeic Acid on Human Cervical Cancer Cells. *Coffee in Health and Disease Prevention.* Elsevier Inc, **2015**. https://doi.org/10.1016/B978-0-12-409517-5.00073-5
25. Rosendahl, A. H.; Perks, C. M.; Zeng, L.; Markkula, A.; Simonsson, M.; Rose, C.; Ingvar, C.; Holly, J. M. P.; Jernström, H. Caffeine and Caffeic Acid Inhibit Growth and Modify Estrogen Receptor and Insulin-like Growth Factor i Receptor Levels in Human Breast Cancer. *Clin. Cancer Res.* **2015,** *21* (8). https://doi.org/10.1158/1078-0432.CCR-14-1748
26. Kabala-Dzik, A.; Rzepecka-Stojko, A.; Kubina, R.; Jastrzębska-Stojko, Ż.; Stojko, R.; Wojtyczka, R. D.; Stojko, J. Migration Rate Inhibition of Breast Cancer Cells Treated by Caffeic acid and Caffeic Acid Phenethyl Ester: An In Vitro Comparison Study. *Nutrients* **2017,** *9* (10), 1–19. https://doi.org/10.3390/nu9101144
27. Veach, D.; Hosking, H.; Thompson, K.; Santhakumar, A. B. Anti-platelet and Anti-Thrombogenic Effects of Shikimic Acid in Sedentary Population. *Food Funct.* **2016,** *7* (8), 3609–3616. https://doi.org/10.1039/c6fo00927a
28. Lee, J.; Nguyen, Q. N.; Park, J. Y.; Lee, S.; Hwang, G. S.; Yamabe, N.; Choi, S.; Kang, K. S. Protective Effect of Shikimic Acid Against Cisplatin-induced Renal Injury: *In Vitro* and *In Vivo* Studies. *Plants* **2020,** *9* (12), 1–21. https://doi.org/10.3390/plants9121681
29. Kang, D. Y.; Sp, N.; Jo, E. S.; Rugamba, A.; Hong, D. Y.; Lee, H. G.; Yoo, J.-S.; Liu, Q.; Jang, K.-J.; Yang, Y. M. The Inhibitory Mechanisms of Tumor PD-L1 Expression by Natural Bioactive Gallic Acid. *Cancers* **2020,** *12* (717), 1–16. https://doi.org/doi:10.3390/cancers12030727

30. Chia, Y. C.; Rajbanshi, R.; Calhoun, C.; Chiu, R. H. Anti-Neoplastic Effects of Gallic Acid, a Major Component of Toona Sinensis Leaf Extract, on Oral Squamous Carcinoma Cells. *Molecules* **2010,** *15* (11), 8377–8389. https://doi.org/10.3390/molecules15118377
31. Xu, Y.; Tang, G.; Zhang, C.; Wang, N.; Feng, Y. Gallic Acid and Diabetes Mellitus: Its Association with Oxidative Stress. *Molecules* **2021,** *26* (23). https://doi.org/10.3390/molecules26237115
32. Tanaka, T.; Tanaka, T.; Tanaka, M. Potential Cancer Chemopreventive Activity of Protocatechuic Acid. *J. Exp. Clin. Med.* **2011,** *3* (1), 27–33. https://doi.org/10.1016/j.jecm.2010.12.005
33. Liu, C. L.; Wang, J. M.; Chu, C. Y.; Cheng, M. T.; Tseng, T. H. In Vivo Protective Effect of Protocatechuic acid on Tert-butyl Hydroperoxide-induced rat Hepatotoxicity. *Food Chem. Toxicol.* **2002,** *40* (5), 635–641. https://doi.org/10.1016/S0278-6915(02)00002-9
34. Xu, X.; Luo, A.; Lu, X.; Liu, M.; Wang, H.; Song, H.; Wei, C.; Wang, Y.; Duan, X. p-Hydroxybenzoic Acid Alleviates Inflammatory Responses and Intestinal Mucosal Damage in DSS-induced Colitis by Activating ERβ Signaling. *J. Funct. Foods* **2021,** *87*. https://doi.org/10.1016/j.jff.2021.104835
35. Oliveira, C.; Cagide, F.; Teixeira, J.; Amorim, R.; Sequeira, L.; Mesiti, F.; Silva, T.; Garrido, J.; Remião, F.; Vilar, S.; Uriarte, E.; Oliveira, P. J.; Borges, F. Hydroxybenzoic acid Derivatives as Dual-target Ligands: Mitochondriotropic Antioxidants and Cholinesterase Inhibitors. *Front. Chem.* **2018,** *6*. https://doi.org/10.3389/fchem.2018.00126
36. Rosiles-Alanis, W.; Zamilpa, A.; García-Macedo, R.; Zavala-Sánchez, M. A.; Hidalgo-Figueroa, S.; Mora-Ramiro, B.; Román-Ramos, R.; Estrada-Soto, S. E.; Almanza-Perez, J. C. 4-Hydroxybenzoic Acid and β-Sitosterol from Cucurbita ficifolia Act as Insulin Secretagogues, Peroxisome Proliferator-Activated Receptor-Gamma Agonists, and Liver Glycogen Storage Promoters: *In Vivo*, *in Vitro*, and *in Silico* Studies. *J. Med. Food* **2022,** *25* (6), 588–596. https://doi.org/10.1089/jmf.2021.0071
37. Kumari, S.; Kamboj, A.; Wanjari, M.; Sharma, A. K. Nephroprotective Effect of Vanillic acid in STZ-induced Diabetic Rats. *J. Diabetes Metab. Disorders* **2021,** *20* (1), 571–582. https://doi.org/10.1007/s40200-021-00782-7
38. Ahmadi, N.; Safari, S.; Mirazi, N.; Karimi, S. A.; Komaki, A. Effects of Vanillic Acid on Aβ1-40-induced Oxidative Stress and Learning and Memory Deficit in Male Rats. *Brain Res. Bull.* **2021,** *170* (February), 264–273. https://doi.org/10.1016/j.brainresbull.2021.02.024
39. Mengdai, X.; Hao, H.; Xiaoxing, M.; Yalun, Z.; Xi, C.; Xiaoqin ,L.; Xiaobo, P.; Zihui, X.; Liangkai C.; Shuang, R.; Wei, Y.; Shuang, L.; Liegang, L. Quercetin-3-O-Glucuronide Alleviates Cognitive Deficit and Toxicity in Aβ1-42 -Induced AD-Like Mice and SH-SY5Y Cells. *Mol. Nutrit. Food Res.* **2021,** *65* (6), e2000660. https://doi.org/10.1002/mnfr.202000660
40. Samrat, B.; Ramesh, P.; Jaehyo, K.; Ho-Sub, L.; Jungwon, S. Quercetin-3-O-Glucuronide Promotes the Proliferation and Migration of Neural Stem Cells. *Neurobiol. Aging* **2017,** *52,* 39–52. https://doi.org/10.1016/j.neurobiolaging.2016.12.024
41. Singeun, K.; Ki-Bae, H.; Kyungae, J.; Hyung J. S. Quercetin-3-O-glucuronide in the Ethanol Extract of Lotus Leaf (Nelumbo nucifera) Enhances Sleep Quantity and Quality in a Rodent Model via a GABAergic Mechanism. *Molecules* **2021,** *26* (10), 3023. https://doi.org/10.3390/molecules26103023

42. Sánchez-González, P. D.; López-Hernández, F. J.; Dueñas, M.; Prieto, M.; Sánchez-López, E.; Thomale, J.; Ruiz-Ortega, M.; López-Novoa, J. M.; Morales, A. I. Differential Effect of Quercetin on Cisplatin-Induced Toxicity in Kidney and Tumor Tissues. *Food Chem. Toxicol.* **2017,** *107* (Pt A), 226–236. https://doi.org/10.1016/j.fct.2017.06.047
43. Clemente-Soto, A. F.; Salas-Vidal, E.; Milan-Pacheco, C.; Sánchez-Carranza, J. N.; Peralta-Zaragoza, O.; González-Maya, L. Quercetin Induces G2 Phase Arrest and Apoptosis with the Activation of p53 in an E6 Expression Independent Manner in HPV Positive Human Cervical Cancer Derived Cells. *Mol. Med. Rep.* **2019,** *19* (3), 2097–2106. https://doi.org/10.3892/mmr.2019.9850
44. Lee W. J.; Hsiao M.; Chang J. L.; Yang S. F.; Tseng T. H.; Cheng C. W.; Chow J. M.; Lin K. H.; Lin Y. W.; Liu C. C.; Lee L. M.; Chien M. H. Quercetin Induces Mitochondrial-derived Apoptosis via Reactive Oxygen Species-mediated ERK Activation in HL-60 Leukemia Cells and Xenograft. *Arch. Toxicol.* **2015,** *89,* 1103–1117. https://doi.org/10.1007/s00204-014-1300-0.
45. Zhang X.; Huang J.; Yu C.; Xiang L.; Li L.; Shi D.; Lin F. Quercetin Enhanced Paclitaxel Therapeutic Effects Towards PC-3 Prostate Cancer Through ER Stress Induction and ROS Production. *OncoTargets Ther.* **2020,** *13,* 513. https://doi.org/10.2147/OTT.S228453
46. Tang, X. L.; Liu, J.; Dong, W.; Li, P.; Li, L.; Hou, J.-C.; Zheng, Y.-Q.; Lin, C.-R.; Ren, J.-G. Protective Effect of Kaempferol on LPS plus ATP-Induced Inflammatory Response in Cardiac Fibroblasts. *Inflammation* **2015,** *38,* 94–101. https://doi.org/10.47808/revistabioagro.v10i1.400
47. Hosseinpour-Niazi, S.; Mirmiran, P.; Fallah-Ghohroudi, A.; Azizi, F. Non-Soya Legume-based Therapeutic Lifestyle Change Diet Reduces Inflammatory Status in Diabetic Patients: A Randomised Cross-over Clinical Trial. *Br. J. Nutr.* **2015,** *114,* 213–219. https://doi.org/10.1017/S0007114515001725
48. Deng, Y.; Ma, J.; Weng, X.; Wang, Y.; Li, M.; Yang, T.; Dou, Z.; Yin, Z.; Shang, J. Kaempferol-3-O-Glucuronide Ameliorates Non-Alcoholic Steatohepatitis in High-Cholesterol-Diet-Induced Larval Zebrafish and HepG2 Cell Models via Regulating Oxidation Stress. *Life* **2021,** *11* (5), 445. https://doi.org/10.3390/life11050445
49. De Araújo, F. M.; Frota, A. F.; de Jesus, L. B.; Cuenca-Bermejo, L.; Ferreira, K. M. S.; Santos, C. C.; Soares, E. N.; Souza, J. T.; Sanches, F. S.; Costa, A. C. S.; Farias, A. A.; de Fatima Dias Costa, M.; Munoz, P.; Menezes-Filho, J. A.; Segura-Aguilar, J.; Costa, S. L.; Herrero, M. T.; Silva, V. D. A. Protective Effects of Flavonoid Rutin Against Aminochrome Neurotoxicity. *Neurotox. Res.* **2023,** *41* (3), 224–241. https://doi.org/10.1007/s12640-022-00616-1
50. Fattouch, S.; Caboni, P.; Coroneo, V.; Tuberoso, C.; Angioni, A.; Dessi, S. Comparative Analysis of Polyphenolic Profiles and Antioxidant and Antimicrobial Activities of Tunisian Pome Fruit Pulp and Peel Aqueous Acetone Extracts. *J. Agric. Food Chem.* **2008,** *56* (3), 1084−1090. https://doi.org/10.1021/jf072409e
51. Liu, S.; Ou, S.; Huang, H. Green Tea Polyphenols Induce Cell Death in Breast Cancer MCF-7 Cells Through Induction of Cell Cycle Arrest and Mitochondrial-Mediated Apoptosis. *J. Zhejiang Univ. Sci. B* **2017,** *18,* 89–98. https://doi.org/10.1631/jzus.B1600022
52. Gupta, D. A.; Bhaskar, D. J.; Gupta, R. K. Green Tea: A Review on Its Natural Anti-oxidant Therapy and Cariostatic Benefits. *Biol. Sci. Pharm. Res.* **2014,** *2,* 8–12.

53. Hara-Terawaki, A.; Takagaki, A.; Kobayashi, H.; Nanjo, F. Inhibitory Activity of Catechin Metabolites Produced by Intestinal Microbiota on Proliferation of HeLa cells. *Biol. Pharm. Bull* **2017,** *40,* 1331–1335. https://doi.org/10.1248/bpb.b17-00127
54. Wang, W.; Liu, X.; Zhang, Z.; Yin, M.; Chen, X.; Zhao, S.; Wu, L Apigenin Induced Apoptosis by Downregulating Sulfiredoxin Expression in Cutaneous Squamous Cell Carcinoma. *Oxid. Med. Cell. Longev.* **2022,** *1.* https://doi.org/10.1155/2022/8172866
55. Ling, J.; Wu, Y.; Zou, X.; Chang, Y.; Li, G.; Fang, M. (-)-Epicatechin Reduces Neuroinflammation, Protects Mitochondria Function, and Prevents Cognitive Impairment in Sepsis-Associated Encephalopathy. *Oxid. Med. Cell. Longev.* **2022,** *2657713.* https://doi.org/10.1155/2022/2657713
56. Yang, Y.; Zhang, T. Antimicrobial Activities of Tea Polyphenol on hytopathogens: A Review. *Molecules* **2019,** *24,* 816. https://doi.org/10.3390/molecules24040816
57. Wang, Z.; Zeng, M.; Wang, Z.; Qin, F.; Chen, J.; He, Z. Dietary Luteolin: a Narrative Review Focusing on its Pharmacokinetic Properties and Effects on Glycolipid Metabolism. *J Agric Food Chem.* **2021,** *69,* 1441–54. https://doi.org/10.1021/acs.jafc.0c08085
58. Kempuraj, D.; Thangavel, R.; Kempuraj, D. D.; Ahmed, M. E.; Selvakumar, G. P.; Raikwar, S. P.; Zaheer, S. A.; Iyer, S. S.; Govindarajan, R.; Chandrasekaran, P. N.; Zaheer, A. Neuroprotective Effects of Flavone Luteolin in Neuroinflammation and Neurotrauma. *BioFactors* **2021,** *47* (2), 190–197. https://doi.org/10.1002/biof.1687
59. Wu L.; Guo T.; Deng R.; Liu L.; Yu Y. Apigenin Ameliorates Insulin Resistance and Lipid Accumulation by Endoplasmic Reticulum Stress and SREBP-1c/SREBP-2 Pathway and SREBP-1c. *J. Basic Clin. Physiol. Pharmacol.* **2018,** *29,* 301–8. https://doi.org/10.1515/jbcpp-20170122
60. Yan, X.; Qi, M.; Li, P.; Zhan, Y.; Shao, H. Apigenin in Cancer Therapy: Anti-Cancer Effects and Mechanisms of Action. *Cell Biosci.* **2017,** *7,* 50. https://doi.org/10.1186/s13578-017-0179-x
61. Jia, B.; Yu, D.; Yu, G.; Cheng, Y.; Wang, Y.; Yi, X.; Li, X.; Wang, Y. Naringenin Improve Hepatitis C Virus Infection Induced Insulin Resistance by Increase PTEN Expression Via p53-dependent Manner. *Biomed. Pharmacother.* **2018,** *103,* 746–754. https://doi.org/10.1016/j.biopha.2018.04.11
62. Wang, Q.; Ou, Y.; Hu, G.; Wen, C.; Yue, S.; Chen, C.; Xu, L.; Xie, J.; Dai, H.; Xiao, H.; Zhang, Y.; Qi, R. Naringenin Attenuates Non-alcoholic Fatty Liver Disease by Downregulating the NLRP3/NF-κB Pathway in mice. *Br. J. Pharmacol.* **2020,** *177* (8), 1806–1821. https://doi.org/10.1111/bph.14938
63. Motallebi, M.; Bhia, M.; Rajani, H. F.; Bhia, I.; Tabarraei, H.; Mohammadkhani, N.; Pereira-Silva, M.; Kasaii, M. S.; Nouri-Majd, S.; Mueller, A. L.; Veiga, F. J. B.; Paiva-Santos, A. C.; Shakibaei, M. Naringenin: A Potential Flavonoid Phytochemical for Cancer Therapy. *Life Sci.* **2022,** *305,* 120752. https://doi.org/10.1016/j.lfs.2022.120752
64. Cui, Y. K.; Hong, Y. X.; Wu, W. Y.; Han, W. M.; Wu, Y.; Wu, C.; Li, G. R.; Wang, Y. Acacetin Ameliorates Cardiac Hypertrophy by Activating Sirt1/AMPK/PGC-1α Pathway. *Eur. J. Pharmacol.* **2022,** *920,* 174858. https://doi.org/10.1016/j.ejphar.2022.174858
65. Zhang, G.; Dong, J.; Lu, L.; Liu, Y.; Hu, D.; Wu, Y.; Zhao, A.; Xu, H. Acacetin Exerts Antitumor Effects on Gastric Cancer by Targeting EGFR. *Front. Pharmacol.* **2023,** *14,* 1121643. https://doi.org/10.3389/fphar.2023.1121643
66. Zielinska, D.; Laparra-Llopis, J. M.; Zielinski, H.; Szawara-Nowak, D.; Giménez-Bastida, J. A. Role of Apple Phytochemicals, Phloretin and Phloridzin, in Modulating

Processes Related to Intestinal Inflammation. *Nutrients* **2019,** *11* (5), 1173. https://doi.org/10.3390/nu11051173

67. Birru, R. L.; Bein, K.; Bondarchuk, N.; Wells, H.; Lin, Q.; Di, Y. P.; Leikauf, G. D. Antimicrobial and Anti-inflammatory Activity of Apple Polyphenol Phloretin on Respiratory Pathogens Associated with Chronic Obstructive Pulmonary Disease. *Front. Cell Infect. Microbiol.* **2021,** *11,* 652944. https://doi.org/10.3389/fcimb.2021.652944
68. Shin, S. K.; Cho, S. J.; Jung, U. J.; Ryu R.; Choi M.S Phlorizin Supplementation Attenuates Obesity, Inflammation, and Hyperglycemia in Diet-induced Obese Mice Fed a High-fat Diet. *Nutrients* **2016,** *8,* 92. https://doi.org/10.3390/nu8020092
69. Khalifa, M. M. A.; Bakr, A. G.; Osman, A. T. Protective Effects of Phloridzin Against Methotrexate-Induced Liver Toxicity in Rats. *Biomed. Pharmacother.* **2017,** *95,* 529–535. https://doi.org/10.1016/j.biopha.2017.08.121.
70. Lei, L. Y.; Wang, R. C.; Pan, Y. L.; Yue, Z. G.; Zhou, R.; Xie, P.; Tang, Z. S. Mangiferin inhibited Neuroinflammation Through Regulating Microglial Polarization and Suppressing NF-κB, NLRP3 Pathway. *Chin. J. Nat. Med.* **2021,** *19* (2), 112–119. https://doi.org/10.1016/S1875-5364(21)60012-2
71. Khoobchandani, M.; Khan, A.; Katti, K. K.; Thipe, V. C.; Al-Yasiri, A. Y.; MohanDoss, D. K. D.; Nicholl, M. B.; Lugão, A. B.; Hans, C. P.; Katti, K. V. Green Nanotechnology of MGF-AuNPs for Immunomodulatory Intervention in Prostate Cancer Therapy. *Sci Rep* **2021,** *11,* 16797 https://doi.org/10.1038/s41598-021-96224-8
72. Wang, Y.; Karmakar, T.; Ghosh, N.; Basak, S.; Gopal Sahoo, N. Targeting Mangiferin Loaded N-succinyl Chitosan-alginate Grafted Nanoparticles Against Atherosclerosis—A Case Study Against Diabetes Mediated Hyperlipidemia in Rat. *Food Chem.* **2022,** *370,* 131376 https://doi.org/10.1016/j.foodchem.2021.131376
73. Liaudanskas, M.; Viškelis, P.; Raudonis, R.; Kviklys, D.; Uselis, N.; Janulis, V. Phenolic Composition and Antioxidant Activity of Malus domestica Leaves. *Sci. World J.* **2014,** 1–10. https://doi.org/http://dx.doi.org/10.1155/2014/306217
74. Granci, V.; Dupertuis, Y. M.; Pichard, C. Angiogenesis as a Potential Target of Pharmaconutrients in Cancer Therapy. *Curr. Opin. Clin. Nutrit. Metab. Care* **2010,** *13* (4), 417–422. https://doi.org/10.1097/MCO.0b013e3283392656
75. Górnaś, P.; Mišina, I.; Olšteine, A.; Krasnova, I.; Pugajeva, I.; Lacis, G.; Siger, A.; Michalak, M.; Soliven, A.; Segliņa, D. Phenolic Compounds in Different Fruit Parts of Crab Apple: Dihydrochalcones as Promising Quality Markers of Industrial Apple Pomace by-products. *Ind. Crops Prod.* **2015,** *74* (November), 607–612. https://doi.org/10.1016/j.indcrop.2015.05.030
76. Barreira, J. C. M.; Arraibi, A. A.; Ferreira, I. C. F. R. Bioactive and Functional Compounds in Apple Pomace from Juice and Cider Manufacturing: Potential use in Dermal Formulations. *Trends Food Sci. Technol.* **2019,** *90* (August), 76–87. https://doi.org/10.1016/j.tifs.2019.05.014
77. Li, X.; Zhang, J.; Gao, W.; Wang, H. Study on chemical composition, anti-Inflammatory and Anti-microbial Activities of Extracts from Chinese pear Fruit (Pyrus Bretschneideri Rehd.). *Food Chem. Toxicol.* **2012,** *50* (10), 3673–3679. https://doi.org/10.1016/j.fct.2012.07.019
78. Kolniak-Ostek, J.; Oszmianski, J. Characterization of Phenolic Compounds in Different Anatomical Pear (Pyrus Communis L.) Parts by Ultra-performance Liquid Chromatography Photodiode Detector-Quadrupole/Time of Flight-mass Spectrometry

(UPLC-PDA-Q/TOF-MS). *Food Res. Int.* **2018,** *113,* 34–43. https://doi.org/10.1016/j.foodres.2018.06.024

79. Matthäus, B.; Özcan, M. Oil Content, Fatty Acid Composition and Distributions of Vitamin-E- active Compounds of Some Fruit Seed Oils. *Antioxidants* **2015,** *4* (1), 124–133. https://doi.org/10.3390/antiox4010124
80. Rodríguez-Cruz, M.; Tovar, A. R.; del Prado, M.; Torres, N. Mecanismos Moleculares de acción de los ácidos Grasos Poliinsaturados y sus Beneficios en la Salud. *Revista de Investigacion Clinica* **2005,** *57* (3), 457–472
81. Ferreira, D. M.; de Oliveira, N. M.; Lopes, L.; Machado, J.; Oliveira, M. B. Potential Therapeutic Properties of the Leaf of Cydonia Oblonga Mill. Based on Mineral and Organic Profiles. *Plants* **2022,** *11* (19), 1–20. https://doi.org/10.3390/plants11192638
82. Ayala Zavala, J. F.; González Aguilar, G. A. Use of Additives to Preserve the Cuality of Fresh-cut Fruits and Vegetables. In *Advances in Fresh-Cut fruits and Vegetables Processing*; Olga, M. B., Robert, S.-F., Eds.; CRC Press Taylor &Francis Group: Boca Raton, FL, **2011**; pp 231–254. ISBN:978-1-4200-7121
83. Herrero, M.; Plaza, M.; Cifuentes, A.; Ibáñez, E.; Green Processes for the Extraction of bioactives from Rosemary: Chemical and Functional Characterization Via Ultraperformance Liquid Chromatography-Tandem Mass Spectrometry and In-Vitro Assays. *J. Chromatogr. A* **2010,** *1217* (16), 2512–2520. https://doi.org/10.1016/j.
84. Galanakis, C. Valorization of Fruit Processing By-Products. Academic Press: Cambridge, MA, USA, 2020. ISBN 9780128171066.
85. Sagar, N. A.; Pareek, S.; Sharma, S.; Yahia, E. M.; Lobo, M. G. Fruit and Vegetable Waste: Bioactive Compounds, Their Extraction, and Possible Utilization. *Compr. Rev. Food Sci. Food Saf.* **2020,** *17,* 512–531. https://doi.org/10.1111/1541-4337.12330
86. Malenica, D.; Kass, M.; Bhat, R. Sustainable Management and Valorization of Agri-Food IndustrialWastes and By-products as Animal Feed: For Ruminants, Non-Ruminants and as Poultry Feed. *Sustainability* **2022,** *15,* 117. https://doi.org/10.3390/su15010117
87. Cakmak, H.; Dekker, M. Optimization of Cellulosic Fiber Extraction from Parsley Stalks and Utilization as Filler in Composite Biobased Films. *Foods* **2022,** *11,* 3932.
88. Vodovic, S.; Tepic Horecki, A.; Vladic, J.; Sumic, Z.; Gavaric, A.; Valkula, A. Chapter 2: Apple in Valorization of Fruit Processing By-Products; Elservier Inc.: Amsterdam, The Netherlands, **2020**. https://doi.org/10.1016/B978-0-12-817106-6.00002-2
89. Lee, E. H.; Kim, Y. J.; Kwon, S. I.; Kim, J. H.; Kang, I. K.; Kim, B. O.; Cho ,Y. J. Functional Properties of Newly Bred Green Ball Apple (Malus pumila Mill.). *Kor. J. Food Preserv.* **2018,** *25,* 837–845 https://doi.org/10.3839/jabc.2019.026
90. Yun, H. J.; Lim, S. Y.; Hur, J. M.; Jeong, J. W.; Yang, S. H.; Kim, D. H. Changes of Functional Compounds in, and Texture Characteristics of, Apples, During Post-irradiation Storage at Different Temperatures. *Kor. J. Food Preserv.* **2007,** *14,* 239–246. https://www.scinapse.io/papers/218764830
91. D'Archivio, M.; Files, C.; Di Benedetto, R.; Gargiulo, R.; Giovannini, C.; Masella, R. Polyphenols, Dietary Sources and Bioavailability. *Annali Dell'Istituto Superiore di Sanita* **2007,** *43,* 348–361. PMID: 18209268
92. Walia M.; Kumar S.; Agnihotria V. K. UPLC-PDA quantification of Chemical Constituents of Two Different Varieties (Golden and Royal) of Apple Leaves and Their Antioxidant Activity. *J. Sci. Food Agric.* **2015**. https://doi.org/10.1002/jsfa.7239

93. Abdul-Ghani, M.; DeFronzo, R.; Inhibition of Renal Glucose Reabsorption: a Novel Strategy for Achieving Glucose Control in type 2 Diabetes Mellitus. *Endocr. Pract.* **2008,** *14* (6), 782–790. https://doi.org/10.4158/ep.14.6.782.
94. Franquin-Trinquier, S.; Maury, C.; Baron, A.; Le Meurlay, D.; Mehinagic, E.; Optimization of the Extraction of Apple Monomeric Phenolics Based on Response Surface Methodology: Comparison of Pressurized Liquid-Solid Extraction and Manual-liquid Extraction. *J. Food Compos. Anal.* **2014,** *34* (1), 56–67. https://doi.org/10.1016/j.jfca.2014.01.005.
95. Grigoras, C. G.; Destandau, E.; Fougére, L.; Elfakir, C.; Evaluation of Apple Pomace Extracts as a Source of Bioactive Compounds. *Ind. Crop. Prod.* **2013,** *49,* 794–804. https://doi.org/10.1016/j.indcrop.2013.06.026.
96. Perussello, C. A.; Zhang, Z.; Marzocchella, A.; Tiwari, B. K.; Valorization of Apple Pomace by Extraction of Valuable Compounds. *Compr. Rev. Food Sci. Food Saf.* **2017,** *16* (5), 776–796. https://doi.org/10.1111/1541-4337.12290.
97. Arain, S.; Sherazi, S. T. H.; Bhanger, M. I.; Memon, N.; Mahesar, S. A.; Rajput, M. T. Prospects of Fatty Acid Profile and Bioactive Composition from Lipid Seeds for the Discrimination of Apple Varieties with the Application of Chemometrics. *Grasas y Aceites* **2012,** *63* (2), 175–183. https://doi.org/10.3989/gya.082811
98. Adebayo, S. E.; Orhevba, B. A.; Adeoye, P. A.; Musa, J. J.; Fase, O. J. Solvent Extraction and Characterization of Oil from African Star Apple (Chrysophyllum albidum) Seeds. *Acad. Res. Int.* **2012,** *3* (2), 178–183. ISSN: 2223-9944
99. Reiland, H.; Slavin, J. Systematic Review of Pears and Health. *Nutrition Today* **2015,** *50* (6), 301–305. https://doi.org/10.1097/NT.0000000000000112
100. Asgary, S.; Sahebkar, A.; Afshani, M. R.; Keshvari, M.; Haghjooyjavanmard, S.; Rafieian-Kopaei, M. Clinical Evaluation of Blood Pressure Lowering, Endothelial Function Improving, Hypolipidemic and Anti-Inflammatory Effects of Pomegranate Juice in Hypertensive Subjects. *Phytother. Res.* **2014,** *28,* 193–199. https://doi.org/10.1002/ptr.4977
101. Yukui, R.; Wenya, W.; Rashid, F.; Qing, L. Fatty Acids Composition of Apple and Pear Seed Oils. *Int. J. Food Prop.* **2009,** *12,* 774–779. https://doi.org/10.1080/10942910802054320
102. Postman, J. Cydonia oblonga: The Unappreciated Quince. *Arnoldia* **2009,** *67* (1), 2–9. ISSN: 0004-2633
103. Sharma, R.; Joshi, V. K.; Rana, J. C. Nutritional Composition and Processed Products of Quince (Cydonia Oblonga Mill.). *Indian J. Nat. Prod. Resour.* **2011,** *2* (3), 354–357.
104. Silva, B. M.; Andrade, P. B.; Ferreres, F.; Domingues, A. L.; Seabra, R. M.; Ferreira, M. A. Phenolic Profile of Quince Fruit (Cydonia oblonga Miller) (pulp and peel). *J. Agric. Food Chem.* **2002,** *50* (16), 4615–4618. https://doi.org/10.1021/jf0203139
105. de Almeida Lopes, M. M.; A. G. Sanches, K. O. de Souza E. O. Silva. Quince *Cydonia oblonga.* In *Exotic Fruits Reference Guide*, Elsevier Inc;.: Amsterdam, Netherlands, **2018**; pp 363–368. ISBN: 978-0-12-803138-4
106. Yıkmış, S.; H. Aksu, B. G. Çöl M. Alpaslan. Thermosonication Processing of Quince (*Cydonia oblonga*) juice: Effects on Total Phenolics, Ascorbic Acid, Antioxidant Capacity, Color and Sensory Properties. *Ciênc. Agrotecnol.* **2019,** *43,* e019919. https://doi.org/10.1590/1413-7054201943019919
107. Yilmaz, E. Karadeniz, F. Effect of Storage on the Bioactive Compounds and Antioxidant Activity of Quince Nectar. *Int. J. Food Sci. Technol.* **2014,** *49,* 718–725. https://doi.org/10.1111/ijfs.12355

108. Cascales, E. V.; García, J. M. R. Characteristics of the Raw Fruit, Industrial Pulp, and Commercial Jam Elaborated with Spanish Quince (Cydonia oblonga Miller). *Emir. J. Food Agric.* **2020,** *32* (8), 623–633. https://doi.org/10.9755/ejfa.2020.v32.i8.2140
109. Herrera-Rocha, K. M.; Rocha-Guzmán, N. E.; Gallegos-Infante, J. A.; González-Laredo, R. F.; Larrosa-Pérez, M.; Moreno-Jiménez, M. R. Phenolic Acids and Flavonoids in Acetonic Extract from Quince (Cydonia oblonga Mill.): Nutraceuticals with Antioxidant and Anti-Inflammatory Potential. *Molecules* **2022,** *27* (2462), 1–14. https://doi.org/10.3390/molecules27082462
110. Ashraf, M. U.; Muhammad, G.; Hussain, M. A.; Bukhari, S. N. A. Cydonia oblonga, M.; A Medicinal Plant Rich in Phytonutrients for Pharmaceuticals. *Front. Pharmacol.* **2016,** *7,* 163. https://doi.org/10.3389/fphar.2016.00163
111. Benzarti, S.; Hamdi, H.; Lahmayer, I.; Toumi, W.; Kerkeni, A.; Belkadhi, K.; Sebei, H. Total Phenolic Compounds and Antioxidant Potential of Quince (Cydonia oblonga Miller) Leaf Methanol Extract. *Int. J. Innov. Appl. Stud.* **2015,** *13,* 518. ISSN: 2028-9324
112. Bernal Rodríguez, C. A. Contribución al Estudio Farmacotécnico del Extracto Estandarizado de Frutos de *Physalis peruviana* L. con Miras a la Obtención de un Producto Fitoterapéutico (tesis de maestría). Departamento de Farmacia, Universidad Nacional de Colombia: Bogotá D.C.; 2012; **2012**. p 193.
113. Naviglio, D.; Scarano, P.; Ciaravolo, M.; Gallo, M. Rapid Solid-liquid Dynamic extraction (RSLDE): A powerful and Greener Alternative to the Latest Solid-liquid Extraction Techniques. *Foods* **2019,** *8* (7), 1–22. https://doi.org/10.3390/foods8070245
114. Azmir, J.; Zaidul, I. S. M.; Rahman, M. M.; Sharif, K. M.; Mohamed, A.; Sahena, F.; Jahurul, M. H. A.; Ghafoor, K.; Norulaini, N. A. N.; Omar, A. K. M. Techniques for extraction of Bioactive Compounds from Plant Materials: A Review. *J. Food Eng.* **2013,** *117* (4), 426–436. https://doi.org/10.1016/j.jfoodeng.2013.01.014
115. De Souza B.; Lúcia, A. Extraction and Characterization of the Biomass and Desmodesmus sp. Microalgae Oil Using Supercritical Carbon Dioxide. Master's Thesis, Universidade Federal Rural do Rio de Janeiro, Instituto de Tecnologia: Rio de Janeiro, Brazil, **2015**.
116. Zygler, A.; Słomi ́nska, M.; Namie ́snik, J. Soxhlet Extraction and New Developments Such as Soxtec. In *Comprehensive Sampling and Sample Preparation*; Elsevier: Amsterdam, The Netherlands, **2012,** 2; pp 65–82. ISBN: 9780123813749.
117. Jacobsen, C.; Sørensen, A. D. M.; Holdt, S. L.; Akoh, C. C.; Hermund, D. B. Source, Extraction, Characterization, and Applications of Novel Antioxidants from Seaweed. *Ann. Rev. Food Sci. Technol.* **2019,** *10,* 541–568. https://doi.org/10.1146/annurev-food-032818-
118. Angela, A.; Meireles, M.; Braga, M.; Leal, P.; Maróstica, M.; Pereira, C.; Takeuchi, T. Low-Pressure Solvent Extraction (Solid-Liquid Extraction, Microwave Assisted, and Ultrasound Assisted) from Condimentary Plants. In *Extracting Bioactive Compounds for Food Products Theory and Applications*; Meireles, M. A. A., Ed.; CRC Press: New York, NY, USA, 2008, **2008**; pp 137–218.
119. Grosso, C.; Valentão, P.; Ferreres, F.; Andrade, P. B. Alternative and Efficient Extraction Methods for Marine-Derived Compounds. *Marine Drugs* **2015,** *13* (5), 3182–3230. https://doi.org/10.3390/md13053182
120. Mussatto, S. I. Generating Biomedical Polyphenolic Compounds from Spent Coffee or Silverskin. In *Coffee in Health and Disease Prevention*, Elsevier: Amsterdam, The Netherlands, **2015**; pp 93–106. ISBN 9780124167162.

121. Kadam, S. U.; Tiwari, B. K.; O'Donnell, C. P. Application of Novel Extraction Technologies for Bioactives from Marine Algae. *J. Agric. Food Chem.* **2013,** *61,* 4667–4675. https://doi.org/10.1021/jf400819p
122. Galanakis, C. M.; Tornbergb, E.; Gekas, V.; Recovery and Preservation of Phenols From Olive Waste in Ethanolic Extracts. *J. Chem. Technol. Biotechnol.* **2010,** *85* (8), 1148–1155. https://doi.org/10.1002/jctb.2413.
123. Prado, J. M.; Vardanega, R.; Debien, I. C. N.; Meireles, M. A. de A.; Gerschenson, L. N.; Sowbhagya, H. B.; Chemat, S.; Conventional Extraction. In *Food Waste Recovery: Processing Technologies and Industrial Techniques.* **2015**. https://doi.org/10.1016/B978- 0-12-800351-0.00006-7.
124. Sowa, A.; Zgórka, G.; Szykuła, A.; Franiczek, R.; Żbikowska, B.; Gamian, A.; Sroka, Z. Analysis of Polyphenolic Compounds in Extracts from Leaves of Some Malus domestica Cultivars: Antiradical and Antimicrobial Analysis of These Extracts. *BioMed. Res. Int.* **2016,** *6705431.* https://doi.org/10.1155/2016/6705431
125. Al-khazraji, S. K. Phytochemical Screening and Antibacterial Activity of the Crude Extract of Cydonia Oblonga Seeds. *Glob. Adv. Res. J. Microbiol.* **2013,** *2,* 137–140
126. Zhang, L.; Rocchetti, G.; Zengin, G.; Ak, G.; Saber, F. R.; Montesano, D.; Lucini, L. The UHPLC-QTOF-MS Phenolic Profiling and Activity of Cydonia oblonga Mill. Reveals a Promising Nutraceutical Potential. *Foods* **2021,** *10,* 1230. https://doi.org/10.3390/foods10061230
127. Strati, I. F.; Oreopoulou, V. Effect of Extraction Parameters on the Carotenoid Recovery From Tomato Waste. *Int. J. Food Sci. Technol.* **2011,** *46* (1), 23–29. https://doi.org/10.1111/j.1365-2621.2010.02496.x
128. Bermejo, D. V.; Luna, P.; Manic, M. S.; Najdanovic-Visak, V.; Reglero, G.; Fornari, T.; Extraction of Caffeine from Natural Matter Using a Bio-renewable Agrochemical Solvent. *Food Bioprod. Process.* **2013,** *91* (4), 303–309. https://doi.org/10.1016/j.fbp.2012.11.007
129. D'Archivio, A. A.; Maggi, M. A.; Ruggieri, F.; Extraction of Curcuminoids by Using Ethyl Lactate and its Optimisation by Response Surface Methodology. *J. Pharmaceut. Biomed. Anal.* **2018,** *149,* 89–95. https://doi.org/10.1016/j.jpba.2017.10.042.
130. Kua, Y. L.; Gan, S.; Morris, A.; Ng, H. K.; Optimization of Simultaneous Carotenes and Vitamin E (tocols) Extraction from Crude Palm Olein Using Response Surface Methodology. *Chem. Eng. Commun.* **2018,** *205* (5), 596–609. https://doi.org/10.1080/00986445.2017.1407760.
131. Villanueva-Bermejo, D.; Reglero, G.; Fornari, T.; Recent Advances in the Processing of Green Tea Biomolecules Using Ethyl Lactate. A Review. *Trends Food Sci. Technol.* **2017,** *62,* 1–12. https://doi.org/10.1016/j.tifs.2016.12.009.
132. Lu, W.; Liu, S. Choline Chloride–based Deep Eutectic Solvents (Ch-DESs) as Promising Green Solvents for Phenolic Compounds Extraction from Bioresources: State-of-the-art, Prospects, and Challenges. *Biomass Convers. Bior.* **2022,** *12* (7), 2949–2962. https://doi.org/10.1007/s13399-020-00753-7
133. Patil, S. S.; Pathak, A.; Rathod, V. K. Optimization and Kinetic Study of Ultrasound Assisted Deep Eutectic Solvent Based Extraction: A Greener Route for Extraction of Curcuminoids from Curcuma longa. *Ultrason. Sonochem.* **2021,** *70,* 105267. https://doi.org/10.1016/j.ultsonch.2020.105267
134. Chemat, F.; Rombaut, N.; Meullemiestre, A.; Turk, M.; Perino, S.; Fabiano-Tixier, A. S.; Abert-Vian, M. Review of Green Food Processing Techniques. Preservation,

Transformation, and Extraction. *Innov. Food Sci. Emerg. Technol.* **2017,** *41,* 357–377. https://doi.org/10.1016/j.ifset.2017.04.016
135. Paniwnyk, L. Applications of Ultrasound in Processing of Liquid Foods: A Review. *Ultrason. Sonochem.* **2017,** *38,* 794–806. https://doi.org/10.1016/j.ultsonch.2016.12.025
136. Jambrak, A. R.; Mason, T. J.; Lelas, V.; Herceg, Z.; Herceg, I. L. Effect of Ultrasound Treatment on Solubility and Foaming Properties of Whey Protein Suspensions. *J. Food Eng.* **2008,** *86* (2), 281–287. https://doi.org/10.1016/j.jfoodeng.2007.10.00
137. Vernès, L.; Vian, M.; Chemat, F. Ultrasound and Microwave as Green Tools for Solid-Liquid Extraction. *Liquid-Phase Extraction* **2019,** *355–374.* https://doi.org/10.1016/B978-0-12-816911-7.00012-8
138. Picot-Allain, C.; Mahomoodally, M. F.; Ak, G.; Zengin, G. Conventional Versus Green Extraction Techniques — a Comparative Perspective. *Curr. Opin. Food Sci.* **2021,** *40,* 144–156. https://doi.org/10.1016/j.cofs.2021.02.009
139. Xia, Q.; Nong-xue, Gao. Ultrasound-assisted Extraction Process of Apple Seed Oil. Analysis of Physical and Chemical Properties. *Chin. J. Oil Crop Sci.* **2007,** *28* (11), 422–426.
140. Belwal, T.; Huang, H.; Li, L.; Duan, Z.; Zhang, X.; Aalim, H.; Luo, Z. Optimization Model for Ultrasonic-assisted and Scale-up Extraction of Anthocyanins from Pyrus Communis 'Starkrimson' Fruit Peel. *In Food Chem.* **2019,** *297.* https://doi.org/10.1016/j.foodchem.2019.124993
141. Badens, E. Mise en Forme de Principes Actifs Pharmaceutiques en Pase Supercritique. *Techniques De l'ingénieur Mise en Forme Des Médicaments, Base Documentaire. TIB611DUO* **2012,** https://doi.org/10.51257/a-v1-chv4010
142. Brunner, G. Supercritical Fluids: Technology and Application to Food Processing. *J. Food Eng.* **2005,** *67* (1–2), 21–33. https://doi.org/10.1016/j.jfoodeng.2004.05.060
143. Uwineza, P. A.; Waśkiewicz, A. Recent Advances in Supercritical Fluid Extraction of Natural Bioactive Compounds From Natural Plant Materials. *Molecules* **2020,** *25* (17). https://doi.org/10.3390/molecules25173847
144. Ishwarya, S. P.; Nisha, P. Headway in Supercritical Extraction of Fragrances and Colors. In *Innovative Food Processing Technologies: A Comprehensive Review,* Elsevier, **2020**. https://doi.org/10.1016/b978-0-08-100596-5.22677-2
145. Khaw, K. Y.; Parat, M. O.; Shaw, P. N.; Falconer, J. R. Solvent Supercritical Fluid Technologies to Extract Bioactive Compounds from Natural Sources: A Review. *Molecules* **2017,** *22* (7). https://doi.org/10.3390/molecules22071186
146. Rozzi, N. L.; Singh, R. K.; Vierling, R. A.; Watkins, B. A. Supercritical Fluid Extraction of Lycopene From Tomato Processing Byproducts. *J. Agric. Food Chem.* **2002,** *50* (9), 2638–2643. https://doi.org/10.1021/jf011001t
147. Chemat, F. Eco-extraction du végétal : Procédés Innovants et Solvants Alternatifs. Dunod, **2011**; 336 p 2011, *Collection L'Usine Nouvelle*. 978-2-10-056543-6. ⟨hal-02804831⟩
148. Da Porto, C.; Decorti, D.; Natolino, A. Water and Ethanol as Co-solvent in Supercritical Fluid Extraction of Proanthocyanidins from Grape Marc: A Comparison and a Proposal. *J. Supercrit. Fluids* **2014,** *87,* 1–8. https://doi.org/10.1016/j.supflu.2013.12.019
149. Daneshvand, B.; Ara, K. M.; Raofie, F. Comparison of Supercritical Fluid Extraction and Ultrasound-assisted Extraction of Fatty Acids from Quince (Cydonia oblonga Miller) seed using Response Surface Methodology and Central Composite Design. *J. Chromatogr. A* **2012,** *1252* (August), 1–7. https://doi.org/10.1016/j.chroma.2012.06.063

150. Ferrentino, G.; Giampiccolo, S.; Morozova, K.; Haman, N.; Spilimbergo, S.; Scampicchio, M. Supercritical Fluid Extraction of Oils From Apple Seeds: Process Optimization, Chemical Characterization and Comparison with a Conventional Solvent Extraction. *Innov. Food Sci. Emerg. Technol.* **2020,** *64,* 102428. https://doi.org/10.1016/j.ifset.2020.102428
151. Zurob, E.; Cabezas, R.; Villarroel, E.; Rosas, N.; Merlet, G.; Quijada-Maldonado, E.; Romero, J.; Plaza, A. Design of Natural Deep Eutectic Solvents for the Ultrasound-assisted Extraction of Hydroxytyrosol from Olive Leaves Supported by COSMO-RS. *Separation and Purification Technology* **2020,** *248* (March), 117054. https://doi.org/10.1016/j.seppur.2020.117054.
152. Nadar, S. S.; Rao, P.; Rathod, V. K. Enzyme assisted Extraction of Biomolecules as an Approach to Novel Extraction Technology: A Review. *Food Res. Int.* **2018,** *108* (2017), 309–330. https://doi.org/10.1016/j.foodres.2018.03.006
153. Marathe, S. J.; Jadhav, S. B.; Bankar, S. B.; Singhal, R. S. Enzyme-Assisted Extraction of Bioactives. In *Food Bioactives*; Puri, M.; Ed.; Springer International Publishing: Cham, Switzerland, 2017; pp 171–201. ISBN 978-3-319-51637-0. https://doi.org/10.1007/978-3-319-51639-4_8
154. Costa, J. R.; Tonon, R. V.; Cabral, L.; Gottschalk, L.; Pastrana, L.; Pintado, M. E. Valorization of Agricultural Lignocellulosic Plant Byproducts Through Enzymatic and Enzyme-Assisted Extraction of High-Value-Added Compounds: A Review. *ACS Sustain. Chem. Eng.* **2020,** *8* (35), 13112–13125. https://doi.org/10.1021/acssuschemeng.0c02087
155. Meini, M. R.; Cabezudo, I.; Boschetti, C. E.; Romanini, D. Recovery of Phenolic Antioxidants from Syrah Grape Pomace Through the Optimization of an Enzymaticmextraction Process. *Food Chem.* **2019,** *283* (January), 257–264. https://doi.org/10.1016/j.foodchem.2019.01.037
156. Wu H.; Zhu J.; Yang L.; Wang R.; Wang C. Ultrasonic-assisted Enzymatic Extraction of Phenolics From Broccoli (Brassica oleracea L. var. italica) Inflorescences and Evaluation of Antioxidant Activity In Vitro. *Food Sci. Technol. Int.* **2015,** *21* (4), 306–319. https://doi.org/10.1177/1082013214536174
157. Li, F.; Mao, Y. D.; Wang, Y. F.; Raza, A.; Qiu, L. P.; Xu, X. Q. Optimization of Ultrasonic-assisted Enzymatic Extraction Conditions for Improving Total Phenolic Content, Antioxidant and Antitumor Activities In Vitro From Trapa Quadrispinosa Roxb. residues. *Molecules* **2017,** *22* (3). https://doi.org/10.3390/molecules22030396

CHAPTER 4

Circular Bioeconomy in Functional Foods and Beverages Development: A Sustainable Holistic Approach

PEDRO HASSIEL SARIÑANA-NÚÑEZ[1],
NURIA ELIZABETH ROCHA-GUZMÁN[1], MARÍA JOSÉ RIVAS-ARREOLA[2],
RUBÉN FRANCISCO GONZÁLEZ-LAREDO[1],
MARTHA ROCÍO MORENO-JIMÉNEZ[1],
JOSÉ ALBERTO GALLEGOS-INFANTE[1], and
SILVIA MARINA GONZÁLEZ-HERRERA[1]

[1]*Laboratorio Nacional CONAHCYT de Apoyo a la Evaluación de Productos Bióticos (LaNAEPBi), Unidad de Servicio, Tecnológico Nacional de México/ I.T. de Durango (TecNM/ITD), Blvd. Felipe Pescador 1830 Ote., C.P. 34080, Durango, Dgo., México*

[2]*University Iberoamericana de Puebla. Blvd. del Niño Poblano 2901, Reserva Territorial Atlixcáyotl, Centro Comercial Puebla, San Andrés Cholula, México*

ABSTRACT

The circular bioeconomy is a key concept in food processing systems, bringing innovation and competitiveness into the food industry. Particularly, when aiming for the revalorization of agro-food by-products, for the conventional and non-conventional extraction of bioactive compounds, and the development of demanded healthy functional foods and beverages from natural sources. In this scenario, holistic and sustainable applications are possible, in pursuit of economic, and social developments, but with environmental protection inter-connected as the fundamentals supporting sustainable development. Therein,

The Food-Energy-Water Triangle for Sustainable Development: Integrative and Synergic Solutions.
Pedro Aguilar-Zárate, Mónica Chávez-González, Ann Rose Abraham, and A. K. Haghi (Eds.)

government participation is crucial in establishing policies and the legal framework needed to warrant productive inputs and higher standards of life. The consumer plays a decisive role in the effective sustainability of food products, as their choices and behaviors can determine the environmental, social, and economic impacts of food production and its consumption. The current consumer trends in processed foods are distinguished by distinctive generations but recognized by the growing demand for eco-friendly products minimally processed, with a low carbon footprint and no genetically modified organisms.

4.1 INTRODUCTION

The circular bioeconomy is a concept that aims to use biological resources in a sustainable and efficient way, minimizing waste and maximizing value. It involves the creation and use of renewable biological resources, such as plants, animals, fish, and other living things, as well as their transformation into food, feed, bio-based goods, and bioenergy. The circular bioeconomy also involves the recycling and reuse of biological materials and nutrients, as well as the restoration of natural capital and ecosystem services.[1–3]

A holistic approach is a way of thinking that considers the whole system, not just its parts. It recognizes the interconnections and interdependencies among different elements and aims to optimize the system, rather than maximizing individual components. The relationship between a holistic approach and the circular economy is that they both share a common vision of creating a more sustainable and resilient society, where economic activities are aligned with ecological limits and human well-being.[4] Therefore, sustainability is crucial, because it is connected to these concepts and is currently thought to be the central idea for the capacity to satisfy the present demands without jeopardizing the ability of future generations to meet their own needs. The relationship between these three concepts is that they all share a common vision of creating a more balanced and harmonious society that respects the environment and human well-being. A holistic approach can help to design and implement circular economy solutions that are aligned with sustainability principles and goals.

The precepts set out in the circular bioeconomy, can have a positive impact on functional foods and beverages development by promoting a holistic and sustainable approach that considers the entire life cycle of the product, from sourcing to consumption and disposal. For example, a circular bioeconomy approach can encourage the use of locally grown ingredients, biodegradable packaging, energy-efficient processing, and innovative recycling methods.

This can benefit both the producers and the consumers of beverages, as well as the society and the planet.

4.1.1 BIOECONOMY AND CIRCULAR ECONOMY IN PROCESS SYSTEMS

The production and application of biological resources, methods, and concepts to the provision of products and services across all economic sectors are referred to as the bioeconomy by the Food and Agriculture Organization of the United Nations (FAO). It encompasses sectors such as agriculture, forestry, fisheries, biotechnology, and bioenergy. Bioeconomy plays a vital role in food processing systems, as it can enhance food quality, safety, and security while reducing environmental impacts and resource use. Bioeconomy can also contribute to innovation and competitiveness in the food industry, by creating new products, markets, and value chains based on biological resources.[5] Furthermore, it can play a significant role in process systems engineering by enhancing the sustainability, efficiency, and resilience of industrial processes.[6] For example, it can enable the development of biorefineries that convert biomass into fuels, chemicals, and materials, or the integration of bioprocesses with conventional processes to reduce waste and emissions. Bioeconomy can also foster innovation and collaboration among different disciplines and stakeholders in process systems engineering.

The circular bioeconomy is a subset of the bioeconomy that seeks to minimize waste and environmental impact by applying the principles of the circular economy such as reducing, reusing, recycling, and recovering resources and biological products. Therefore, the revalorization of agro-industrial wastes for the development of value-added products matches the precepts that have been put forward in developing of healthy functional foods and beverages from natural sources.

4.1.2 SUSTAINABILITY

The term sustainability is a complex and elusive concept, given that it is an extremely important issue since it has to do nothing with less than the possibilities of humankind´s survival on this planet. The excessive use of non-renewable resources (i.e., soil, biodiversity, fuels) and renewable resources (i.e., water, forest, clean air), in addition to the lack of application of corrective strategies, denotes an uncertain future. It follows that such a complex topic does not have a simple and straightforward treatment, especially considering that sustainability is not a goal but a process. Some authors[7] such as Caniglia et al. agree with this idea, stating that action-oriented research for sustainability conceives sustainability as a process and has increasingly focused on how change can be created, enacted, and facilitated. On the other hand, Norström et al.[8] refer that human domination of the biosphere has led to substantial advances in human well-being and economic development. However, at the same time, it threatens the planetary conditions that sustain societal well-being and prosperity and stresses the emerging challenges, such as water scarcity, food security, and biodiversity loss, which are intractable, interconnected, and influenced by several factors at different scales and complex feedback mechanisms. In the last century, human beings have effectively managed the earth's resources, cutting down much of the existing forests, deteriorating the atmosphere (i.e., courtesy of organochlorine chemicals), polluting rivers, and altering the climate. In addition, they have caused the extinction of thousands of species.

Economic development, social development, and environmental protection are interwoven and support one another as the three pillars of sustainable development. The idea of sustainability in three dimensions (i.e., economic, social, environmental) arose from the concept of the "Triple Bottom Line," coined by Elkington in 1994, where he conceived it as a way of operationalizing the corporate social liability.[9]

Pondering on current events, the COVID-19 pandemic caused a slowdown in global economic growth, which led to some environmental improvements, such as a decrease in greenhouse gas emissions. However, the economic rebound eventually led to unmitigated higher pollutant emissions. In this period (2020–2023), employment and economic concerns were paramount in the recovery of the economy, neglecting or eliminating any hint of concern for the environment. Several governments have already proposed a policy of economic growth that includes rolling back or removing some environmental regulations.[10] Similarly, Barbier and Burgess[11] stated that global carbon dioxide (CO_2) emissions decreased drastically after the pandemic. However, the trend in recent years has been an increase in emissions due to the growing burning of fossil fuels. In this sense, research should support in all areas, national authorities, civil society organizations, private companies, as well as indigenous peoples.[7]

4.1.3 SYMBIOTIC APPROACH BETWEEN HOLISTIC AND SUSTAINABLE APPLICATIONS

Holistic and sustainable applications in a circular bioeconomy scenario are those that consider the environmental, social, and economic impacts of using biological resources for various purposes. These applications aim to minimize waste, reduce greenhouse gas emissions, enhance biodiversity, and improve human well-being. Figure 4.1 illustrates a symbiotic approach between holistic and sustainable applications, highlighting the production of biodegradable packaging developed from agricultural waste or algae; the transformation of organic waste into biogas, fertilizer, or compost; and the development of cosmetics, pharmaceuticals, or textiles, and other related activities that ultimately support the local or regional food system.

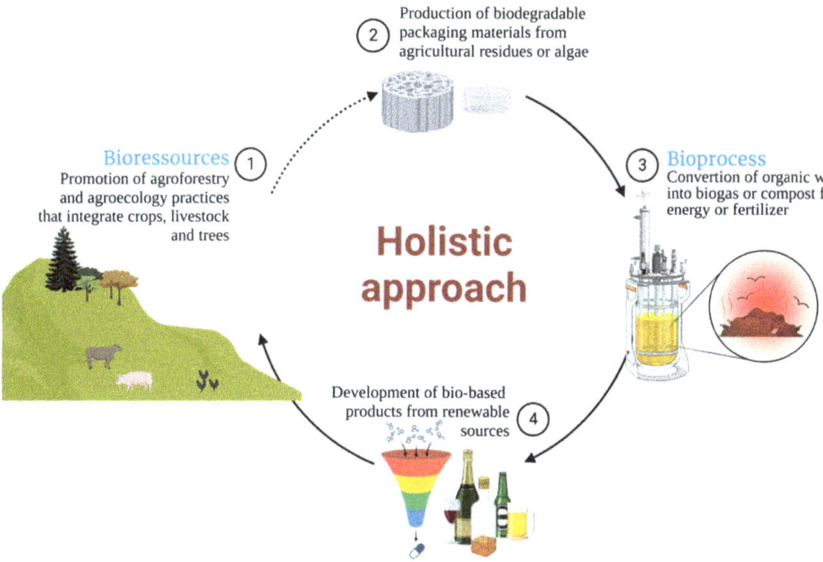

FIGURE 4.1 Holistic and sustainable applications in a circular bioeconomy scenario.

In summary, these holistic and sustainable approaches support local and regional food systems with the consequent reduction of food miles, increase food security, and allow the revalorization of agro-industrial waste for the development of value-added products.

4.2 GOVERNANCE IN THE CIRCULAR BIOECONOMY AND SUSTAINABILITY

Governments around the world, having a greater impact on the environment, are addressing sustainability by encouraging industries to comply with sustainability policies since industry, due to its large scale, is one of the largest producers of pollutants. Some types of government offer greater participation than others, but all have some form of citizen participation. Thus, Fig. 4.2 depicts the main elements involved in some forms of governance. On the other hand, local governments work with broader stakeholders in decision-making in a state or national context. A variety of organizations work to assist local governments on sustainability issues.

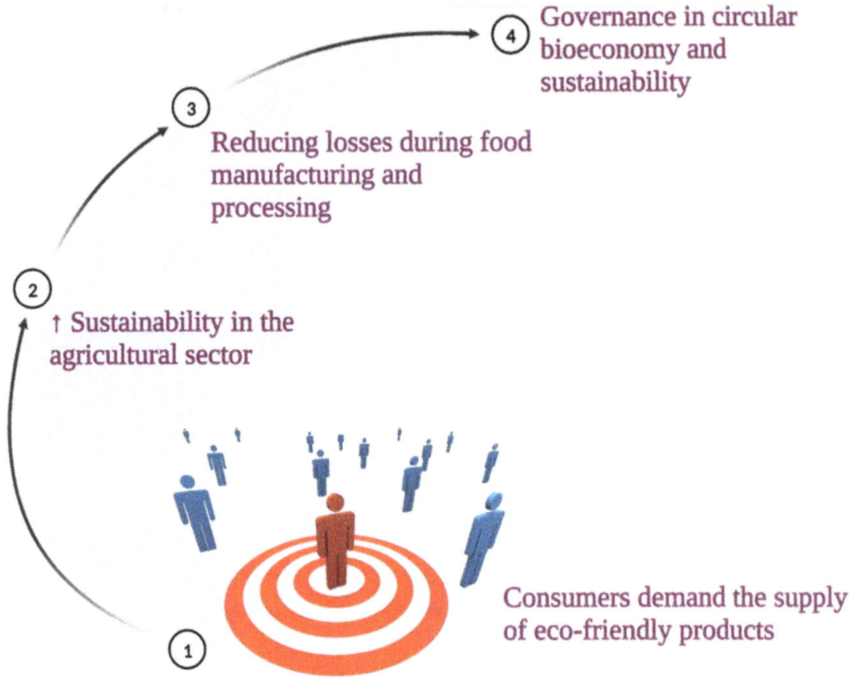

FIGURE 4.2 Main elements involved in some forms of governance in circular bioeconomy and sustainability.

Considering this, some authors such as Chen et al.[12] remark that the two main ways in which industries can benefit from government subsidy are through a "direct cost" official assistance (e.g., subsidies and tax concessions) or through the "market" (e.g., a domestic government taxing foreign-made products so heavily that they are not competitive with local products). An advantage for the industry is that consumers are willing to pay more for products that are seen as friendly to the environment, as sustainable or eco-friendly products, in addition to governments making sustainability a political issue.[13] Undoubtedly, governments play a crucial role in sustainability by establishing policies and the legal framework needed to preserve productive inputs and quality of life. However, doing so could be perceived as expensive and risky to industrial competitiveness. With no doubt, these increased expenses could lead to both immediate and long-term challenges for the sector, particularly if proper preparation work has not been done.[14] Therefore, the role of government and industry go hand in hand as they are interconnected, having the same goal, to establish a sustainable process.

For the above reasons, governance in the circular bioeconomy and sustainability is a complex and multidimensional topic that involves different actors, interests, values, and goals. Governance refers to the policies, technologies, organizations, behaviors, cultures, and markets that shape and enable the transition to a circular bioeconomy[15,16] Its challenges include balancing the trade-offs and synergies between different biomass uses, ensuring social justice and equity, fostering innovation and collaboration, and enhancing transparency and accountability.[17] Thus, governance in the circular bioeconomy refers to the processes and institutions that shape the production, consumption, and recycling of biological resources and products. It involves multiple actors and levels of decision-making, from local to global, and requires coordination and collaboration among different sectors and stakeholders. Governance in the circular bioeconomy aims to balance economic, environmental, and social objectives while ensuring the sustainability and resilience of natural systems and human well-being.

In conclusion, governance in the circular bioeconomy and sustainability is a crucial issue that requires a holistic and participatory approach. A circular bioeconomy can offer multiple benefits for the environment, society, and economy, but it also poses significant challenges that need to be addressed. By applying the principles of circularity and bioeconomy, for both ourselves and the world, we can build a future that may be more robust and sustainable.

4.2.1 SUSTAINABILITY IN THE AGRICULTURAL SECTOR

Sustainability in the agricultural sector is the ability to produce food and other agricultural products without compromising the natural resources and ecosystems that support them. It involves balancing the environmental, social, and economic aspects of farming and food systems, such as soil health, water quality, biodiversity, climate change, animal welfare, human health, and livelihoods. Sustainability in the agricultural sector is essential for ensuring food security, reducing poverty, and protecting the planet for future generations. To illustrate, agricultural lands are multifunctional and provide a variety of cultural, supporting, and regulating ecosystem services, in addition to food, fodder, and fuel.

Given that agriculture and related industries are the backbone of the economies of many developed and developing countries, which directly or indirectly account for nearly 82% of the world's population, the interest in agricultural sustainability dates to the environmental concerns that first emerged in the 1950s and 1960s.[18] This implies that sustainability in this sector would increase human prosperity. Improved production technologies must be introduced, adopted, and developed to achieve this goal.[19] However, the adoption of such input-intensive agricultural practices often leads to severe soil nutrient depletion and causes a linear increase in crop residue production in modern agricultural practices. Besides, farmers often burn these crop residues on site, which generates serious environmental impacts.[20] Burning crop leftovers have severe detrimental effects on both human health and the environment. Sustainable management of crop residues to reduce the harmful impact on the environment and humans is a serious challenge for scientists and policymakers. Since these approaches frequently fail to directly boost farm income, farmers are frequently unwise to implement any corrective actions for proper crop residue management. Therefore, it is crucial to choose adequate crop residue management techniques that not only protect the environment but also boost farm profitability.[21,22]

4.2.2 SUSTAINABILITY IN FOOD PROCESSING AND MANUFACTURING

Food businesses are still under pressure from outside parties to adopt sustainable practices in their operations and supply chains, including customers, regulators, and non-governmental groups.[23] According to some researchers, a typical food item is handled 33 times on average before a customer touches it in a supermarket.[24] The study at the process level has generally been disregarded, even though this has given rise to a significant number of supply-side studies that have focused on food waste and losses at specific levels of the supply chain (i.e., wholesalers, distributors, and retailers).[25] Food losses can occur for a variety of causes during the manufacturing process, including overproduction, defective products, product and container damage, inadequate cleaning methods, and low compliance.[26] Similarly, product spills, damage, and contamination may be caused by, for instance, operator neglect, improper handling techniques, and equipment failure leading to insufficient package seals. Additionally, food loss can happen because of inadequate compliance with any ingredient or product that falls short of requirements for quality, appearance, taste, or flavor.[27]

Recently, studies conducted by Dora et al.[28] in 47 food processing companies in Belgium showed that on average, the amount of food wasted in the companies represents 2.1% of the total production. This translates to an avcragc loss of 1,011 tons based on an average production of 48,839 tons in all subsectors. Additionally, these researchers discovered that more than 10% of the businesses reported significant losses because of product changes (13%) or human error (10.9%); however, they also found that excessive losses were primarily attributable to product flaws (8.7%) and purchase agreements (6.4%).

Another important type of waste, which has a great potential for revalorization, is the underutilized comestibles. For instance, it is possible to create bioactive hydrolysates using the protein by-products of animal processing.[29] Currently underused is wheat bran, a by-product of wet milling wheat. It has dietary fiber, B vitamins, minerals, and protein. The protein content alone equates to the annual loss of 15.5 million tons of high-quality wheat protein. The idea of extracting the protein for usage as a dietary additive and for the creation of bioactive peptides is, therefore, gaining popularity.[30] Another underutilized resource is waste from the production of fruit juice. Pectin, hemicelluloses, cellulose, and lignin are examples of dietary fiber from apple pomace. Flavonols, phenolic acids, dihydrochalcones, and anthocyanins are examples of phenolic chemicals that can be removed and reincorporated into the food chain.[31] Wastewater from olive mills can be used to recover valuable phenolic compounds with antioxidant characteristics for the olive and palm oil industries.[32]

Therefore, the relationship between sustainability and the potential uses of underutilized comestibles is an issue of great relevance for human development and biodiversity conservation. To clarify, underutilized comestibles are those foods that have nutritional or medicinal value, but are not adequately utilized for various reasons, such as lack of knowledge, lack of access, preference for other products, or poor promotion. These foods can contribute to improving food security, health, and well-being of populations, as well as reducing the environmental impact of agricultural production. Examples of underutilized comestibles include native species used as "quelites" in Mexico, ultra-processed products containing low-cost, high-energy ingredients, and wild or cultivated species for medicinal purposes in Venezuela.

In summary, sustainability implies satisfying current needs without compromising those of future generations, which requires a rational and equitable use of natural resources. Underutilized comestibles can offer a sustainable alternative to diversify diets, take advantage of local resources, rescue traditional knowledge, and encourage innovation. However, they also face several challenges, such as habitat loss, genetic erosion, competition with other crops, lack of regulation or certification, and low social or cultural acceptance. It is, therefore, necessary to promote actions that encourage the recognition, valorization, and sustainable use of underutilized comestibles, as well as to generate scientific and technological information that supports their potential as sources of food and health.

4.2.3 THE ROLE OF CONSUMERS IN THE SUSTAINABILITY OF FOOD PRODUCTS

The consumer plays a crucial role in the sustainability of food products, as their choices and behaviors can influence the environmental, social, and economic impacts of food production and consumption. The consumer can contribute to sustainability by choosing foodstuffs that are produced in a way that minimizes the use of natural resources, reduces greenhouse gas emissions, preserves biodiversity, respects animal welfare and human rights, and supports local communities. The consumer can also adopt sustainable behaviors such as reducing food waste, eating a balanced and healthy diet, and avoiding overconsumption of animal products. By doing so, the consumer can help to ensure the availability and accessibility of food for current and future generations, as well as to improve their own health and well-being.

The methods used to produce, distribute, and consume food have an impact on how sustainability is approached. According to the European Environment Agency,[33] this requires a delicate shift in perspective regarding the entire food system, mainly to improve its fundamental understanding and to discover possibilities for creating meaningful activities to promote the transition. All actors with the power to influence the system, including the consumer, must be involved in this regard. The success of sustainable food products may possibly be directly influenced by the consumer, just like in any other market sector.[34] As a result, starting with the strategies of institutions and businesses, the consumer can switch from being a passive receiver to playing an active role and being able to affect the market trend through their purchasing decisions.[35] Aschemann-Witzel et al.[36] indicate that it is possible to change the consumer behavior by providing information and knowledge, stressing that this could be effectively embodied in food labeling. However, it is noted that awareness of the problem in general is required, as well as warning individuals of the fact that it may occur in one's places and not only in other people's homes.

4.2.3.1 CURRENT CONSUMER TRENDS

One of the current consumer trends in processed foods is the growing demand for minimally processed and without genetically modified organisms. Furthermore, consumers are becoming increasingly conscious of the negative effects that ultra-processed foods, which are rich in calories, sugar, salt, fat, and additives, have on their health and the environment. According to a comprehensive review by Zhang and Giovannucci,[37] ultra-processed foods are associated with increased risks of mortality, cardiovascular disease, obesity, diabetes, cancer, and other diseases. Therefore, customers are calling for the provision of goods that not only do no damage but also have advantages like lowering disease risk factors and promoting healthy aging through better health and mental well-being.[38] As a result, a lot of customers are searching for substitutes that are more like nature intended and have fewer artificial components. Fresh fruits and vegetables, nuts, seeds, eggs, milk, cheese, yogurt, and whole grains are a few examples of minimally processed foods.

The preference for eco-friendly items has increased in addition to the aforementioned factors. Customers relate them to goods that minimize or improve environmental effects, such as lowering waste, conserving energy, or promoting biodiversity. Because of this and their desire to change the world, consumers are becoming more and more demanding of eco-friendly goods and shopping experiences. Some of the current consumer trends in eco-friendly products are listed in Table 4.1.

TABLE 4.1 Main Characteristics of the Products Recognized as Eco-friendly.

Type	Motivation for the choice	References
Natural, non-toxic products	The maintenance of health and safety by avoiding the intake of products containing dangerous chemicals or synthetic materials is of increased concern. Additionally, buyers are increasingly favoring goods made of natural, organic, or recycled materials.	[39]
Zero waste	Products with no or little waste are more in demand, and refillable or biodegradable packaging is preferred.	[40]

TABLE 4.1 *(Continued)*

Type	Motivation for the choice	References
Closed loop manufacturing	Customers value products with a longer lifespan and are supportive of companies that collect and recycle their used products.	[41,42]
Brands are getting political	Consumers expect firms to show their dedication to sustainability through their messages and activities to gain their confidence and loyalty.	[43]
Makers and hyper-local merchandise	Customers help small businesses to use nearby resources and materials to make their products; Thus, lowering their carbon footprint, and giving back to the neighborhood.	[44]

The integration of these approaches has established the significance of the consumer's role in the development of new foods and beverages: the factors influencing consumer acceptance center on the innovation of traditional products; the expanding market for new healthy foods and beverages; and consumers' willingness to pay for innovations in the production of food products.[45] Thus, in food and beverage development, in addition to considering the specific needs of industrialists and developers, it is necessary to consider the customers to whom the food product may be focused.[46] Consequently, the customer archetype is an important factor in this adventure and in these preferences, it states that products derived from natural, eco-friendly, and preferably minimally processed products are currently in demand in the food market.[47] Hence, what was really moving the market in the COVID period was the behavior that consumers acquired at that time according to their age.

Figure 4.3 shows the varying preferences and demands that different generations have and the impact on the development of foods and beverages[45,48] Baby boomers, those born between 1946 and 1964, are worried about chronic diseases and aging and are looking for products that will improve their health and fitness.[49] The generation known as Generation X, which was born between 1965 and 1980, is more skeptical and autonomous and seeks out items that are practical, genuine, and natural. Since they are more diversified and adventurous than previous generations, millennials—those born between 1981 and 1996—demand goods that are creative, individualized, and sustainable. Born after 1997, millennials are increasingly digital and social, demanding products that are open, moral, and interesting.

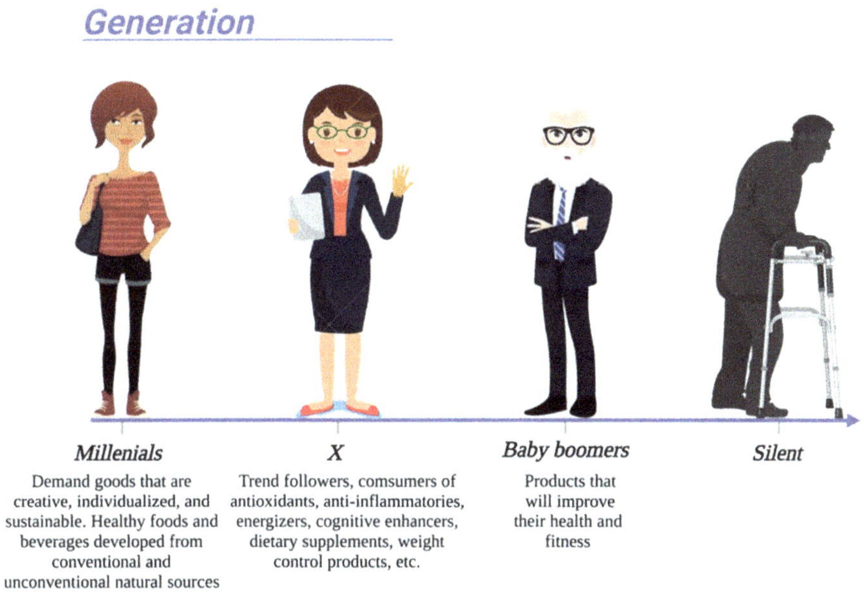

FIGURE 4.3 Attitudes of consumers according to the generation to which they belong for the choice of healthy products.

4.3 REVALORIZATION OF SOURCES FOR SUSTAINABLE PROCESSING OF FUNCTIONAL FOODS

Due to its potential to impact positively on health and its link to the utilization of raw materials of natural origin, functional nutrition has emerged as one of the key directions for a healthy lifestyle and sustainable food production.[50] Because of this, it has drawn a lot of interest from consumers and the food businesses, who are concerned with economic growth and human welfare.[51] Some organizations such as Health Canada define functional foods as "similar in appearance to a conventional food, consumed as part of the regular diet, with demonstrated physiological benefits in reducing the risk of chronic disease beyond basic nutritional functions," this definition is extrapolated to the concept of functional beverage.

Through the revalorization of agro-industrial by-products, efforts are being concentrated to generate healthy and environmentally friendly products. These patterns demonstrate how customers are becoming more conscious of and accountable for their buying decisions, and they expect brands to follow suit. Green products are beneficial to consumers' health and satisfaction as well as the environment. Many of these goods, which fall under the category of functional foods, are also known for their health advantages.

This trend is centered on the demands of consumers who demand accessible products that they consider healthier and eco-friendly. However, not having scientific support in the selection of these natural sources for the development of new products can be counterproductive. In a scenario where several contemporary diseases are recognized as multifactorial, they can hardly be associated with the excessive consumption of natural products. Although they may be traditionally recognized as safe foods, there is already enough scientific evidence indicating that they can promote health problems. For example, we have the case of tea, which is a popular beverage with various health advantages, including lowering the risk of cardiovascular disease, diabetes, neurodegeneration, and cancer. These bioactivities have been connected to polyphenols, which are found in tea and can scavenge free radicals and shield cells from oxidative damage[52–54] Tea consumption due to its polyphenol composition has also been recognized for their modulating effect on the microbiota with a significant impact on digestion, immunity, metabolism, and mood.[55,56]

Despite these beneficial effects, the form of consumption may promote adverse effects mainly associated with its alkaloid content. To exemplify, some people may feel lightheaded after drinking tea, especially if they do so in big quantities or first thing in the morning. This might be brought on by the fact that tea contains caffeine, which affects blood pressure[57,58] and dehydrates people.[59] The caffeine naturally present in tea can stimulate the nervous system and cause nervousness, anxiety, sleep disorder, and cognition.[60] A part of the population that consumes this type of beverage, especially those predisposed to migraine or tension headaches, may experience headaches after drinking tea. It has been reported that caffeine has the potential to constrict or dilate blood vessels in the brain and activate pain receptors. In addition, it has been indicated that the consumption of caffeine during pregnancy may pose some dangers to both the mother and the fetus, with a negative correlation between caffeine consumption and newborn birth weight and height,[61] and miscarriage. The heart rate and blood pressure of the fetus may be affected by caffeine as it crosses the placenta.[62] It is important to emphasize that these effects have been observed at high consumption doses. Thus, in COVID times, it was important to raise the alert given the increased demand for the supply of "functional" foods and beverages, regularly associated with natural products often abundant in alkaloids. Meanwhile, the consumer perception of caffeine-rich sources may differ depending on several variables, including health consciousness, cultural preferences, individual preferences, and price sensitivity. Caffeinated products may be viewed favorably by some users as improving alertness, mood, and performance, while negatively by others as creating addiction, insomnia, anxiety, and cardiovascular issues.[63] Additionally, some consumers might favor organic caffeine sources like green tea or yerba mate over man-made ones like energy drinks or tablets.

Another industry, on the other hand, is searching for natural substitutes with a low level of alkaloids but maintaining the quality of recognized bioactives like polyphenols in their composition. The selection of unconventional sources that were originally ingested by the population and that currently have ethnobotanical or scientific proof of prospective health impacts on the population has led to a return to the rescue of knowledge. Also, according to consumer interests, it was related to ethical and sustainable aspects in the production of food and beverages. Therefore, it is important to highlight business strategies, as in recent years they have been based on the circular economy and the use of agro-industrial waste from those sources rich in bioactive compounds. As a result, different conventional and non-conventional sources have been considered for the development of healthy foods and beverages, especially based on the importance of the chemical composition of these undervalued matrices. Thus, these non-conventional sources are important from a socioeconomic point of view.

4.3.1 NON-CONVENTIONAL SOURCES FOR THE RECOVERY OF BIOACTIVE COMPOUNDS

There is a growing interest in obtaining bioactive compounds from agro-industrial residues. Among them are little explored sources such as those derived from pruning processes of fruit trees such as guava or berries; producing charcoal, from large volumes of foliar resources; using wasted pods and acorns, sources rich in polyphenolic compounds, including mesquite and oak trees. Therefore, it is important to establish, in a holistic and sustainable approach, those conventional and non-conventional extraction methods that allow a higher yield of bioactives with greater functional properties.

As previously mentioned, revalorization is the flagship of the circular bioeconomy, and a holistic and sustainable approach is mandatory. The shift to the bioeconomy era will be aided by valorizing agri-food leftovers to create bio-based platform chemicals. The revalorization of non-conventional sources to obtain bioactive compounds tries to recover and make use of valuable substances from typically discarded or underutilized natural resources.[64]

The term "non-conventional sources" refers to organisms that are not typically exploited or recognized for their therapeutic benefits. These sources include microalgae, fungi, insects, marine organisms, agricultural waste, and agroforestry. Bioactive substances, such as antioxidants, antimicrobials, anti-inflammatories, and anticancer drugs, have positive effects on living systems with favorable impacts on human health. It is possible to create new uses in the food, pharmaceutical, cosmetic, and biotechnology industries by extracting and purifying these substances from unconventional sources.

As previously established in this chapter, consumer perception of these sources is generally favorable because they are seen as natural, sustainable, and innovative alternatives. However, some customers might be worried about the security, usefulness, and caliber of these goods, particularly if they are not ruled or standardized. Thus, revalorization is a strategy to recover valuable components from food by-products and waste streams and use them to produce functional foods, but it must be escorted by the supervision of regulatory agencies in each country.

Some of the trends in source revalorization for sustainable processing of functional foods and beverages are using biotechnology to enhance the extraction and modification of bioactive compounds from plant and animal sources, such as enzymes, peptides, polyphenols, and carotenoids. It comprises the developing of novel processing methods that minimize the environmental impact and energy consumption of source revalorization, as supercritical fluid extraction, microwave-assisted extraction, and ultrasound-assisted extraction. It also involves exploring new sources of functional ingredients from underutilized or unconventional food materials, such as algae, insects, microorganisms, and marine by-products, and incorporating source revalorization into circular economy models that aim to reduce waste generation and maximize resource efficiency in the food sector.[65,66]

Again, tea is the classical example to extrapolate the revalorization of foliar resources derived from the production of fruits or charcoal. In the conventional methods, the extraction of polyphenols and other bioactive chemical groups has contemplated the use of different extraction solvents such as water, hydroalcoholic solutions, acetone, methanol, and isopropanol, among others. The selection of these solvents depends on the polarity of the chemical group to be obtained. Also, as a conventional method, the temperature has been used through boiling and Soxhlet, even though in the case of polyphenol extraction, this chemical group will be affected by its sensitivity to high temperatures. To make the extraction of bioactive compounds of phenolic nature more efficient, the pharmaceutical, nutraceutical, and food industries have made enormous efforts to replace these conventional techniques, firstly to decrease the volume of contaminants when solvents are used and secondly to increase the yields of extracts rich in bioactive compounds. Figure 4.4 shows some innovative technologies currently considered as non-conventional methods, but they have proven to be safe, fast, energy-saving, eco-friendly, and significantly increasing the extraction yields of target bioactive compounds. These emerging technologies include ultrasound, microwaves, pressurized liquids, electric pulses, and supercritical fluids.[67]

4.3.1.1 USE OF BIOACTIVE INGREDIENTS IN THE DEVELOPMENT OF FUNCTIONALIZED FOOD

A functionalized food is a type of food product that claims additional benefits for health promotion or disease prevention, besides providing nutrients and energy. Functionalized foods can be made from natural, whole foods including fruits, vegetables, legumes, grains, nuts, seeds, shellfish, fermented foods, spices, and herbs that are high in vitamins, minerals, antioxidants, and other health-promoting substances. Functionalized meals can also be altered or fortified with additional ingredients or supplements to improve their health benefits, such as probiotics, fiber, or vitamins. Functionalized foods are popular in many countries and may provide protection against disease, prevent nutrient deficiencies, and support proper growth and development.

Global initiatives to encourage healthy eating have failed frequently due to consumer opposition to changing their preferences. This situation gives an opportunity to encourage better lifestyles using functional items without significantly altering consumer behavior by the inclusion of bioactive ingredients in highly consumed meals, such as pasta.[68] The incorporation of extracts or bioactive compounds produced using green technology to starch reduces their digestibility by boosting their resistance to enzymatic digestion, which also reduces glycemic response.[69] Another application is reported by Wu et. al.[70] who indicate that whey proteins and isolated whey protein fractions can fully interact with different food phenolic sources and forms in aqueous solutions and at various pH levels, creating complexes and conjugates that protect and supply food phenols, while improving their bioaccessibility and bioavailability. These authors indicate that when whey proteins and dietary phenols are combined, novel functional whey protein components can be produced at both the molecular and combination levels.

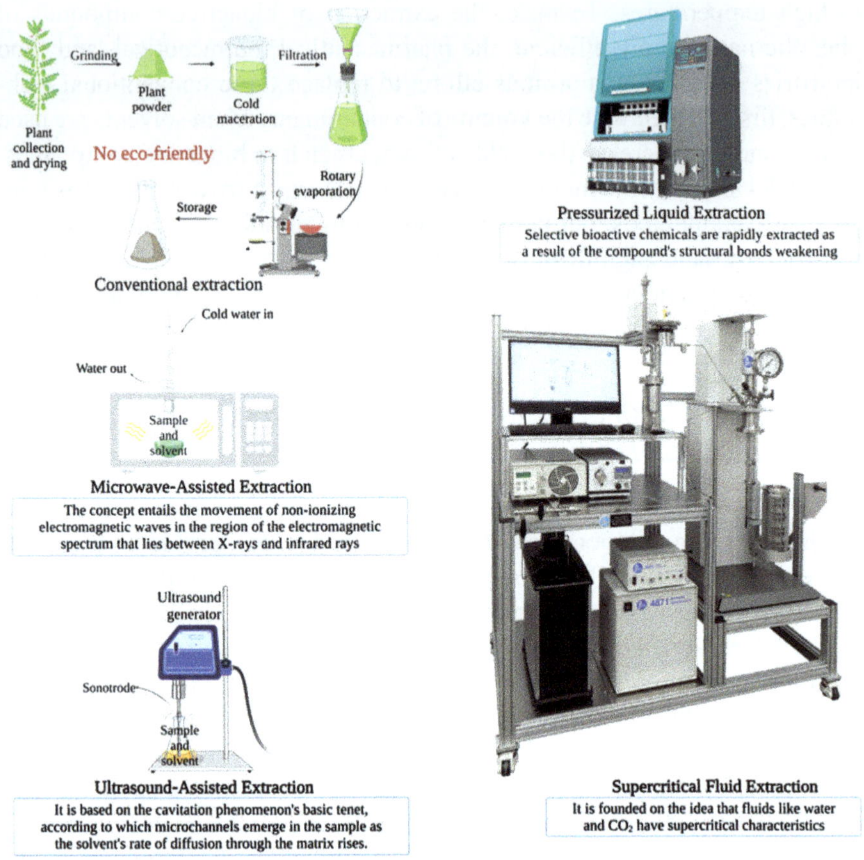

FIGURE 4.4 Conventional and non-conventional processing technologies for the extraction of bioactive compounds.

Studies by Zokaityte et al.[71] indicate that beverages enriched with bioactive-rich sources are more acceptable to consumers because they induce positive emotions and have desirable nutraceutical properties. Such beverages, which are prepared in a sustainable and environmentally friendly way, have a great potential to position themselves in the market.

4.3.2 FOOD PROCESSING WITH SUSTAINABLE FUNCTIONAL PROPERTIES

Food processing with sustainable functional properties is a type of food processing that aims to preserve or enhance the nutritional, sensory, and health benefits of food products while minimizing the environmental impact of the processing methods. The value of the global market for functional foods was $178 billion in 2019, $258.8 billion in 2020, and $281.4 billion in 2021, respectively. Forecasts include a value of $530 billion in 2028 and a compound annual growth rate of roughly 10% according to Fortune Business Insights.[72]

As already mentioned previously, changing lifestyles, rising affluence, and awareness of the health advantages of functional foods are the primary drivers of the increase in demand for healthy and nourishing diets.[73] To achieve a "clean label" designation, food companies have adopted technologies and improved their processes to reduce energy consumption, use less plastic for packaging materials, use more naturally derived ingredients in formulations, and reduce the use of synthetic ingredients.[74] Some of the characteristics of this type of food processing are the use of natural or renewable resources as raw materials, additives, or packaging materials; the reduction of energy consumption, water usage, waste generation, and greenhouse gas emissions during the processing stages; the application of modern technologies to enhance the quality and safety of food items without affecting their functional features, such as high-pressure processing, pulsed electric fields, ultrasound, or cold plasma; the incorporation of bioactive compounds such as antioxidants, probiotics, prebiotics, or functional peptides into food products to enhance their health-promoting effects; and the development of new food products that meet the specific needs and preferences of consumers in terms of taste, texture, appearance, convenience, and shelf life.[75] Finally, an important application of bioactive compounds obtained by green technologies is related to their antimicrobial activities, which is why they are being used in the manufacture of packaging to extend the shelf life of foods.[76–78]

4.4 CONCLUSIONS

In this chapter, we have emphasized the role of functional foods and beverages played in the idea of sustainability. The reduction of residues on the bioactive chemicals recovered from food waste is an essential component of sustainability. It is typically performed as a drying operation during the recovery of bioactive compounds from food waste to concentrate these compounds, using the least amount of solvent. Numerous studies have been conducted to determine the best methods for utilizing and valorizing food waste to recover the most content of bioactive components while having the least negative environmental effects. These tactics can be used regardless of the source of the bioactive molecule, whether it comes from a food waste recovery or is naturally present in the food matrix.

KEYWORDS

- **consumer**
- **holistic**
- **sustainability**
- **eco-friendly**
- **foods**

REFERENCES

1. Kardung, M.; Cingiz, K.; Costenoble, O.; Delahaye, R.; Heijman, W.; Lovric, M.; van Leeuwen, M.; M´Barek, R.; van Meijl, H.; Piotrowski, S.; Ronzon, T.; Sauer, J.; Verhoog, D.; Verkerk, P. J.; Vrachioli, M.; Wesseler, J. H. H.; Zhu, B. X. Development of the Circular Bioeconomy: Drives and Indicators. *Sustainability* **2021,** *13,* 413.
2. Venkatesh, G. Circular Bio-economy-paradigm for the Future: Systematic Review of Scientific Journal of Scientific Journal Publications from 2015 to 2021. *Circ. Econ. Sustain.* **2021,** *2,* 231–279.
3. Dahiya, D.; Sharma, H.; Rai, A. K.; Singh-Nigam, P. Application of Biological System and Processes Employing Microbes and Algae to Reduce, Recycle, Reuse (3Rs) for the Sustainability of Circular Bioeconomy. *AIMS Microbiol.* **2022,** *8* (1), 83–102.
4. Velenturf, A. P. M.; Purnell, P. Principles for a Sustainable Circular Economy. *Sustain. Prod. Consump.* **2021,** *27,* 1437–1457.
5. Lokko, Y.; Heijde, M.; Schebesta, K.; Scholtès, P.; Van Montagu, M.; Guacca, M. Biotechnology and the Bioeconomy-towards Inclusive and Sustainable Industrial Development. *New Biotechnol.* **2018,** *40,* 5–10.
6. Egea, F. J.; Torrente, R. G.; Aguilar, A. An Efficient Agro-Industrial Complex in Almería (Spain): Towards and Integrated and Sustainable Bioeconomy Model. *New Biotechnology* **2018,** *40,* 103–112.
7. Caniglia, G.; Luederitz, C.; von Wirth, T.; Fazey, I.; Martín-López, B.; Hondrila, K.; König, A.; von Wehrden, H.; Schäpke, N. A.; Laubichler, M. D.; Lang, D. J. A Pluralistic and Integrated Approach to Action-oriented Knowledge for Sustainability. *Nat. Sustain.* **2021,** *4,* 93–100.
8. Norström, A. V.; Cvitanovic, C.; Löf, M. F.; West, S.; Wyborn, C.; Balvarena, P.; Bednarek, A. T.; Bennett, E. M.; Biggs, R.; de Bremond, A.; Campbell, B. M.; Canadell, J. B.; Carpenter, S. R.; *et al.* Principles for Knowledge Co-production in Sustainability Research. *Nat. Sustain.* **2020,** *3,* 182–190.

9. Elkington, J. Towards the Sustainable Corporation: Win-Win-Win Business Strategies for Sustainable Development. *California Management Review* **1994,** *36* (2), 90–100.
10. Sarkis, J. Supply Chain Sustainability: Learning from the COVID-19 Pandemic. *Int. J. Operat. Prod. Manag.* **2020,** *41* (1), 63–73.
11. Barbier, E. B.; Burgess, J. C. Sustainability and Development After COVID-19. *World Dev.* **2020,** *135,* 105082.
12. Chen, J.-Y.; Dimitrov, S.; Pun, H. The Impact of Government Subsidy on Supply Chains' Sustainability Innovation. *Omega* **2018,** *86,* 42–58.
13. Adams, A. T. Global Consumers are Willing to Put Their Money Where Their Heart is. 2014. Retrieved from: https://www.businesswire.com/news/home/20140616006605/en/Global-Consumers-Are-Willing-to-Put-Their-Money-Where-Their-Heart-is-When-It-Comes-to-Goods-and-Services-from-Companies-Committed-to-Social-Responsibility
14. Wilkinson, A.; Hill, M.; Gollan, P. The Sustainability Debate. *Int. J. Operat. Prod. Manag.* **2001,** *21* (12), 1492–1502.
15. Dietz, T.; Börner, J.; Förster, J. J.; von Braun, J. Governance of the Bioeconomy: A Global Comparative Study of National Bioeconomy Strategies. *Sustainability* **2018,** *10,* 3190.
16. D´Amato, D.; Veijonaho, S.; Toppinen, A. Towards Sustainability? Forest-based Circular Bioeconomy Business Models in Finnish SMEs. *Forest Policy Econ.* **2020,** *110,* 101848.
17. Fallah-Shayan, N.; Mohabbati-Kalejahi, N.; Alavi, S.; Alli-Zahed, M. Sustainable Development Goals (SDGs) as a Framework for Corporate Social Responsibility (CSR). *Sustainability* **2022,** *14,* 1222.
18. Pretty, J.; Bharucha, Z. P. Sustainable Intensification in Agricultural Systems. *Ann. Botany* **2014,** *114* (8), 1571–1596.
19. Kar, S.; Pramanick, B.; Brahmachari, K.; Saha, G.; Mahapatra, B.; Saha, A.; Kumar, A. Exploring the Best Tillage Option in Rice Based Diversified Cropping Systems in Alluvial Soil of Eastern India. *Soil Tillage Res.* **2021,** *205,* 104761.
20. Sarkar, S.; Skalicky, M.; Hossain, A.; Brestic, M.; Saha, S.; Garai, S.; Ray, K.; Brahmachari, K. Management of Crop Residues for Improving Input use Efficiency and Agricultural Sustainability. *Sustainability* **2020,** *12* (23), 9808.
21. Valkama, E.; Kunypiyaeva, G.; Zhapayev, R.; Karabayev, M.; Zhusupbekov, E.; Perego, A.; Schillaci, C.; Sacco, D.; Moretti, B.; Grignani, C.; Acutis, M. Can Conservation Agriculture Increase Soil Carbon Sequestration? A Modelling Approach. *Geoderma* **2020,** *369,* 114298.
22. Raza, M. H.; Abid, M.; Yan, T.; Ali-Naqvi, A. S.; Akhtar, S.; Faisal, M. Understanding Farmers' Intentions to Adopt Sustainable Crop Residue Management Practices: A Structural Equation Modeling Approach. *J. Clean. Prod.* **2019,** *227,* 613–623.
23. Adams, D.; Donovan, J.; Topple, C. Achieving Sustainability in Food Manufacturing Operations and Their Supply Chains: Key Insights From a Systematic Literature Review. *Sustain. Prod. Consump.* **2021,** *28,* 1491–1499.
24. Kantor, L. S.; Lipton, K.; Manchester, A.; Oliveira, V. Estimating and Addressing America's food. *Food Rev.* **1997,** *20* (1), 2–12.
25. Sheahan, M.; Barrett, C. B. Food Loss and Waste in Sub-Saharan Africa: A Critical Review. *Food Policy* **2017,** *70,* 1–12.
26. Halloran, A.; Clement, J.; Kornum, N.; Bucatariu, C.; Magid, J. Addressing Food Waste Reduction. *Food Policy* **2014,** *49,* 294–301.

27. Dora, M.; Van-Goubergen, D.; Molnar, A.; Gellynck, X.; Kumar, M. Adoptability of Lean Manufacturing Among Small and Medium Food Processing Enterprises. *62nd IIE Annual Conference and Expo* **2012,** 806–814.
28. Dora, M.; Wesana, J.; Gellynck, X.; Seth, N.; Dey, B.; De-Steur, H. Importance of Sustainable Operations in Food Loss: Evidence from the Belgian Food Processing Industry. *Ann. Operat. Res.* **2020,** *290,* 47–72.
29. Martínez-Alvarez, O.; Chamorro, S.; Brenes, A. Protein Hydrolysates from Animal Processing By-Products as a Source of Bioactive Molecules with Interest in Animal Feeding: A Review. *Food Res. Int.* **2015,** *73,* 204–212.
30. Balandrán-Quintana, R. R.; Mercado-Ruiz, J. N.; Mendoza-Wilson, A. M. Wheat Bran Proteins: A Review of Their Uses and Potential. *Food Rev. Int.* **2015,** *31* (3), 279–293.
31. Rabetafika, H. N.; Bchir, B.; Blecker, C.; Richel, A. Fractionation of Apple by-products as Source of New Ingredients: Current Situation and Perspectives. *Trends Food Sci. Technol.* **2014,** *40,* 99–114.
32. El-Abbassi, A.; Kiai, H.; Hafidi, A. Phenolic Profile and Antioxidant Activities of Olive Mill Wastewater. *Food Chem.* **2012,** *132,* 406–412.
33. EEA. Food in Green Light: A Systems Approach to Sustainable Food. *EEA Rep.* **2017,** *16,* 64–95.
34. Notarnicola, B.; Sala, S.; Anton, A.; McLaren, S. J.; Saouter, E.; Sonesson, U. The Role of Life Cycle Assessment in Supporting Sustainable Agri-food. *J. Clean. Prod.* **2017,** *140,* 399–409.
35. Peano, C.; Merlino, V. M.; Sottile, F.; Borra, D.; Massaglia, S. Sustainability for Food Consumers: Which Perception?. *Sustainability* **2019,** *11* (21), 5955.
36. Aschemann-Witzel, J.; De-Hooge, I.; Amani, P.; Bech-Larsen, T.; Oostindjer, M. Consumer-related Food Waste: Causes and Potential for Action. *Sustainability* **2015,** *7* (6), 6457–6477.
37. Zhang, Y.; Giovannucci, E. L. Ultra-processed Foods and Health: a Comprehensive Review. *Crit. Rev. Food Sci. Nutrit.* **2022,** 1–13.
38. Augustin, M. A.; Riley, M.; Stockmann, R.; Bennett, L.; Kahl, A.; Lockett, T.; Osmond M.; Sanguansri, P.; Stonehouse, W.; Zajac, I.; Cobiac, L. Role of Food Processing in Food and Nutrition Security. *Trends Food Sci. Technol.* **2016,** *56,* 115–125.
39. Bom, S.; Jorge, J.; Ribeiro, H. M.; Marto, J. A step Forward on Sustainability in the Cosmetic Industry: A Review. *J. Clean. Prod.* **2019,** *225,* 270–290.
40. Boz, Z.; Korhonen, V.; Sand, C. K. Consumer Considerations for the Implementation of Sustainable Packaging: A Review. *Sustainability* **2020,** *12,* 2192.
41. De Man, J. C.; Stranhagen, J. O. An Industry 4.0 Research Agenda for Sustainable Business Models. *Procedia CIRP* **2017,** *63,* 721–726.
42. Vanapalli, K. R.; Sharma, H. B.; Ranjan, V. P.; Samal, B.; Bhattacharya, J.; Dubey, B. K.; Goel, S. Challenges and Strategies for Effective Plastic Waste Management During and Post COVID-19 Pandemic. *Sci. Total Environ.* **2021,** *750,* 141514.
43. Iglesias, O.; Markovic, S.; Bagherzadeh, M.; Singh, J. J. Co-creation: a Key Link Between Corporate Social Responsibility, Customer Trust, and Customer Loyalty. *J. Bus. Ethics* **2020,** *163,* 151–166.
44. Nandi, S.; Sarkis, J.; Hervani, A. A.; Helms, M. M. Redesigning Supply Chains Using Blockchain-Enabled Circular Economy and COVID-19 Experiences. *Sustain. Prod. Consump.* **2021,** *27,* 10–22.

45. Guiné, R. P. F.; Florenca, S. G.; Barroca, M. A.; Anjos, O. The Link Between the Consumer and the Innovations in Food Product Development. *Foods* **2020,** *9,* 1317.
46. Wu, W.; Zhang, A.; van Klinken, R. D.; Schrobback, P.; Muller, J. M. Consumer Trust in Food and the Food System: A Critical Review. *Foods* **2021,** *10* (10), 2490.
47. Smiglak-Krajewska, M.; Wojciechowska-Solis, J. Consumer Versus Organic Products in the COVID-19 Pandemic: Opportunities and Barriers to Market Development. *Energies* **2021,** *14,* 5566.
48. Küster-Boluda, I.; Vidal-Capilla, I. Consumer Attitudes in the Election of Functional Foods. *Span. J. Mark.-ESIC* **2017,** *21* (S1), 65–79.
49. LeRouge, C. M.; Tao, D.; Ohs, J.; Lach, H. W.; Jupka, K.; Wray, R. Challenges and Opportunities with Empowering Baby Boomers for Personal Health Information Management using Consumer Health Information Technologies: An Ecological Perspective. *AIMS Public Health* **2014,** *1* (3), 160–181.
50. Ogori, A. F.; Amove, J.; Aduloju, P.; Sardo, G.; Okpala, C. O.; Bono, G.; Korzeniowska, M. Functional and Quality Characteristics of Ginger, Pineapple, and Turmeric Juice Mix as Influenced by Blend Variations. *Foods* **2021,** *10* (3), 525.
51. Putnik, P.; Bezuk, I.; Barba, F. J.; Lorenzo, J. M.; Polunić, I.; Bursać, D. K. Chapter 5 - Sugar Reduction: *Stevia rebaudiana* Bertoni as a Natural Sweetener. *Agri-Food Industry Strat. Healthy Diets Sustai.***2020,** 123–152.
52. Zaveri, N. T. Green Tea and its Polyphenolic Catechins: Medicinal uses in Cancer and Noncancer Applications. *Life Sci.* **2006,** *78,* 2073–2080.
53. Hayat, K.; Iqbal, H.; Malik, U.; Bilal, U.; Mushtaq, S. Tea and its Consumption: Benefits and Risks. *Crit. Rev. Food Sci. Nutrit.* **2015,** *55* (7), 939–959.
54. Aguilera, Y.; Martin-Cabrejas, M. A.; González-de Mejía, E. Phenolic Compounds in Fruits and Beverages Consumed as Part of the Mediterranean Diet: Their Role in Prevention of Chronic Diseases. *Phytochem. Rev.* **2016,** *15,* 405–423.
55. Ma, H.; Zhang, B.; Hu, Y.; Liu, W. J.; Qin, R.; Lv, S.; Wang, S. Correlation Analysis of Intestinal Redox State with the Gut Microbiota Reveals the Positive Intervention of Tea Polyphenols on Hyperlipidemia in High Fat Diet Fed Mice. *J. Agric. Food Chem.* **2019,** *67* (26), 7325–7335.
56. Zhang, Y.; Cheng, L.; Liu, Y.; Zhang, R.; Wu, Z.; Cheng, K.; Zhang, X. Omics Analyses of Intestinal Microbiota and Hypothalamus Clock Genes in Circadian Disturbance Model Mice Fed with Green Tea Polyphenols. *J. Agric. Food Chem.* **2022,** *70* (6), 1890–1901.
57. Rogers, P. J.; Smith, J. E.; Heatherley, S. V.; Pleydell-Pearce, C. W. Time for Tea: Mood, Blood Pressure and Cognitive Performance Effects of Caffeine and Theanine Administered Alone and Together. *Psychopharmacology* **2008,** *195,* 569–577.
58. Hodgson, J. M.; Puddey, I. B.; Woodman, R. J. Effects of Black Tea on Blood Pressure: A Randomized Controlled Trial. *The Archives of Internal Medicine* **2012,** *172* (2), 186–188.
59. Ruxton, C. H. S. The Impact of Caffeine on Mood, Cognitive Function, Performance and Hydration: A Review of Benefits and Risks. *Nutrit. Bull.* **2008,** *33* (1), 15–25.
60. Bertasi, R. A. O.; Humeda, Y.; Bertasi, T. G. O.; Zins, Z.; Kimsey, J.; Pujalte, G. Caffeine Intake and Mental health in College Students. *Cureus* **2021,** *13* (4), e14313.
61. Gür, E. Y.; Cil, M. A.; Apay, S. E. The Caffeine Consumption Among Turkish Pregnant Women. *Niger. J. Clin. Pract.* **2022,** *25,* 1507–1516.

62. Nojima, T.; Naito, H.; Kosaki, Y.; Osako, T.; Tanaka, K.; Murata, A.; Nakao, A. Caffeine Intoxication in Pregnancy; A Case Report. *Arch. Acad. Emerg. Med.* **2019,** *7* (1), e67.
63. Samoggia, A.; Riedel, B. Consumers' Perceptions of Coffee Health Benefits and Motives for Coffee Consumption and Purchasing. *Nutrients* **2019,** *11* (3), 653.
64. Otoni, C. G.; Henriette, M. C.; Azeredo, M. C.; Matos, B. D.; Beaumont, M.; Correa, D. S.; Rojas, O. J. The Food-Materials Nexus: Next Generation Bioplastics and Advanced Materials from Agri-Food Residues. *Adv. Mater.* **2021,** *33,* 2102520.
65. Cádiz-Gurrea, M. L.; Villegas-Aguilar, M. C.; Leyva-Jiménez, F. J.; Pimentel-Moral, S.; Fernández-Ochoa, A.; Alañón, M. E.; Segura-Carretero, A. Revalorization of Bioactive Compounds from Tropical Fruit by-products and Industrial Applications by Means of Sustainable Approaches. *Food Res. Int.* **2020,** *138*, part B, 109786.
66. García-Perez, P.; Xiao, J.; Munekata, P. E. S.; Lorenzo, J. M.; Barba, F. J.; Rajoka, M. S. R.; Barros, L.; Mascoloti-Sprea, R.; Amaral, J. S.; Prieto, M. A.; Simal-Gandara, J. Revalorization of Almond By-products for the Design of Novel Functional Foods: An Updated Review. *Foods* **2021,** *10,* 1823.
67. Raghunath, S.; Budaraju, S.; Taghi-Gharibzahedi, S. M.; Koubaa, M.; Roohinejad, S.; Mallikarjunan, K. Processing Technologies for the Extraction of Value-Added Bioactive Compounds From Tea. *Food Engineer. Rev.* **2023,** *15,* 276–308.
68. Carpentieri, S.; Larrea-Wachtendorff, D.; Donsi, F.; Ferrari, G. Functionalization of Pasta Through the Incorporation of Bioactive Compounds from Agri-food By-products: Fundamentals, Opportunities, and Drawbacks. *Trends Food Sci. Technol.* **2022,** *122,* 49–65.
69. Wu, T.-Y.; Sun, N.-N.; Chan, Z.; Chen, C.-J.; Wu, Y.-C.; Chau, C.-F. Enhancement of Digestion Resistance and Glycemic Control of Corn Starch Through Conjugation with Gallic Acid and Quercetin Using the Free Radical Grafting Method. *Processes* **2022,** *10,* 2610.
70. Wu, G.; Hui, X.; Gong, X.; Tran, K. N.; Stipkovits, L.; Mohan, M. S.; Brennan, M. A.; Brennan, C. S. Functionalization of Bovine Whey Proteins by Dietary Phenolics from Molecular-level Fabrications and Mixture-Level Combinations. *Trends Food Sci. Technol.* **2021,** *110,* 107–119.
71. Zokaityte, E.; Lele, V.; Starkute, V.; Zavistanaviciute, P.; Cernauskas, D.; Klupsaite, D.; Ruzauskas, M.; Alisauskaite, J.; Baltrusaityté, A.; Dapsas, M.; Siriakovaite, K.; Trunce, S.; Guiné, R. P. F.; Viskelis, P.; Steibliene, V.; Bartkiene, E. Antimicrobial, Antioxidant, Sensory Properties and Emotions Induced for the Consumers of Nutraceutical Beverages Developed from Technological Functionalized Food Industry By-products. *Foods* **2020,** *9,* 1620.
72. Fortune Business Insights. Functional Foods. https://www.fortunebusinessinsigh (Accessed on: 18 April 2023).
73. Granato, D.; Carocho, M.; Barros, L.; Zabetakis, I.; Mocan, A.; Tsoupras, A.; Gomes-Cruz, A.; Pimentel, T. C. Implementation of Sustainable Development Goals in the Dairy Sector: Perspectives on the use of Agro-industrial Side-streams to Design Functional Foods. *Trends Food Sci. Technol.* **2022,** *124,* 128–139.
74. Noguerol, A. T.; Pagán, M. J.; García-Segovia, P.; Varela, P. Green or clean? Perception of Clean Label Plant-based Products by Omnivorous, Vegan, Vegetarian and Flexitarian Consumers. *Food Res. Int.* **2021,** *149,* 110652.
75. Arshad, R. N.; Abdul-Malek, Z.; Roobab, U.; Nawaz-Rahjha, M. M. A.; Jambrak, A. R.; Qureshi, M. I.; Khan, N.; Lorenzo, J. M.; Aadil, R. M. Nonthermal Food Processing: A

Step Towards a Circular Economy to Meet the Sustainable Development Goals. *Food Chem. X* **2022,** *16,* 100516.

76. Getachew, A. T.; Ahmad, R.; Park, J.-S.; Chun, B.-S. Fish Skin Gelatin based Packaging Films Functionalized by Subcritical Water Extract from Spent Coffee Ground. *Food Packag. Shelf Life* **2021,** *29,* 100735.
77. Salgado-Ramos, M.; Tabasso, S.; Calcio-Gaudino, E.; Moreno, A.; Mariatti, F.; Cravotto, G. An Innovative, Green Cascade Protocol for Grape Stalk Valorization with Process Intensification Technology. *Appl. Sci.* **2022,** *12,* 7417.
78. Jyske, T.; Liimatainen, J.; Tienaho, J.; Brännström, D.; Kunnas, S.; Halmemies, E.; Nakayama, E.; Kilpeläinen, P.; Ora, A.; Kaseva, J.; Hellström, J.; Marjomäki, V. S.; Karonen, M.; Fikushima, K. Inspired by Nature: Fiber Networks Functionalized with Tannic Acid and Condensed Tannin-rich Extracts of Norway Spruce Bark Show Antimicrobial Efficacy. *Front. Bioengineer. Biotechnol.* **2023,** *11,* 1171908.

CHAPTER 5

Current Perspectives on Sustainable Food Product Formulation by Microalgae Cultivation

MARÍA GORETY CONTRERAS-HERNÁNDEZ[1,2],
CORDERO-SOTO ITZA NALLELY[1,3],
ALFREDO DE JESÚS MARTÍNEZ-ROLDÁN[2,4],
MARÍA DOLORES JOSEFINA RODRÍGUEZ-ROSALES[1,2],
SERGIO VALLE-CERVANTES[1,2], and OLGA MIRIAM RUTIAGA-QUIÑONES[3]

[1]*Tecnológico Nacional de México/Instituto Tecnológico de Durango, Departamento de Ingenierías Química y Bioquímica, Felipe Pescador 1803 Ote, Colonia Nueva Vizcaya, Durango, Durango, México*

[2]*Tecnológico Nacional de México/Instituto Tecnológico de Durango. Maestría en Sistemas Ambientales, Departamento de Ingenierías Química y Bioquímica, Felipe Pescador 1803 Ote, Colonia Nueva Vizcaya, Durango, Durango, México*

[3]*Laboratorio Nacional CONAHCYT de Apoyo a la Evaluación de Productos Bióticos (LaNAEPBi), Unidad de Servicio, Tecnológico Nacional de México/ I.T. de Durango (TecNM/ITD)*

[4]*CONAHCYT-Tecnológico Nacional de México/Instituto Tecnológico de Durango. Maestría en Sistemas Ambientales, Departamento de Ingenierías Química y Bioquímica, Felipe Pescador 1803 Ote, Colonia Nueva Vizcaya, Durango, Durango, México*

ABSTRACT

Nowadays, people are interested in food bioproducts that positively impact health. Therefore, the food market has been searching for natural and organic

The Food-Energy-Water Triangle for Sustainable Development: Integrative and Synergic Solutions.
Pedro Aguilar-Zárate, Mónica Chávez-González, Ann Rose Abraham, and A. K. Haghi (Eds.)

products with nutritional attributes that meet consumers' expectations. Microalgae emerge as a solution in this regard. These unicellular photosynthetic microorganisms offer numerous advantages for sustainable food production. On the one hand, their easy cultivation under controlled conditions in photobioreactors reduces the need for land and freshwater, facilitating a circular bioeconomy. On the other hand, microalgae have a high growth rate and productivity, allowing for the production of large quantities of biomass compared to conventional crops. Among the natural active biological products used as food ingredients are lipids (omega-3 and 6) rich in essential fatty acids, pigments (chlorophyll, astaxanthin, and phycocyanin) with antioxidant attributes, phenolic compounds, vitamins, and exopolysaccharides with functional and nutritional properties. All these compounds are employed in the food industry as ingredients in colorants, beverages, additives, and protein formulations, among others. However, critical factors for successful food production from these microorganisms include growth system strategies, substrate use, and cultivation conditions. Photobioreactors have been extensively explored in terms of stability and control parameters such as temperature, pH, nutrient distribution, reflected light, and aeration. Other important parameters include the development of efficient extraction techniques for byproducts, regulations, and standards for using microalgae in food. The technical and economic aspects of largescale production continue to challenge researchers as they are crucial for process viability.

5.1 INTRODUCTION

Overpopulation and the negative impacts of conventional farming and livestock rearing on the environment force the introduction of sustainable alternative ingredients on the food market. Microalgae can serve as an alternative source of proteins due to their easy cultivation as well as low land and resource requirements.[1,3]

Microalgae are unicellular photosynthetic microorganisms found in various aquatic ecosystems, offering numerous advantages for developing sustainable and environmentally friendly food production strategies.[4,5] They have garnered attention in recent years as a promising source of bioactive compounds and valuable biomolecules in the food industry.

The potential of microalgae in this regard is high because of their unique ability to convert solar energy and carbon dioxide (CO_2) into biomass through photosynthesis. This makes them a sustainable source of nutrients and bioactive compounds as their cultivation does not directly compete with land used for conventional agriculture.[1,6]

These aquatic organisms can produce various compounds, including lipids, proteins, carbohydrates, vitamins, pigments, polysaccharides, and phenolic compounds. Among these, they contain approximately 40–70% protein, 12–30% carbohydrates, 4–20% lipids, 8–14% carotene, and considerable amounts of the vitamins B_1, B_2, B_3, B_6, B_{12}, E, K, and D.[7] These biomolecules possess functional and nutritional properties (https://phytobloom.com/), making them valuable ingredients for formulating healthy and functional foods.[8]

The versatility of the microalgae (*Chlorella vulgaris*, *Nannochloropsis* sp., *Arthrospira platensis* (spirulina), *Scenedesmus* sp., *Isochrysis* sp., *Dunaliella salina*, *Haematococcus pluvialis*) allows their use in various food products. For example, microalgae lipids are a potential source of polyunsaturated fatty acids, rich in omega-3 fatty acids, and beneficial for the cardiovascular system and the brain.[9,13] The pigments present in microalgae, such as chlorophyll, astaxanthin, and phycocyanin, are used as natural colorants in foods and beverages, replacing synthetic additives.[14,16] Another notable aspect is the ability of microalgae to produce phenolic compounds, vitamins, and exopolysaccharides with antioxidant, antimicrobial, and anti-inflammatory properties, which are used in the form of capsules or added to foods, improving their quality and shelf-life while promoting consumer health and well-being.[17,18]

The submerged fermentation (SmF) of microalgae can be carried out in closed systems such as photobioreactors, reducing water consumption and minimizing the emission of greenhouse gases.[19,21]

Microalgae can use nutrients from agricultural and industrial waste and byproducts, such as agroindustrial residues and CO_2 produced in various industrial processes, contributing to a circular and sustainable economy.[6]

Microalgae represent a valuable and sustainable source of compounds, biomolecules, and products used in food. Their ability to convert solar energy and inorganic resources into nutrient-rich biomass and bioactive compounds makes them a good resource for the food industry. Harnessing microalgae in food products offers nutritional, functional, and environmental benefits, promoting more sustainable production and consumption. The following section provides an overview of the current perspectives on the sustainable application of microalgae in the food industry.

This chapter presents a description of microalgae (section 2); circular economy and cultivation strategies (section 3); products and biomolecules obtainable from microalgae (section 4); and commercially viable microalgae (section 5). Finally, section 6 presents the conclusions and perspectives.

5.2 MICROALGAE

Microalgae are a diversified group of eukaryotic and prokaryotic photosynthetic microorganisms that can show single-cell morphology and arrangements where small colonies, or coenobia, are presented. The prokaryotic category includes cyanobacteria and prochlorophytes, which are classified based on pigmentation. The eukaryotic category includes *Chlorophyta*, *Rhodophyta*, *Cryptophyta*, *Ochrophyta*, *Myzozoa*, *Haptophyta*, and *Euglenozoa*.[22,23] These microorganisms can be motionless, have flagella, or be mobile.[24]According to some authors, there are estimates of thousands of different species distributed around the world.

Microalgae can be found in water bodies (such as oceans, seas, lagoons, and rivers), in soil, snow, glaciers, or in ecological relationships with plants, fungi, and animals; they can also be symbionts inside other organisms.[4] As photosynthetic microorganisms, microalgae share many metabolic pathways with plants,[24] making them a potential source of many products and molecules that are generally obtained from plants and crops. Nevertheless, the implementation of such bioprocesses is still under development.[5]

Many of these compounds can be applied in the pharmaceutical and food industries, or the biomass can be used as a nutraceutical, prebiotic, or dietary supplement.[25] Few strains can be produced on a commercial scale (Nutraceuticals; Nutraingredients), either for the sale of the whole biomass or for the extraction of high-value molecules such as astaxanthin, phycobilin, lutein, carotenoids, polysaccharides, and omega 3 and 6 fatty acids.[25]

The production of microalgal biomass in liquid fermentations (SmFs) with technological devices, similar to the fermenter developed for bacteria and yeasts, makes microalgae a potential source of high-value compounds. Nevertheless, many physical, chemical, and biological aspects need to be considered to ensure correct growth and, consequently, a high productivity of the process.[21,28,29]

5.3 CIRCULAR ECONOMY AND CULTIVATION STRATEGIES

Population growth and the COVID-19 pandemic have significantly affected the production, distribution, and consumption of food (United Nations, 2021), leading to a global demand for food that will be challenging to meet by 2050 (FAO, 2021). This has resulted in adverse environmental impacts caused by the outdated food supply system based on traditional crops, including increased greenhouse gas emissions, excessive pesticide use, and the excessive use of land and freshwater resources.[1,6]

Currently, new ways of obtaining food are being sought to deal with food demand and consumption, supporting the circular economy based on the sustainable use of biological resources, minimizing waste, and maximizing waste use. Microalgae have become a promising approach in the food industry due to their high nutrient content and speed of growth.[2,8]

Compared to traditional crops, microalgae do not require herbicides or pesticides, have no soil condition requirements, and are not restricted by seasons. The use of microalgae in food involves taking advantage of their nutrient-rich biomass to develop supplements, feed ingredients, and substitutes for conventional ingredients. This promotes a sustainable approach, reduces the dependency on limited natural resources, and contributes to a circular economy. Since byproducts or residues from other sectors can be used for their cultivation, greenhouse gas emissions are reduced.[1,30]

However, with the advancement of microalgal biotechnology as a food field, issues related to microalgae-based foods still need to be solved. Challenges such as production costs, low product acceptance, and unknown safety are the bottlenecks that restrict practical application. Therefore, the joint efforts of specialists in various fields are necessary to promote the development of microalgae-based foods.[6]

Several strategies for microalgae cultivation have been developed based on submerged fermentation; however, some of the main factors limiting bioreactor design are the cultivation requirements. These have been optimized under different operating conditions using artificial and natural light, as well as a wide range of nutrients and CO_2 concentrations.[31]

5.3.1 PHOTOBIOREACTORS

Microalgae and cyanobacteria can be cultured by submerged fermentation, but there is a necessity to add light to the culture as they are photosynthetic microorganisms. Also, their cell walls are not as strong as those of bacteria and yeasts, and therefore, the shear stress must be low. Both factors need to be considered in the design of suitable bioreactors.[32] The most suitable bioreactors are therefore photobioreactors (PBRs), considering not only the light supply to the culture but also the distribution of the light inside the culture.[19]

According to several authors, there are some aspects to be considered when designing PBRs, mainly when the organisms are autotrophs:

1. High mass transfer to ensure an efficient CO_2 supply and the removal of O_2 produced via photosynthesis.
2. Mixing to avoid the stratification of the culture, circumvent biomass sedimentation, and ensure light exposition of the cells.
3. Precise control of the temperature, pH, and CO_2 concentration.[29]

Today, PBRs can be configured in various ways, but light distribution and low shear stress are mandatory in almost all configurations. Normally, the construction material is transparent, such as acrylic, glass, or PVC, to ensure the entrance of light to the culture. The classification can be made considering different operational parameters, but the most useful characterization is related to the shape and agitation regime (regulating light entrance). Considering this, there are three main groups: agitation by an airlift device (the mixing pattern is defined and the mass transfer is efficient), mixing carried out by bubbling (low shear stress and the option to grow sensitive cells), and mixing using a mechanical mixer (stirring propels and pumps).[32]

Considering the shape, the PBR can be classified as a column, annular, flat plate/flat panel, torus, dome, thin-layer, or tubular (horizontal, inclined, and vertical configurations), with a specific configuration for the production of specific strains.[19,21,29,33,36] Table 5.1 shows the advantages and disadvantages of some configurations for microalgal biomass production.

TABLE 5.1 Comparison of Different PBRs' Configurations.[2,19,20,35]

Configuration	Advantages	Disadvantages
Raceway ponds	Low construction costs.	Higher land extensions are required.
	Easy operation.	Low illuminated area.
	Simple scaling (modular).	Inefficient mixing.
	Temperature control is not required.	High risk of contamination.
	Low energy consumption.	Poor CO_2 supply.
Tubular	Efficient illumination and good light distribution.	Mandatory to include a temperature control.
	High photosynthetic efficiency.	Necessity to add a degasser.
	Low contamination risk.	High construction costs.
	Flexibility to make different configurations (serpentine, manifold, or helical).	Limitation in the length and diameter of the tubes.
		Poor mixing, especially in longer tubes.
		High energy consumption for pumped systems.
Column	Efficient illumination and good light distribution.	Mandatory to include a temperature control.
	Feasibility to include an airlift device.	Limitation in the diameter of the tube (8 inches).
	Low contamination risk.	The light distribution is inefficient for bigger diameters.
	Low shear stress.	

TABLE 5.1 *(Continued)*

Configuration	Advantages	Disadvantages
Flat plate/flat panel	Efficient illumination and good light distribution. Simple scaling (modular). High biomass productivity. Low shear stress.	Higher land extensions are required. Mandatory to include a temperature control. Photoinhibition risk is high for diluted cultures.
Thin-layer	Low construction costs. Good light distribution. High biomass productivity.	Higher land extensions are required. High contamination risk. High energy consumption because the pump has a deficient CO_2 supply. High evaporation rate.

Generally, raceways and thin-layer systems are employed for the culture of microalgae with a high growth rate or with selective growing conditions, such as *Dunaliella salina*, *Spirulina/Arthrospira*, and *Chlorella*, reaching even commercial scale[14] and allowing the cultivation of *Nannochloropsis* and *Phaeodactylum tricornutum.*[32] *Dunaliella* can be produced in such systems because it grows at high salinity levels, whereas *Spirulina* has an optimal pH above 10 and tolerates high alkalinity, decreasing the possibility of contamination by other microalgae. *Chlorella* can be cultured in open systems because it has a high growth rate, ensuring its prevalence inside the PBR.[37] For *Nannochloropsis* and *Phaeodactylum*, the contamination risk is also low.[38,39]

In contrast, closed PBRs allow the production of any strain of microalgae, even in the presence of organic carbon sources (growing mixotrophic or heterotrophic) or genetically modified microalgae.[32] For heterotrophic cultures, sterilization of the PBR is mandatory, which considerably increases the construction costs and reduces the material used to only glass. In sterilizable systems, the production of *Schizochytrium* and *Crypthecodinium* is feasible; these microalgae are economically viable sources of unsaturated fatty acids.[16] The final product, the specific operational factors of the productive process, and the metabolic characteristics and necessities of the microalgae govern the selection of the PBR.[14,19,21,37,39]

5.4 PRODUCTS AND BIOMOLECULES

Microalgae are an excellent source of high-quality carbohydrates, lipids, and proteins, and some species can accumulate or produce some high-value molecules (bioactive compounds) such as carotenoids, chlorophylls, antioxidants, vitamins, fatty acids, polyunsaturated fatty acids, antivirals, antifungals, and immunostimulants.[37,40,43] Table 5.2, and Figure 5.1 show the most common species of microalgae and some of their potential products.

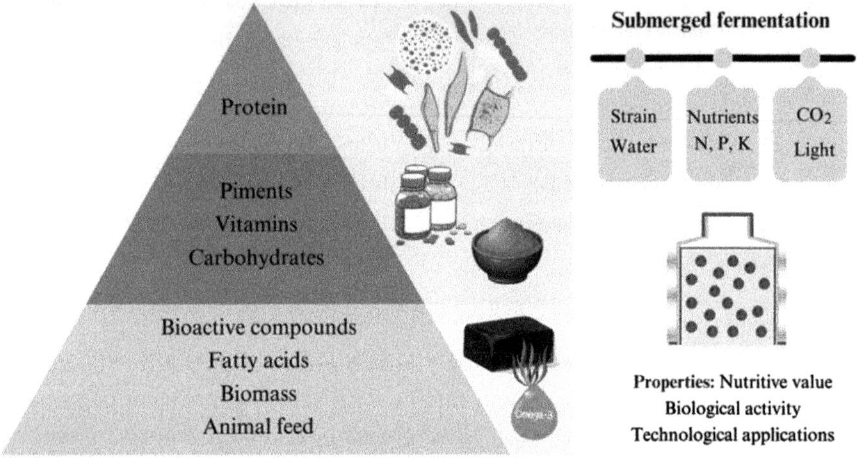

FIGURE 5.1 Main bioproducts and biomolecules detected in microalgae and cyanobacteria with sustainable food potential.

TABLE 5.2 High-Value Products Produced by Microalgae and Cyanobacteria.[14–16,42,50,92–95]

Species/Genus	Product/compound	Sector of application
Arthrospira/ Spirulina	Phycocyanin	Cosmetics, healthy food, and nutraceuticals
	Biomass, Vitamin B12	Cosmetics, antioxidants, and pharmaceuticals
Chlorella	Biomass, carbohydrate fraction, and polysaccharides	Healthy food, nutraceuticals, and animal feed
	Tocopherol, ascorbic acid, and polyunsaturated fatty acids	Cosmetics, nutraceuticals, and healthy food
Dunaliella	Carotenoids (mainly β-carotene)	Healthy food, nutraceuticals, animal feed, poultry feed, and aquaculture
Haematococcus	Carotenoids (mainly astaxanthin)	Healthy food, nutraceuticals, animal feed, and aquaculture
Schizochytrium	Docosahexaenoic acid	Milk formulas for babies, healthy food, nutraceuticals, animal feed, and aquaculture
	Eicosapentaenoic acid	
Crypthecodinium cohnii	Docosahexaenoic acid	Milk formulas for babies, healthy food, nutraceuticals, animal feed, and aquaculture
Nannochloropsis	Biomass	Nutraceuticals, animal feed, and aquaculture
	Polyunsaturated fatty acids	

The diversity of high-value products that can be obtained from microalgal biomass has caused considerable interest in microalgal biomass production. Today, most of the biomass produced and consumed around the world is single-cell protein biomass because of its excellent elemental composition and health benefits for humans and animals.[37,40] Recently, it has been proposed to use microalgal biomass enriched with some specific metabolites, such as carotenoids, polyunsaturated fatty acids, vitamins, and antioxidants, in animal feed.[40,43,44]

5.4.1 BIOACTIVE COMPOUNDS

5.4.1.1 PIGMENTS

Microalgae share many characteristics of the photosynthetic apparatus with plants, and many molecules used by plants to harvest light, such as chlorophylls, carotenoids, and xanthophylls, are also present in microalgae.[24] Nevertheless, the specific amounts depend on the strains and the culturing conditions; some of the compounds produced by microalgae are astaxanthin, β-carotene, lutein, canthaxanthin, fucoxanthin, and lycopene, all of which have benefits for human and animal health.[40,41,43,45] Regarding cyanobacteria, in addition to chlorophylls and carotenoids, they contain protein pigments called phycobilins (including phycocyanin, phycoerythrin, and allophycocyanin), which stimulate the immune system and cause beneficial metabolic modifications.[46,47]

5.4.1.2 CAROTENOIDS AND XANTHOPHYLLS

Carotenoids and xanthophylls are molecules composed of carbon and hydrogen; they are generally symmetric and contain several double bonds in their central structure. The section of the succession of alternate double bounds allows the molecule to absorb electromagnetic radiation.[48] In the photosynthetic apparatus, carotenoids and xanthophylls have two functions. They act as accessory pigments to support light absorption by the chlorophylls and as antioxidants, stabilizing the reactive oxygen species produced during the luminous phase of photosynthesis.[24] The largest difference between carotenoids and xanthophylls is the absence of oxygen in the carotenoids (e.g., lycopene, γ-carotene, and β-carotene), whereas the xanthophylls contain oxygen (e.g., lutein and astaxanthin).[48]

Some species show a high production of specific pigments (chlorophyll, carotenoids, or even xanthophylls) during the exponential growth phase.[45,49]

Nevertheless, for secondary pigments, such as the majority of carotenoids and xanthophylls, overproduction occurs during the stationary phase or by the employment of an induction strategy. This requires the development of processes in two stages: one for biomass production (under mixotrophic or even heterotrophic conditions) and one with high irradiation, nutrient limitation, nutrient starvation, increased salinity, or the addition of an inductor.[45,50,51] Table 5.3 shows examples of pigments produced by microalgae and the characteristics of the PBRs used.

TABLE 5.3 Pigment Production by Microalgae and Cyanobacteria.

Microalga	Pigment	PBR and operational conditions	Reference
Dunaliella salina	β-carotene	Tubular de 55 L	[96]
		Semicontinuous operation	
		Continuous operation (turbidostate)	[97]
		Flat panel 2.5 L	
		Flat panel (34 L)	[98]
		Indoors	
Muriellopsis	Lutein	Continuous operation Tubular de 55L	[99]
		Raceway (800 L)	[100]
Scenedesmus almeriensis	Lutein	Continuous operation Column (2 L)	[101]
		Outdoor cultivation	[102]
		Raceways (11800 L)	
		Thin layer (2400 L)	
		Indoor cultivation	[103]
		Column (1.25L)	
Haematococcus pluvialis	Astaxanthin	Outdoor cultivation (25 000 L)	[104]
		Indoor cultivation	[105]
		Column (10 L)	
Chlorella sorokiniana	Astaxanthin	Addition of an inductor	[106]
		External loop airlift	
Chlorella zofingiensis	Astaxanthin	Heterotrophic culture	[107]
Chromochloris zofingiensis	Ketocarotenoids, Astaxanthin, Canthaxanthin, and Adonixanthin	Indoors	[108]
		Columns (1 L)	

5.4.1.3 VITAMINS

Vitamins are essential organic compounds for the proper functioning of the body. Although they are needed in small amounts, they play a crucial role in numerous physiological functions and are vital to health. Notably, vitamins cannot be synthesized in adequate amounts by the body and must therefore be obtained through the diet or, in some cases, through supplements.[52]

Microalgae are an excellent source of vitamins with antioxidant properties since they can synthesize and accumulate vitamins inside their cells, which can be used as ingredients or food supplements, in powder or tablet form, or as dietary supplements.[8,17]

Several microalgae are known for their ability to produce vitamins and have applications in the commercial food industry. One example is *Chlorella* sp., which is widely used in the production of food supplements and is rich in vitamins B_{12}, B complex, B_2 (riboflavin), and B_3 (niacin).[53] *Arthrospira* sp. is considered an excellent source of nutrients and contains a wide range of vitamins, such as vitamin A (β-carotene), K, E, and B complex vitamins (B_1 (thiamine), B_2, and B_6).[53,54] *Dunaliella* sp. is known for its high content of β-carotene, a precursor to vitamin A and a powerful antioxidant. However, this genus can also produce the vitamins C, E, B_2, B_{12}, B_9, and B_3.[55,56] Other species of microalgae can be commercial sources of vitamins, such as *Isochrysis galbana*, which can be used in the production of the vitamins B_1, B_2, B_3, B_5, B_6, B_{12}, A, E, and biotin, and *Euglena gracilis* Z, which contains provitamin A and the vitamins C and E.[52] *Nannochloropsis oculata* (*Eustigmatophyceae*) is also suitable for the production of vitamin E (α-tocopherol).[57]

Obtaining vitamins from microalgae is a promising research field, with the opportunity for the sustainable production of vitamins through renewable nutrient sources such as wastewater or CO_2 from industrial plants. These vitamins can be used as ingredients in food, which completes the resource use cycle, contributes to sustainability, and reduces dependence on traditional vitamin sources.[8,58]

5.4.1.4 PHENOLIC COMPOUNDS

Phenolic compounds are bioactive compounds with potential benefits for human health.[59] These include antioxidant, anti-inflammatory, immunomodulatory, antimicrobial, and antitumor activities.[60] Microalgae can produce phenolic compounds such as phenolic acids, lignans, flavonoids, flavones, isoflavones, flavonols, stilbenes, and anthocyanins.[52,61] The production of these compounds can be stimulated by controlled abiotic stress conditions (high salinity, intense light, nitrogen deprivation, high temperature, and ultraviolet radiation)[22] and depends on the type of microalgae, culture conditions, and environmental stimuli.[58] Phenolic compounds are extracted and purified using solvent-extraction techniques.[62] Some microalgae have shown a more remarkable ability to synthesize and accumulate phenolic compounds than others. For example, Morowvat and Ghasemi (2016)[18] reported that the microalgae *Dunaliella salina*, *Fischerella ambigua*, *Nostoc muscorum*, *Oocystis pusilla*, and *Scenedesmus rubescens* produce gallic acid. Another recent study showed that *Chlorococcum* sp. MC1 and *Scenedesmus* sp. B2-2 produce high concentrations of carotenoids and phenolic compounds and have a greater antioxidant capacity during stress, which justifies their largescale cultivation.[63]

In a previous study, *Tetraselmis suecica* showed high contents of protocatechuic acid, caffeic acid, caffeoyl glucoside, dimethoxyflavone, p-coumaroyl tyrosine, apigenin-O-rhamnosyl rutinoside, and hexosyl-methyl-quercetin.[64] *Chlorella vulgaris* produces high amounts of gallic, caffeic, p-coumaric, and ferulic acids,[62] whereas *Spirulina platensis* presents a high level of radical scavenging activity and contains apigenin (4′, 5, 7-trihydroxy flavone).[65] *Scenedesmus* sp., with its high antioxidant activity, contains the flavonols quercetin and rutin.[66] The phenolic compounds produced by microalgae have various potential applications in the food and pharmaceutical industries, but their largescale production is still a developing field, and further studies are required to optimize the production methods, enhance the yields, and improve their properties and applications.

5.4.1.5 FATTY ACIDS

Fatty acids (FAs) are essential lipids with specific chains (CH_3–$CH_{2(n)}$–COOH) in their chemical structure.[67] Polyunsaturated fatty acids (PUFAs) are necessary to produce nutraceuticals and are essential in the human diet as a functional food. Although fish and shellfish are the main sources of PUFA, current issues such as overfishing and global warming have created the need to find alternatives.[68]

Microalgae are a promising source of healthy fatty acids and lipids. Of these, fatty acids correspond to the largest lipid fraction, and, in certain species, PUFAs account for 25–60% of the total lipids.[69] Microalgae can accumulate large amounts of lipids in their cells, including eicosapentaenoic acid (EPA), docosahexaenoic acid (DHA), which consists of omega-3 and omega-3-rich γ-linolenic acid (GLA), and arachidonic acid (ARA) with omega-6.[70,72]

Microalgae are mainly cultivated to produce DHA as a feed additive (Martek Biosciences Corporation, USA). Among the main reported microalgae with PUFA production potential are *Spirulina*, *Chlorella*, *Dunaliella salina*, and *Aphanizomenon flosaquae*, which are already on the market as food products in the form of powders, tablets, drink extracts, and liquor.[9,72] The heterotrophic microalga *Crypthecodinium cohnii* is recognized as a DHA producer since the content can reach up to 65% of the total PUFA present in its lipid profile.[73] Heterotroph microalgae, including the genera *Thraustochytrium*, *Schizochytrium*, and the dinoflagellate *Crypthecodinium cohnii*, contain large amounts of DHA and EPA in their cells, along with small amounts of other PUFAs. Being nonpathogenic and nontoxigenic and having a practical use in human nutrition, these species represent the most important alternative sources of DHA/EPA.[74,75]

Microalgae are some of the most promising sources of fatty acids due to their ability to produce biomass. However, the selection of suitable strains is crucial. Since they must be nontoxic, fast-growing, and with a high content of fatty acids, the culture of microalgae must be adjusted to the maximum efficiency for the production of biomass and lipids, taking into account the proportion of fatty acids and the culture costs.[70,71]

5.4.1.6 EXOPOLYSACCHARIDES

Microalgae polysaccharides are complex and heterogeneous heteropolysaccharide macromolecules formed of xylose, glucose, and galactose as well as considerable amounts of monosaccharides such as fucose, methylated sugars, rhamnose, and iduronic acid, sometimes also glucuronic acid and sulfate groups.[76] They can not only be used as fibers, namely, prebiotics,[77] but also as thickeners and gelling additives in the food industry.[78] In addition, they have antioxidant, antibacterial and antiviral, antitumor, antihyperlipidemic, and anticoagulant activities.[79]

Microalgae produce large amounts of bioactive polysaccharides, divided into storage and cell wall-related polysaccharides.[58] Exopolysaccharides (EPS) are extracellular polysaccharides that can be excreted into the environment or bound to the cell walls.[78] The functions of EPS are the protection of cells against biotic and abiotic stress, mainly against dehydration or the toxic substances.[80]

The chemical structures and quantities of EPS depend on the species of the microalgae and the culture conditions.[81] In a previous study, *Botryococcus braunii* CCALA 220 produced 666 mg g^{-1} of exopolysaccharides at varying nitrogen concentrations and light/dark cycles.[82] *Nostoc flageliforme*,[83] *Nostoc* sp.,[84] *Porphyridium sordidum*, *Porphyridium purpureum*,[85] espacio and *Parachlorella*[86] can increase EPS production by modifying their growth conditions (pH, light, and culture medium). Other microalgae (such as *Chlorella stigmatophora*, *Isochrysis* sp., *Phaeodactylum tricornutum*, and *Dictyosphaerium chlorelloides*) produce EPS (around 20 g L^{-1}) based on optimization.[81,87]

The production of EPS from microalgae faces significant challenges in obtaining biomass via the photosynthetic pathway, such as increased production costs due to the design and operation of the PBR. However, different configurations and techniques are still being tested to improve this aspect.[88]

5.5 COMMERCIALLY VIABLE MICROALGAE

Globally, several companies have specialized in microalgal biomass production to obtain various final products such as polyunsaturated astaxanthin, fatty acids, and carotenoids. They can be used in poultry feed, as food additives, as colors and antioxidants in cosmetics and food, as components of infant formulas, and in beverages and healthy foods.[15] Today, the commercialization of carotenoids produced by microalgal biomass is a large market with highly developed technology.[89] In 2022, carotenoids had a market of close to 2 billion dollars, with an estimated annual growth rate of approximately 4%, potentially reaching 2.7 billion dollars in 2027. The increase in the sales of carotenoids is based only on a few molecules, mainly astaxanthin, β-carotene, β-apo-8-carotenal, β-apo-8-carotenal-ester, canthaxanthin, lutein, lycopene, and zeaxanthin.[90] However, the consumption of other carotenoids in the near future is likely.

The carotenoid market focuses its efforts on increasing the fraction of the carotenoids obtained from natural sources, highlighting microalgae and cyanobacteria.[89] Organically produced compounds realize a higher price than those obtained via chemical synthesis, and there are estimations of 300–3000 USD per kg of β-carotene and of 2500–10,000 USD per kg of astaxanthin. Ketocarotenoids, such as astaxanthin or canthaxanthin, are among the most expensive carotenoids on the market.[91] Recently, the high market prices of carotenoids have stimulated the development of different companies around the world that produce power biomass, oleoresins, or extracts with a successful marketing strategy. Some of those companies are described below, in addition to their main products.

The company with the largest PBR is Cyanotech®, located on the Hawaiian archipelago, which produces spirulina biomass in raceways under outdoor conditions. Cyanotech® also produces microalgal biomass to extract astaxanthin in a two-stage process. In the first stage, high amounts of biomass are produced in the laboratory, followed by an induction phase in raceways. Some of the products offered by Cyanotech® are BioAstin Hawaiian astaxanthin, Hawaiian spirulina, and the recently developed BioAstin® water-dispersible 1% and 2% astaxanthin powder, which is used in food coloring and is free of isomers (https://www.cyanotech.com/).

In Portugal, the company Algicel, Biotecnologia e Investigação®, offers Azora Astaxanthin®, extracted from *Haematococcus pluvialis*. According to the information on their website, the biomass is produced in vertical flat plate PBRs under laboratory conditions to avoid contamination (https://algicel.pt/es/).

In the Czech Republic, Algamo S.R.O. offers biomass of *Haematococcus pluvialis* with a high amount of astaxanthin, produced in a vertical bag PBR operated under indoor conditions. Some of the products produced by this company are Algamo Organic Dried Biomass® (microalgal biomass with an astaxanthin content of 5%), Algamo Organic Oleoresin® (an oleoresin with an astaxanthin content of 5% or 10%), astaxanthin capsules, and Algamo CWD®, which is a red water-soluble powder (liposomal encapsulated astaxanthin). Algamo CWD® can be used in water-based cosmetics, drinks, supplements, and food, among others, to obtain a brilliant red color and to enrich the product with antioxidants (https://www.algamo.cz/). In Austria, there is the company BDI-BioLife Science GmbH®, which offers ASTAFIT® CWD25 and CWD50 microencapsulated astaxanthin (this procedure increases the shelf-life by avoiding oxidation) to be used in health drinks, premixes, or powder mixtures (https://www.bdi-biolifescience.com/de/).

The company Algatech®, located in the Arava Desert in Israel, commercializes products that contain biomass from *Haematococcus pluvialis*, *Phaeodactylum tricornutum*, *Porphyridium cruentum*, *Nannochloropsis*, and *Euglena gracilis*. Biomass production is performed in a horizontal tubular PBR operated outdoors. Some of the products are FucoVital™ (fucoxanthin), BioGlena™ (betaglucan from Euglena), and a huge variety of formulas that include astaxanthin, such as AstaPure®Max, AstaPure® Arava, and AstaPure®Astaxanthin (https://www.algatech.com/).

5.6 CONCLUSIONS AND PERSPECTIVES

Microalgae are a promising option for obtaining foodgrade products, considering the type of strain and essential nutrients required. Their growth capacity, high productivity, and variety of beneficial compounds make them essential sources of nutrients and biomolecules in the food industry. By harnessing their properties, sustainable development is promoted. The technoeconomic viability of investing in using microalgae as an alternative source for food production is an unconventional and innovative approach. However, their cultivation still has limitations, and further exploration is needed. Microalgae cultivation can contribute to food security.

KEYWORDS

- **submerged fermentation**
- **photobioreactors**
- **circular economy**
- **bioactive compounds**
- **biomolecules**

REFERENCES

1. Ciani, M.; Lippolis, A.; Fava, F.; Rodolfi, L.; Niccolai, A.; Tredici, M. R. Microbes: Food for the Future. *Foods* **2021,** *10* (5). https://doi.org/10.3390/foods10050971.
2. Nagarajan, D.; Varjani, S.; Lee, D. J.; Chang, J. S. Sustainable Aquaculture and Animal Feed from Microalgae – Nutritive Value and Techno-Functional Components. *Renew. Sustain. Energy Rev.* **2021,** *150*, 111549. https://doi.org/10.1016/J.RSER.2021.111549.
3. D'Auria, G.; Nitride, C.; Ferranti, P. Microalgae to Contrast the Climate Change: A Novel Food and Feed Ingredient with Technological Applications. In *Reference Module in Food Science.* 2023; pp 146–163. https://doi.org/10.1016/B978-0-12-823960-5.00024-X.
4. Khan, A. K.; Kausar, H.; Jaferi, S. S.; Drouet, S.; Hano, C.; Abbasi, B. H.; Anjum, S. An Insight into the Algal Evolution and Genomics. *Biomolecules* **2020,** *10* (11), 1524. https://doi.org/10.3390/biom10111524.

5. Martínez-Roldán, A. J.; Ibarra-Berumen, J. Employment of Wastewater to Produce Microalgal Biomass as a Biorefinery Concept. In *Microalgae Biotechnology for Development of Biofuel and Wastewater Treatment*; Springer Singapore: Singapore, **2019**; pp 487–504. https://doi.org/10.1007/978-981-13-2264-8_19.
6. Chen, C.; Tang, T.; Shi, Q.; Zhou, Z.; Fan, J. The Potential and Challenge of Microalgae as Promising Future Food Sources. *Trends Food Sci. Technol.* **2022,** *126*, 99–112. https://doi.org/10.1016/J.TIFS.2022.06.016.
7. Batista, A. P.; Gouveia, L.; Bandarra, N. M.; Franco, J. M.; Raymundo, A. Comparison of Microalgal Biomass Profiles as Novel Functional Ingredient for Food Products. *Algal. Res.* **2013,** *2* (2), 164–173. https://doi.org/10.1016/J.ALGAL.2013.01.004.
8. Levasseur, W.; Perré, P.; Pozzobon, V. A Review of High Value-Added Molecules Production by Microalgae in Light of the Classification. *Biotechnol. Adv.* **2020,** *41*, 107545. https://doi.org/10.1016/J.BIOTECHADV.2020.107545.
9. Moradi-Kheibari, N.; Ahmadzadeh, H.; Lyon, S. R. Correlation of Total Lipid Content of *Chlorella Vulgaris* With the Dynamics of Individual Fatty Acid Growth Rates. *Front. Mar. Sci.* **2022,** *9*, 1–11. https://doi.org/10.3389/fmars.2022.837067.
10 . Hulatt, C. J.; Wijffels, R. H.; Bolla, S.; Kiron, V. Production of Fatty Acids and Protein by *Nannochloropsis* in Flat-Plate Photobioreactors. *PLoS One* **2017,** *12* (1), 1–17. https://doi.org/10.1371/journal.pone.0170440.
11. Gentscheva, G.; Nikolova, K.; Panayotova, V.; Peycheva, K.; Makedonski, L.; Slavov, P.; Radusheva, P.; Petrova, P.; Yotkovska, I. Application of *Arthrospira Platensis* for Medicinal Purposes and the Food Industry: A Review of the Literature. *Life* **2023,** *13* (3), 845. https://doi.org/10.3390/life13030845.
12. González-Falfán, K. A.; Guerrero-Barajas, C.; Badillo-Corona, J. A.; Fernández-Linares, L. C. Production of Metabolites from *Scenedesmus* Sp. and a Microalgal Consortium Cultured in Unconventional Media. *Ciencias Mar.* **2021,** *47* (2), 89–103. https://doi.org/10.7773/cm.v47i2.3138.
13. Sun, Z.; Wang, X.; Liu, J. Screening of *Isochrysis* Strains for Simultaneous Production of Docosahexaenoic Acid and Fucoxanthin. *Algal Res.* **2019,** *41*, 101545. https://doi.org/10.1016/J.ALGAL.2019.101545.
14. Mobin, S.; Alam, F. Some Promising Microalgal Species for Commercial Applications: A Review. *Energy Procedia* **2017,** *110*, 510–517. https://doi.org/10.1016/J.EGYPRO.2017.03.177.
15. Vilakazi, H.; Olasehinde, T. A.; Olaniran, A. O. Chemical Characterization, Antiproliferative and Antioxidant Activities of Polyunsaturated Fatty Acid-Rich Extracts from *Chlorella* Sp. S14. *Molecules* **2021,** *26* (14), 4109. https://doi.org/10.3390/molecules26144109.
16. Winwood, R. J. Recent Developments in the Commercial Production of DHA and EPA Rich Oils from Micro-Algae. *OCL* **2013,** *20* (6), D604. https://doi.org/10.1051/ocl/2013030.
17. Zhou, L.; Li, K.; Duan, X.; Hill, D.; Barrow, C.; Dunshea, F.; Martin, G.; Suleria, H. Bioactive Compounds in Microalgae and Their Potential Health Benefits. *Food Biosci.* **2022,** *49*, 101932. https://doi.org/10.1016/j.fbio.2022.101932.
18. Morowvat, M. H.; Ghasemi, Y. Evaluation of Antioxidant Properties of Some Naturally Isolated Microalgae: Identification and Characterization of the Most Efficient Strain. *Biocatal. Agric. Biotechnol.* **2016,** *8*, 263–269. https://doi.org/10.1016/J.BCAB.2016.09.010.

19. Acién, F. G.; Molina, E.; Reis, A.; Torzillo, G.; Zittelli, G. C.; Sepúlveda, C.; Masojídek, J. Photobioreactors for the Production of Microalgae. In *Microalgae-Based Biofuels and Bioproducts*; Elsevier, **2017**; pp. 1–44. https://doi.org/10.1016/B978-0-08-101023-5.00001-7.
20. Chang, J. S.; Show, P. L.; Ling, T. C.; Chen, C. Y.; Ho, S. H.; Tan, C. H.; Nagarajan, D.; Phong, W. N. Photobioreactors. In *Current Developments in Biotechnology and Bioengineering*; Elsevier, **2017**; pp. 313–352. https://doi.org/10.1016/B978-0-444-63663-8.00011-2.
21. Martínez-Roldán, A. J.; Cañizares-Villanueva, R. O. Photobioreactors: Improving the Biomass Productivity. In *Microalgae and Other Phototrophic Bacteria: Culture, Processing, Recovery and New Products First Edit*; Torres-Bustillos, L., Ed.; Nova Science Publishers, **2015**; pp 145–170.
22. Hamidi, M.; Safarzadeh Kozani, P.; Safarzadeh Kozani, P.; Pierre, G.; Michaud, P.; Delattre, C. Marine Bacteria versus Microalgae: Who Is the Best for Biotechnological Production of Bioactive Compounds with Antioxidant Properties and Other Biological Applications? *Mar. Drugs* **2020,** *18* (1), 1–38. https://doi.org/10.3390/md18010028.
23. Patras, D.; Moraru, C. V.; Socaciu, C. Bioactive Ingredients from Microalgae: Food and Feed Applications. *Bull. Univ. Agric. Sci. Vet. Med. Cluj-Napoca. Food Sci. Technol.* **2019,** *76* (1), 1. https://doi.org/10.15835/buasvmcn-fst:2018.0018.
24. Barsanti, L.; Gualtieri, P. In *Algae: Anatomy, Biochemistry, and Biotechnology*, 2nd Ed.; CRC Press, **2014**.
25. Camacho, F.; Macedo, A.; Malcata, F. Potential Industrial Applications and Commercialization of Microalgae in the Functional Food and Feed Industries: A Short Review. *Mar. Drugs* **2019,** *17* (6), 312. https://doi.org/10.3390/md17060312.
26. Nutraceuticals, P. Parry Nutraceuticals. https://www.parrynutraceuticals.com/.
27. NutraIngredients. Ocean Nutrition Canada. https://www.nutraingredients.com/Suppliers/Ocean-Nutrition-Canada.
28. Fernandes, B. D.; Mota, A.; Teixeira, J. A.; Vicente, A. A. Continuous Cultivation of Photosynthetic Microorganisms: Approaches, Applications and Future Trends. *Biotechnol. Adv.* **2015,** *33* (6), 1228–1245. https://doi.org/10.1016/j.biotechadv.2015.03.004.
29. Zittelli, G. C.; Rodolfi, L.; Bassi, N.; Biondi, N.; Tredici, M. R. Photobioreactors for Microalgal Biofuel Production. In *Algae for Biofuels and Energy*; Springer Netherlands: Dordrecht, **2013**; pp 115–131. https://doi.org/10.1007/978-94-007-5479-9_7.
30. Zittelli, G. C.; Silva Benavides, A. M.; Silovic, T.; Ranglová, K.; Masojídek, J.; Cicchi, B.; Faraloni, C.; Touloupakis, E.; Torzillo, G. Productivity and Nutrient Removal by the Microalga *Chlamydopodium Fusiforme* Grown Outdoors in BG-11 and Piggery Wastewater. *Front. Mar. Sci.* **2022,** *9*, 1–18. https://doi.org/10.3389/fmars.2022.1043123.
31. Almomani, F.; Hosseinzadeh-Bandbafha, H.; Aghbashlo, M.; Omar, A.; Joo, S. W.; Vasseghian, Y.; Karimi-Maleh, H.; Shiung Lam, S.; Tabatabaei, M.; Rezania, S. Comprehensive Insights into Conversion of Microalgae to Feed, Food, and Biofuels: Current Status and Key Challenges towards Implementation of Sustainable Biorefineries. *Chem. Eng. J.* **2023,** *455*, 140588. https://doi.org/10.1016/J.CEJ.2022.140588.
32. Martínez-Roldán, A. J.; Cañizares–Villanueva, R. O. Effect of the Nitrogen-Source Modification and CO_2-Addition on Growth and Lipids Composition in *Nannochlorospis* Sp. In *Nannochloropsis: Biology, Biotechnological Potential and Challenges*, First Ed.; Jan, M., Kazik, P., Eds.; **2015**; pp 279.

33. Alfredo de Jesús, M. R.; Brenda Paloma, G. L. Growth of *Spirulina* Maxima in an Optimized Concentric Internal Tube Photobioreactor. *Iran. J. Biotechnol.* **2022,** *20* (3), 44–53. https://doi.org/10.30498/ijb.2022.273070.2985.
34. Sirohi, R.; Kumar Pandey, A.; Ranganathan, P.; Singh, S.; Udayan, A.; Kumar Awasthi, M.; Hoang, A. T.; Chilakamarry, C. R.; Kim, S. H.; Sim, S. J. Design and Applications of Photobioreactors- a Review. *Bioresour. Technol.* **2022,** *349*, 126858. https://doi.org/10.1016/j.biortech.2022.126858.
35. Torzillo, G.; Chini Zittelli, G. Tubular Photobioreactors. In *Algal Biorefineries: Volume 2: Products and Refinery Design*; Springer International Publishing, 2015; pp 187–212.
36. Zhou, W.; Lu, Q.; Han, P.; Li, J. Microalgae Cultivation and Photobioreactor Design. In *Microalgae Cultivation for Biofuels Production*; Elsevier, **2020**; pp 31–50. https://doi.org/10.1016/B978-0-12-817536-1.00003-5.
37. Richmond, A. In *Handbook of Microalgal Culture*; Richmond, A., Ed.; Blackwell Publishing Ltd: Oxford, UK, **2003**. https://doi.org/10.1002/9780470995280.
38. Silva-Benavides, A.; Torzillo, G. Productivity and Biochemical Composition of *Phaeodactylum Tricornutum* (Bacillariophyceae) Cultures Grown Outdoors in Tubular Photobioreactors and Open Ponds. *Biomass Bioenergy* **2013,** *54*, 115–122.
39. Torzillo, G.; Giannelli, L.; Martínez-Roldán, A. J.; Verdone, N.; De Filippis, P.; Scarsella, M.; Bravi, M.; Martínez-Roldan, A. J.; Verdone, N.; De Filippis, P.; Scarsella, M.; Bravi, M. Microalgae Culturing in Thin-Layer Photobioreactors. *Chem. Eng. Trans.* **2010,** *20*, 265–270. https://doi.org/10.3303/CET1020045.
40. Flórez-Miranda, L.; Martínez-Roldán, A. J.; Cañizares-Villanueva, R. O. Microalgae in Livestock Nutrition. In *Microalgae and Other Phototrophic Bacteria: Culture, Processing, Recovery and New Products*; Torres-Bustillos, L., Ed.; Nova Science Publishers: USA, **2016**; pp 87–110.
41. Borowitzka, M. A. Dunaliella : Biology, Production, and Markets, In *Handbook of Microalgal Culture*; Richmond, A., Hu, Q., Eds.; John Wiley & Sons, Ltd: Oxford, UK, **2013**; pp 359–368. https://doi.org/10.1002/9781118567166.ch18.
42. Cañizares-Villanueva, R. O.; Rojo-Gómez, E.; Arroyo-Sánchez, B. I.; Martínez-Roldán, A. *Chlorella* as a Source of Antioxidant Compounds. In *Chlorella and its Health Benefits*; Marutholi, M., Ed.; **2015**. https://doi.org/10.52305/XXUW5187.
43. Ritu, J. R.; Ambati, R. R.; Ravishankar, G. A.; Shahjahan, M.; Khan, S. Utilization of Astaxanthin from Microalgae and Carotenoid Rich Algal Biomass as a Feed Supplement in Aquaculture and Poultry Industry: An Overview. *J. Appl. Phycol.* **2023,** *35* (1), 145–171. https://doi.org/10.1007/s10811-022-02817-9.
44. Madeira, M. S.; Cardoso, C.; Lopes, P. A.; Coelho, D.; Afonso, C.; Bandarra, N. M.; Prates, J. A. M. Microalgae as Feed Ingredients for Livestock Production and Meat Quality: A Review. *Livest. Sci.* **2017,** *205*, 111–121. https://doi.org/10.1016/j.livsci.2017.09.020.
45. Flórez-Miranda, L.; Cañizares-Villanueva, R. O.; Melchy-Antonio, O.; Martínez-Jerónimo, F.; Flores- Ortíz, C. M. Two Stage Heterotrophy/Photoinduction Culture of Scenedesmus Incrassatulus : Potential for Lutein Production. *J. Biotechnol.* **2017,** *262*, 67–74. https://doi.org/10.1016/j.jbiotec.2017.09.002.
46. Fernández-Rojas, B.; Hernández-Juárez, J.; Pedraza-Chaverri, J. Nutraceutical Properties of Phycocyanin. *J. Funct. Foods* **2014,** *11* (C), 375–392. https://doi.org/10.1016/J.JFF.2014.10.011.

47. Romay, C.; Gonzalez, R.; Ledon, N.; Remirez, D.; Rimbau, V. C-Phycocyanin: A Biliprotein with Antioxidant, Anti-Inflammatory and Neuroprotective Effects. *Curr. Protein Pept. Sci.* **2003,** *4* (3), 207–216. https://doi.org/10.2174/1389203033487216.
48. Rodriguez-Amaya, D. B. Structures and Analysis of Carotenoid Molecules. Subcell. Biochem. **2016**, 79, 71–108. https://doi.org/10.1007/978-3-319-39126-7_3.
49. Bazarnova, J.; Smyatskaya, Y.; Shlykova, A.; Balabaev, A.; Đurović, S. Obtaining Fat-Soluble Pigments—Carotenoids from the Biomass of *Chlorella* Microalgae. *Appl. Sci.* **2022,** *12* (7). https://doi.org/10.3390/app12073246.
50. Martínez-Roldán, A. J.; Perales-Vela, H. V.; Cañizares-Villanueva, R. O.; Torzillo, G. Physiological Response of *Nannochloropsis* Sp. to Saline Stress in Laboratory Batch Cultures. *J. Appl. Phycol.* **2014,** *26* (1), 115–121. https://doi.org/10.1007/s10811-013-0060-1.
51. Barone, M. E.; Parkes, R.; Herbert, H.; McDonnell, A.; Conlon, T.; Aranyos, A.; Fierli, D.; Fleming, G. T. A.; Touzet, N. Comparative Response of Marine Microalgae to H_2O_2-Induced Oxidative Stress. *Appl. Biochem. Biotechnol.* **2021,** *193* (12), 4052–4067. https://doi.org/10.1007/s12010-021-03690-x.
52. Del Mondo, A.; Sansone, C.; Brunet, C. Insights into the Biosynthesis Pathway of Phenolic Compounds in Microalgae. *Comput. Struct. Biotechnol. J.* **2022,** *20*, 1901–1913. https://doi.org/10.1016/J.CSBJ.2022.04.019.
53. Edelmann, M.; Aalto, S.; Chamlagain, B.; Kariluoto, S.; Piironen, V. Riboflavin, Niacin, Folate and Vitamin B_{12} in Commercial Microalgae Powders. *J. Food Compos. Anal.* **2019,** *82*, 103226. https://doi.org/10.1016/J.JFCA.2019.05.009.
54. Karkos, P. D.; Leong, S. C.; Karkos, C. D.; Sivaji, N.; Assimakopoulos, D. A. *Spirulina* in Clinical Practice: Evidence-Based Human Applications. *Evid. Based Complement. Altern. Med.* **2011,** *2011*, 1–4. https://doi.org/10.1093/ecam/nen058.
55. Ljubic, A.; Thulesen, E. T.; Jacobsen, C.; Jakobsen, J. UVB Exposure Stimulates Production of Vitamin D3 in Selected Microalgae. *Algal Res.* **2021,** *59*, 102472. https://doi.org/10.1016/J.ALGAL.2021.102472.
56. Udayan, A.; Arumugam, M.; Pandey, A. Nutraceuticals From Algae and Cyanobacteria. In *Algal Green Chemistry*; Elsevier, **2017**; pp 65–89. https://doi.org/10.1016/B978-0-444-63784-0.00004-7.
57. Durmaz, Y. Vitamin E (α-Tocopherol) Production by the Marine Microalgae *Nannochloropsis Oculata* (Eustigmatophyceae) in Nitrogen Limitation. *Aquaculture* **2007,** *272* (1–4), 717–722. https://doi.org/10.1016/J.AQUACULTURE.2007.07.213.
58. Zhou, L.; Li, K.; Duan, X.; Hill, D.; Barrow, C.; Dunshea, F.; Martin, G.; Suleria, H. Bioactive Compounds in Microalgae and Their Potential Health Benefits. *Food Biosci.* **2022,** *49*, 101932. https://doi.org/10.1016/j.fbio.2022.101932.
59. Finkel, T. Signal Transduction by Reactive Oxygen Species. *J. Cell Biol.* **2011,** *194* (1), 7–15. https://doi.org/10.1083/jcb.201102095.
60. Schieber, M.; Chandel, N. S. ROS Function in Redox Signaling and Oxidative Stress. *Curr. Biol.* **2014,** *24* (10), R453–R462. https://doi.org/10.1016/J.CUB.2014.03.034.
61. Sansone, C.; Brunet, C. Promises and Challenges of Microalgal Antioxidant Production. *Antioxidants* **2019,** *8* (7), 199. https://doi.org/10.3390/antiox8070199.
62. Wan Mahmood, W. M. A.; Lorwirachsutee, A.; Theodoropoulos, C.; Gonzalez-Miquel, M. Polyol-Based Deep Eutectic Solvents for Extraction of Natural Polyphenolic Antioxidants from *Chlorella Vulgaris*. *ACS Sustain. Chem. Eng.* **2019,** *7* (5), 5018–5026. https://doi.org/10.1021/acssuschemeng.8b05642.

63. León-Vaz, A.; León, R.; Vigara, J.; Funk, C. Exploring Nordic Microalgae as a Potential Novel Source of Antioxidant and Bioactive Compounds. *N. Biotechnol.* **2023,** *73*, 1–8. https://doi.org/10.1016/J.NBT.2022.12.001.
64. Haoujar, I.; Cacciola, F.; Abrini, J.; Mangraviti, D.; Giu, D.; Oulad, Y.; Majdoub, E.; Kounnoun, A.; Miceli, N.; Taviano, M. F.; Mondello, L.; Rigano, F.; Senhaji, N. S. The Contribution of Carotenoids, Phenolic Compounds, and Flavonoids to the Antioxidative Properties of Marine Microalgae Isolated from Mediterranean Morocco. *Molecules* **2019,** *24*, 4037. https://doi.org/10.3390/moléculas24224037.
65. Bellahcen, T. O.; AAmiri, A.; Touam, I.; Hmimid, F.; Amrani, A. El; Cherif, A.; Cherki, M. Evaluation of Moroccan Microalgae: *Spirulina Platensis* as a Potential Source of Natural Antioxidants. *J. Complement. Integr. Med.* **2020,** *17* (3). https://doi.org/10.1515/jcim-2019-0036.
66. Bulut, O.; Akın, D.; Sönmez, Ç.; Öktem, A.; Yücel, M.; Öktem, H. A. Phenolic Compounds, Carotenoids, and Antioxidant Capacities of a Thermo-Tolerant *Scenedesmus* Sp. (Chlorophyta) Extracted with Different Solvents. *J. Appl. Phycol.* **2019,** *31* (3), 1675–1683. https://doi.org/10.1007/s10811-018-1726-5.
67. Barta, D. G.; Coman, V.; Vodnar, D. C. Microalgae as Sources of Omega-3 Polyunsaturated Fatty Acids: Biotechnological Aspects. *Algal Res.* **2021,** *58*, 102410. https://doi.org/10.1016/J.ALGAL.2021.102410.
68. Colombo, S. M.; Rodgers, T. F. M.; Diamond, M. L.; Bazinet, R. P.; Arts, M. T. Projected Declines in Global DHA Availability for Human Consumption as a Result of Global Warming. *Ambio* **2020,** *49* (4), 865–880. https://doi.org/10.1007/s13280-019-01234-6.
69. Ferreira, S. P.; De Souza-Soares, L.; Vieira Costa, J. A. Revisão: Microalgas: Uma Fonte Alternativa Na Obtenção de Ácidos Gordos Essenciais Review: Microalgae: An Alternative Source to Obtain Essential Fatty Acids. *Rev. Ciências Agrárias* **2013,** *36* (3), 275–287.
70. Vaz, B. da S.; Moreira, J. B.; Morais, M. G. de; Costa, J. A. V. Microalgae as a New Source of Bioactive Compounds in Food Supplements. *Curr. Opin. Food Sci.* **2016,** *7*, 73–77. https://doi.org/10.1016/J.COFS.2015.12.006.
71. Perdana, B. A.; Chaidir, Z.; Kusnanda, A. J.; Dharma, A.; Zakaria, I. J.; Syafrizayanti; Bayu, A.; Putra, M. Y. Omega-3 Fatty Acids of Microalgae as a Food Supplement: A Review of Exogenous Factors for Production Enhancement. *Algal Res.* **2021,** *60*, 102542. https://doi.org/10.1016/J.ALGAL.2021.102542.
72. Noer Abyor Handayania, D. A. and H. Potential Production of Omega Fatty Acids from Microalgae. *Int. J. Pharm. Sci. Rev. Res.* **2015,** *34* (2), 210–215. https://doi.org/10.4172/scientificrep.
73. Ratledge, C.; Streekstra, H.; Cohen, Z.; Fichtali, J. Downstream Processing, Extraction, and Purification of Single Cell Oils. In *Single Cell Oils*; Elsevier, **2010**; pp 179–197. https://doi.org/10.1016/B978-1-893997-73-8.50013-X.
74. Fedorova-Dahms, I.; Marone, P. A.; Bailey-Hall, E.; Ryan, A. S. Safety Evaluation of Algal Oil from *Schizochytrium* Sp. *Food Chem. Toxicol.* **2011,** *49* (1), 70–77. https://doi.org/10.1016/J.FCT.2010.09.033.
75. Lupette, J.; Benning, C. Human Health Benefits of Very-Long-Chain Polyunsaturated Fatty Acids from Microalgae. *Biochimie* **2020,** *178*, 15–25. https://doi.org/10.1016/j.biochi.2020.04.022.
76. Gloaguen, V.; Ruiz, G.; Morvan, H.; Mouradi-Givernaud, A.; Maes, E.; Krausz, P.; Strecker, G. The Extracellular Polysaccharide of *Porphyridium* Sp.: An NMR Study of

Lithium-Resistant Oligosaccharidic Fragments. *Carbohydr. Res.* **2004,** *339* (1), 97–103. https://doi.org/10.1016/j.carres.2003.09.020.

77. Gouda, M.; Tadda, M. A.; Zhao, Y.; Farmanullah, F.; Chu, B.; Li, X.; He, Y. Microalgae Bioactive Carbohydrates as a Novel Sustainable and Eco-Friendly Source of Prebiotics: Emerging Health Functionality and Recent Technologies for Extraction and Detection. *Front. Nutr.* **2022,** 9. https://doi.org/10.3389/fnut.2022.806692.
78. Liu, L.; Pohnert, G.; Wei, D. Extracellular Metabolites from Industrial Microalgae and Their Biotechnological Potential. *Mar. Drugs* **2016,** *14* (10), 1–19. https://doi.org/10.3390/md14100191.
79. Jacob-Lopes, E.; Maroneze, M. M.; Deprá, M. C.; Sartori, R. B.; Dias, R. R.; Zepka, L. Q. Bioactive Food Compounds from Microalgae: An Innovative Framework on Industrial Biorefineries. *Curr. Opin. Food Sci.* **2019,** *25*, 1–7. https://doi.org/10.1016/J.COFS.2018.12.003.
80. Morais, M. G.; Santos, T. D.; Moraes, L.; Vaz, B. S.; Morais, E. G.; Costa, J. A. V. Exopolysaccharides from Microalgae: Production in a Biorefinery Framework and Potential Applications. *Bioresour. Technol. Rep.* **2022,** *18*, 101006. https://doi.org/10.1016/j.biteb.2022.101006.
81. Halaj, M.; Matulová, M.; Capek, P. Structural Features of Biologically Active Extracellular Polysaccharide Produced by Green Microalgae *Dictyosphaerium Chlorelloides*. *Int. J. Biol. Macromol.* **2022,** *214*, 152–161. https://doi.org/10.1016/j.ijbiomac.2022.05.197.
82. Cepák, V.;Přibyl, P. Light Intensity and Nitrogen Effectively Control. *Genet. Plant Physiol.* **2018,** *8*, 24–37. https://doi.org/http://www.ifrg-bg.com.
83. Han, P.; Sun, Y.; Jia, S.; Zhong, C.; Tan, Z. Effects of Light Wavelengths on Extracellular and Capsular Polysaccharide Production by *Nostoc Flagelliforme*. *Carbohydr. Polym.* **2014,** *105*, 145–151. https://doi.org/10.1016/j.carbpol.2014.01.061.
84. Ge, H.; Xia, L.; Zhou, X.; Zhang, D.; Hu, C. Effects of Light Intensity on Components and Topographical Structures of Extracellular Polysaccharides from the Cyanobacteria *Nostoc* Sp. *J. Microbiol.* **2014,** *52* (2), 179–183. https://doi.org/10.1007/s12275-014-2720-5.
85. Medina-Cabrera, E. V.; Rühmann, B.; Schmid, J.; Sieber, V. Optimization of Growth and EPS Production in Two *Porphyridum* Strains. *Bioresour. Technol. Rep.* **2020,** *11*, 100486. https://doi.org/10.1016/j.biteb.2020.100486.
86. Sasaki, M.; Takagi, A.; Sasaki, D.; Nakamura, A.; Asayama, M. Characteristics and Function of an Extracellular Polysaccharide from a Green Alga *Parachlorella*. *Carbohydr. Polym.* **2021,** *254*, 117252. https://doi.org/10.1016/j.carbpol.2020.117252.
87. Michaud, P. Polysaccharides from Microalgae, What's Future? *Adv. Biotechnol. Microbiol.* **2018,** *8* (2), 1–2. https://doi.org/10.5772/51572.This.
88. Franco-Morgado, M.; Amador-Espejo, G. G.; Pérez-Cortés, M.; Gutiérrez-Uribe, J. A. Microalgae and Cyanobacteria Polysaccharides: Important Link for Nutrient Recycling and Revalorization of Agro-Industrial Wastewater. *Appl. Food Res.* **2023,** *3* (1), 100296. https://doi.org/10.1016/j.afres.2023.100296.
89. Karmakar, A.; Das, A. K.; Ghosh, S.; Sil, P. C. Carotenoids as Coloring Agents. In *Carotenoids: Structure and Function in the Human Body*; Springer International Publishing: Cham, **2021**; pp 189–207. https://doi.org/10.1007/978-3-030-46459-2_6.
90. McWilliams, A. Global Carotenoids Market Research Report to 2022. Glob. Mark. Carotenoids, **2018**.

91. Yaqoob, S.; Riaz, M.; Shabbir, A.; Zia-Ul-Haq, M.; Alwakeel, S. S.; Bin-Jumah, M. Commercialization and Marketing Potential of Carotenoids. In *Carotenoids: Structure and Function in the Human Body*; Springer International Publishing: Cham, **2021**; pp 799–826. https://doi.org/10.1007/978-3-030-46459-2_27.
92. El-Chaghaby, G.; Rashad, S.; Abdel-Kader, S. F.; Rawash, E. S. A.; Moneem, M. A. Assessment of Phytochemical Components, Proximate Composition and Antioxidant Properties of *Scenedesmus Obliquus*, *Chlorella Vulgaris* and *Spirulina Platensis* Algae Extracts. *Egypt. J. Aquat. Biol. Fish.* **2019,** *23* (4), 521–526. https://doi.org/10.21608/EJABF.2019.57884.
93. Flórez-Miranda, L.; Martínez Roldán, A. J.; Cañizares-Villanueva, R. O. In *Microalgae in Livestock Nutrition*, **2016**.
94. Martínez-Roldán, A. J.; Cañizares–Villanueva, R. O. Effect of the Nitrogen-Source Modification and CO_2-Addition on Growth and Lipids Composition in *Nannochlorospis* Sp. In *Nannochloropsis: Biology, Biotechnological Potential and Challenges*, First Ed.; Jan, M., Kazik, P., Eds.; **2017**; p 279.
95. Safafar, H.; Wagenen, J. Van; Møller, P.; Jacobsen, C. Carotenoids, Phenolic Compounds and Tocopherols Contribute to the Antioxidative Properties of Some Microalgae Species Grown on Industrial Wastewater. *Mar. Drugs* **2015,** *13* (12), 7339–7356. https://doi.org/10.3390/MD13127069.
96. García-González, M.; Moreno, J.; Manzano, J. C.; Florencio, F. J.; Guerrero, M. G. Production of *Dunaliella Salina* Biomass Rich in 9-Cis-β-Carotene and Lutein in a Closed Tubular Photobioreactor. *J. Biotechnol.* **2005,** *115* (1), 81–90. https://doi.org/10.1016/j.jbiotec.2004.07.010.
97. Del Campo, J. A.; Rodríguez, H.; Moreno, J.; Vargas, M. Á.; Rivas, J.; Guerrero, M. G. Lutein Production by *Muriellopsis* Sp. in an Outdoor Tubular Photobioreactor. *J. Biotechnol.* **2001,** *85* (3), 289–295. https://doi.org/10.1016/S0168-1656(00)00380-1.
98. Khadim, S. R.; Singh, P.; Singh, A. K.; Tiwari, A.; Mohanta, A.; Asthana, R. K. Mass Cultivation of *Dunaliella Salina* in a Flat Plate Photobioreactor and Its Effective Harvesting. *Bioresour. Technol.* **2018,** *270*, 20–29. https://doi.org/10.1016/j.biortech.2018.08.071.
99. Blanco, A. M.; Moreno, J.; Del Campo, J. A.; Rivas, J.; Guerrero, M. G. Outdoor Cultivation of Lutein-Rich Cells of *Muriellopsis* Sp. in Open Ponds. *Appl. Microbiol. Biotechnol.* **2007,** *73* (6), 1259–1266. https://doi.org/10.1007/s00253-006-0598-9.
100. Cavieres, L.; Bazaes, J.; Marticorena, P.; Riveros, K.; Medina, P.; Sepúlveda, C.; Riquelme, C. Pilot-Scale Phycoremediation Using *Muriellopsis* Sp. for Wastewater Reclamation in the Atacama Desert: Microalgae Biomass Production and Pigment Recovery. *Water Sci. Technol.* **2021,** *83* (2), 331–343. https://doi.org/10.2166/wst.2020.576.
101. Sánchez, J. F.; Fernández, J. M.; Acién, F. G.; Rueda, A.; Pérez-Parra, J.; Molina, E. Influence of Culture Conditions on the Productivity and Lutein Content of the New Strain *Scenedesmus Almeriensis*. *Process Biochem.* **2008,** *43* (4), 398–405. https://doi.org/10.1016/j.procbio.2008.01.004.
102. Morillas-España, A.; Lafarga, T.; Gómez-Serrano, C.; Acién-Fernández, F. G.; González-López, C. V. Year-Long Production of *Scenedesmus Almeriensis* in Pilot-Scale Raceway and Thin-Layer Cascade Photobioreactors. *Algal Res.* **2020,** *51*. https://doi.org/10.1016/j.algal.2020.102069.

103. Molino, A.; Mehariya, S.; Iovine, A.; Casella, P.; Marino, T.; Karatza, D.; Chianese, S.; Musmarra, D. Enhancing Biomass and Lutein Production From *Scenedesmus Almeriensis*: Effect of Carbon Dioxide Concentration and Culture Medium Reuse. *Front. Plant Sci.* **2020,** *11*. https://doi.org/10.3389/fpls.2020.00415.
104. Olaizola, M. Commercial Production of Astaxanthin from *Haematococcus Pluvialis* Using 25,000-Liter Outdoor Photobioreactors. *J. Appl. Phycol.* **2000,** *12* (3–5), 499–506. https://doi.org/10.1023/a:1008159127672.
105. Torzillo, G.; Goksan, T.; Faraloni, C.; Kopecky, J.; Masojídek, J. Interplay between Photochemical Activities and Pigment Composition in an Outdoor Culture of *Haematococcus Pluvialis* during the Shift from the Green to Red Stage. *J. Appl. Phycol.* **2003,** *15* (2/3), 127–136. https://doi.org/10.1023/A:1023854904163.
106. Yadavalli, R.; Ratnapuram, H.; Peasari, J. R.; Reddy, C. N.; Ashokkumar, V.; Kuppam, C. Simultaneous Production of Astaxanthin and Lipids from *Chlorella Sorokiniana* in the Presence of Reactive Oxygen Species: A Biorefinery Approach. *Biomass Convers. Biorefin.* **2022,** *12* (3), 881–889. https://doi.org/10.1007/s13399-021-01276-5.
107. Wang, Y.; Peng, J. Growth-Associated Biosynthesis of Astaxanthin in Heterotrophic *Chlorella Zofingiensis* (Chlorophyta). *World J. Microbiol. Biotechnol.* **2008,** *24* (9), 1915–1922. https://doi.org/10.1007/s11274-008-9692-8.
108. Minyuk, G.; Sidorov, R.; Solovchenko, A. Effect of Nitrogen Source on the Growth, Lipid, and Valuable Carotenoid Production in the Green Microalga *Chromochloris Zofingiensis*. *J. Appl. Phycol.* **2020,** *32* (2), 923–935. https://doi.org/10.1007/s10811-020-02060-0.

CHAPTER 6

Biostimulation of Grains and Plant Seeds Using Probiotic Microorganisms

STEPHANY NEFERTARI CHÁVEZ-GARCÍA, DORA ELISA CRUZ-CASAS, ERIK DE LEÓN-JUÁREZ, ÁLVARO JAVIER PASTRANA PASTRANA, STEFFANNY SÁNCHEZ PORTILLO, and ADRIANA CAROLINA FLORES-GALLEGOS

Food Research Department, School of Chemistry, Autonomus University of Coahuila, Saltillo, Coahuila, México

ABSTRACT

Plants are a food source that is increasingly in demand in our current society due to various factors such as their high nutritional value or sustainability. For this reason, we are currently seeking to produce these plants from their grains and seeds with green technology. Biostimulation is a technique with several advantages that promotes the effective development of the plant. It is already known that the use of microorganisms is a useful, cheap, and potential tool for plant growth. However, there are microorganisms, such as probiotics, which are known to have multiple beneficial applications for consumer health. However, there are only a few reports that discuss how probiotics improve plant growth as well as the nutritional quality of the plant, and they also have positive effects on the consumer. Therefore, it is important to know the role of the probiotic biostimulation method, describe what this method consists of, and compare it with others. The beneficial biostimulation effects of chitin on facilitating plant growth and improving stress tolerance have been found. As a biostimulant, plant growth-promoting bacteria have been widely used in horticultural plants to promote their growth. Plant growth-promoting bacteria can improve plant nutrient availability,

The Food-Energy-Water Triangle for Sustainable Development: Integrative and Synergic Solutions.
Pedro Aguilar-Zárate, Mónica Chávez-González, Ann Rose Abraham, and A. K. Haghi (Eds.)

promote the release of plant hormones, and enhance abiotic stress resistance by releasing carriers, organic acids, and enzymes. In addition to explaining the mechanism by which probiotics promote plant growth, microorganisms, including bacteria, yeast, fungi, and microalgae, have been proven to have biostimulation activity. Many plant-associated microbes can synthesize plant hormones such as auxin, ethylene, and cytokinins that have crucial, multifaceted roles in plants. Moreover, some bacteria can produce abscisic acid and gibberellins; inoculation with these microbes helps plants to resist stresses.

6.1 INTRODUCTION

Sprouting is a bioprocessing technique used by the food industry where the nutrients contained in the seeds are used to give rise to the growth of a sprout (Zhang et al., 2020; Samec et al., 2018). The best use of nutrients comes from the enzymatic activity that begins with the soaking of the seed, unfolding the larger nutrients into simple molecules such as amino acids, simple sugars, and other nutritional compounds (Liu et al., 2019), in addition to reducing the content of antinutritional components present in grains and seeds, making these foods more digestible and with better sensory attributes (Ohanenye et al., 2020) and considering them as a rich source of phenolic compounds, vitamins, and minerals (Mekky et al., 2020). The current trend in the consumption of healthy foods has led to an increase in the production of sprouts, which are considered a functional food with beneficial properties for the consumer. Today's research is focused on the study of the effects of the germination process as part of biostimulation in plants, including changes in nutritional composition and phytochemical profile. Therefore, research has focused on studying the effects that germination has on the nutritional composition and phytochemical profile (Galieni et al., 2020).

The germination process consists of three stages: soaking, enzyme activation, and growth. During each of its phases, there is a series of factors that influence the growth and nutritional content of the plant. Soaking, also known as imbibition, has been reported as one of the most important stages for the reduction of antinutritional components, in addition to the fact that the substance in which the seed is soaked can influence the metabolism of the seed (Vötterl et al., 2019). One of the substances in which favorable changes have been seen in the development of the plant are probiotics, which are live microorganisms that, when administered in the right amount, can provide a health benefit to the host.[4] This strategy would allow probiotics to be considered biostimulatory substances, in addition to developing a food of excellent nutritional quality with beneficial effects on consumer health (Świeca et al., 2019). The objective of this work was to study the effect of sprouting on biostimulation, nutritional quality, and the probiotic effect of sprouts.

6.2 PROBIOTICS

The most widely accepted definition of probiotics is the one suggested by the Food and Agriculture Organization of the United Nations/World Health Organization in 2002. It defines them as "live microorganisms which, when administered in an appropriate dosage, provide health benefits to the host."

The beneficial effects that probiotics have shown to have on the health of the consumer include immunological, antiviral, antidiabetic, and anticancer activity; they restore the microbial balance in the human intestine, prevent and treat diarrhea caused by various factors, intestinal diseases, and irritable bowel syndrome, among others.[3]

The most commercialized probiotic microorganisms are those belonging to the genera *Lactobacillus*, *Bifidobacteria*, *Lactococcus*, *Streptococcus*, and *Enterococcus*, in addition to some strains of *Bacillus* and *Saccharomyces*.[4] However, LABs are considered the most important probiotic species. They are gram-positive, catalase-negative, aerotolerant, fastidious, strictly fermentative, and nonspore-forming organisms, among others. They are mainly isolated from fermented dairy products (yogurt, kefir, and cheese) and silage (cabbage).[1] Probiotics are also characterized by producing compounds of high importance such as bacteriocins, hydrogen peroxide, diacetyl, lactic acid, acetic acid, aromatic compounds, and exopolysaccharides, as well as proteolytic and lipolytic capacity.[6] Probiotics belong to a multibillion-dollar industry in the food industry and are highly consumed worldwide as food supplements. In the food industry, to be considered a probiotic microorganism, it must be metabolically stable and active in the food, survive in the digestive tract, have beneficial effects in the host intestine, and be transferred to the consumer.[2,6]

In other areas, the importance of probiotics has also increased, such as the agricultural industry. These microorganisms have been identified as plant growth promoters. In recent years, the use of microorganisms to stimulate plant growth has become more involved. They are an environmentally friendly alternative.[5]

Probiotic microorganisms have multiple applications in addition to those beneficial to human health. It is essential to apply these beneficial microorganisms in other areas, such as agriculture, where sustainable production can be achieved and the consumption of these plants would be beneficial for the consumer.

6.3 APPLICATIONS

6.3.1 HEALTH BENEFITS

The probiotic bacteria promote some benefits for host health when provided as live supplements. Many of the effects of probiotics depend on their interaction with dendritic cells; this is possible due to their capacity to polarize the adaptive immune response. The maturity of these cells and their cytosine production depend on the probiotic strain; some examples of this are the stimulation and synthesis of IL-10 and IL-12 cytosines, which produce an immune response.[26] Probiotics can also prevent the overgrowth of dangerous pathogenic bacteria by improving the epithelial barrier function or increasing the resistance of the gut to invasions.[28] Specific strains have specific effects on the host's health, including neurological and endocrinological effects, antimicrobial activities, and the production of specific bioactive substances, as long as they are the ones described previously in this text.[27]

Some studies have shown differences in the microbiota between healthy and allergic patients, and despite that, few studies have found positive results using probiotics to treat allergies. A recent research has shown that a combination of *Bifidobacterium breve* M-16V with other probiotics significantly reduces the Th2 cytosine production.[32]Other studies claim that the administration of *Lactobacillus rhamnosus* GG during the last 4 weeks of pregnancy and the 3 months after birth improved the immune protection provided by breastmilk.[26] Talking about breastmilk, it contains an important quantity of nondigestible carbohydrates, which stain the gastrointestinal tract as a substratum to microbiota like Bifidobacterium. These bacteria affect the microbiota in the first year of life and have been observed to have the highest concentration in healthy children in comparison to overweight children.[33] (WHO, 2013.)

6.3.2 AGRICULTURE

The use of probiotics on plants is not a human invention; plants are always accompanied by microorganisms known as Plant Probiotic Microorganisms (PPMs), and they have a symbiotic association with plants.[29] These microbes support the vegetable host by stimulating growth, increasing yield, reducing pathogen infections, and reducing abiotic/biotic and salt stress.[30] Some of the most studied are the microorganisms that establish an endosymbiotic relationship with the plant, like *Fusarium oxysporum*, *Alternaria sp.*, *Phoma*, *Aspergillus spp.*, *Penicillium oxalicum*, and *Penicillium simplicissimum*, which have demonstrated their growth-promoting capacity. This positions probiotics as an alternative to the use of chemical fertilizers.[31] Fungi are used as probiotics; lactic acid bacteria are mostly used in fermented foods and can be considered biocontrollers thanks to their activity against pathogens and their generally recognized as safe (GRAS) status.[35] Probiotics are located in soil too, and because the microflora of soil mediates many processes in plants, improving soils with probiotics is an alternative to fertilization with no adverse effects.[36]

6.4 BIOSTIMULANTS IN PLANTS

One of the most challenging aspects of agriculture today is increasing crop production to provide food for the world's population. Chemicals, such as pesticides and fertilizers, increase crop yields. However, this practice has significant implications for production costs and raises environmental concerns about water, land, and air resources. For agricultural productivity to be achieved, new approaches that promote more sustainable production are needed. An alternative approach in agriculture is biostimulants.[14]

The term "biostimulant" is used in agriculture to describe a formulation of components, substances, or microbes that stimulate growth, yield, and stress tolerance in plants and crops.[18] A biostimulant can be classified into five main groups: seaweed and plant extracts, humic substances, hydrolyzed proteins and nitrogen-containing compounds, microorganisms, and inorganic compounds with biostimulant action. Each of them has a specific effect on plants.[9]

Biostimulants are substances capable of increasing crop growth or tolerance to adverse environmental conditions, reducing disease incidence, increasing soil nutrient availability, improving plant nutrient use efficiency, and promoting the degradation and humification of organic substances in soils. The use of biostimulants has been increasingly integrated into agricultural production systems such as seed treatment, plant growth, and harvesting in order to enhance plant productivity by altering physiological processes. It is difficult to identify the mechanisms of action or biochemical pathways involved in the physiological processes induced by biostimulants. In part, this is because the raw materials used in biostimulant production are heterogeneous and mixtures are complex. However, a hypothesis is that plant metabolism is affected by biostimulants, increasing growth potential, development, and stress response.[19] On the other hand, a projected USD 4.14 billion by 2025 is the projected growth rate for the global biostimulants market. The use of these components in agriculture is trending and becoming increasingly important. However, biostimulants cannot solve all agricultural problems with a one-size-fits-all approach. Depending on the climatic conditions, the type of crop, and the dosage, the efficacy of these compounds can vary.[7]

6.4.1 HYDROPONIC METHOD

The growth of the world's population primarily occurs in urban areas, which means that agricultural soil will compete with urban centers for land and water, which will mean more food needs to be produced on a smaller area of land in the future. Hydroponics is the science of growing crops without the need for soil. It seeks to generate an optimal environment for plant growth through an optimal ratio of nutrients and climate control. Hydroponic systems are efficient in the use of water and nutrients because they allow their reuse, which contributes to a reduction in the consumption of water and fertilization compared to conventional agriculture. In addition, hydroponic systems do not present any problems when it comes to diseases, pests, or weeds that originate in the soil.[10]

There are numerous crops that can be grown hydroponically, including beans, corn, cucumbers, lettuce, tomatoes, carrots, rice, pepper, spinach, strawberries, and potatoes. Hydroponics has become more and more flexible due to the development of new materials and equipment. Examples are the wick system, drip irrigation, aeroponics, and nutrient film technique. Hydroponic systems offer technological opportunities in regions with poor agronomic conditions, such as deserts and infertile lands, with the advantage that hydroponics is highly productive and allows process automation. However, the high initial cost of hydroponic systems is the main disadvantage of this crop production model. Constant control is required because many pathogens can grow in hydroponic conditions due to nutrient concentrations, the most common microorganisms being *Fusarium*, *Phytophthora*, and *Pythium*. The use of probiotics as a biocontrol alternative has emerged in recent years. Traditional chemical and physical disinfection procedures can be replaced by this method. Moreover, probiotics could improve the crop's nutraceutical properties and productivity. However, problems arising from a lack of understanding of the mechanism that prevents damage by plant pathogens need to be addressed.[11,16]

6.4.2 PROBIOTIC METHODS

In nature, there are specific microbial communities associated with plants, both externally and internally within their tissues, which are known as the plant microbiome. These communities include rhizosphere and endosphere microorganisms. Plant probiotic microorganisms (PPMs) are microorganisms, usually bacteria, that promote plant growth through symbiotic or free-living relationships. This approach shows promise as an alternative to chemical fertilizers and pesticides.[12]

Plant growth-promoting bacteria (PGPB) are one of the most common PPM groups in soil. PGPB are constituted by different microorganisms (bacteria, fungi, yeasts, and microalgae) that can grow in, on, or around plant tissues, which promote growth, enhance resistance to abiotic stress, or increase plant nutrient availability through direct or indirect mechanisms. Direct mechanisms associated with nutrient enhancement include biological nitrogen fixation, phosphate solubilization, and hormone synthesis. Indirect mechanisms are related to plant defense mechanisms against phytopathogens. Although nitrogen accounts for approximately 80% of the earth's atmosphere, neither plants nor animals have an easy way to obtain sufficient nitrogen for their growth. This situation results because molecular nitrogen is chemically very stable and therefore unusable for most biological organisms. It must therefore be fixed before it can be assimilated. Microbial processes therefore play a crucial role in supplying the nutrients needed for growth. Only a few prokaryotes can carry out biological nitrogen fixation (BNF), among which are the genera *Azospirillum*, *Azotobacter*, *Bacillus*, and *Nitrobacter*. Phosphorus is the second most relevant nutrient for plants after nitrogen. Genera such as *Pseudomonas*, *Bacillus*, *Paenibacillus*, *Deftia*, *Azotobacter*, *Klebsiella*, and *Flavobacterium* have been reported as efficient phosphorus solubilizers. Phytohormone biosynthesis could be involved in plant pathogenesis. A wide spectrum of bacteria has the ability to produce phytohormones such as auxins, cytokinins, and gibberellins. Likewise, some microbial strains are capable of producing siderophores, organic compounds capable of sequestering ferric iron Fe^{3+}, which increases the availability of the nutrient to the plant. Some endophytic strains are *Micrococcus* and *Streptomyces*. On the other hand, lactic acid bacteria are extensively employed in probiotic and fermented products and can be regarded as biocontrollers as they possess the ability to combat pathogenic microorganisms. Overall, understanding the roles of different bacteria in nutrient cycling and hormone production can lead to more sustainable and efficient agricultural practices.[8,12,13]

6.4.3 OTHER SUSTAINABLE METHODS

Other techniques to improve soil quality, growth, and food production are as follows: bioremediation and compost.

Bioremediation: Soil contamination by heavy metals generates problems related to damage to soil quality and fertility, leading to reduced production. An alternative to addressing this technology through biological means is the use of microorganisms. Microorganisms possess inherent mechanisms to degrade a variety of contaminant components and remove them from the environment. Microorganisms have advantages compared to traditional methods of contaminant control. It is a simpler technique, easy to implement, economical, and environmentally friendly because it does not produce additional toxic waste.[15] Compost: Composting is a biochemical process based on the action of aerobic microorganisms with the ability to degrade organic matter into a more stable material. The product obtained seeks to improve physical (e.g., porosity and moisture retention capacity), chemical (e.g., pH and cation exchange capacity), and microbiological properties that favor a physicochemical and biological balance for proper plant–soil interaction.[17]

6.5 PROBIOTIC MECHANISMS IN THE GROWTH OF PLANTS

The soil is one of the nonrenewable natural resources that is used as a medium for the development of food crops, providing sustenance for humans and animals. Its main function is to be the main source of substrates such as organic and mineral components, water retention, and the basis of the nutritional contribution for microbial growth to provide a suitable environment for plant development.[37] The microorganisms in the phytomicrobiome provide many benefits to the plant for the acquisition of nutrients, improvement in soil fertility, release, and modulation of metabolites such as hormones, antimicrobial compounds, and other compounds that suppress the development of phytopathogens. Bacteria, fungi, and other microorganisms present in the soil are responsible for the decomposition of organic matter and the transformation of chemical substances into nutrients used by the plant. Some of the microorganisms work through a symbiosis with the plant, modifying its metabolism and causing a greater availability of micronutrients in the soil, which are made available for use by the plant, in addition to playing an important role in tolerance to drought, soil acidity and salinity, and control of the development of pathogenic microorganisms that limit plant growth.[38] There are a number of microorganisms that have been studied for their nitrogen-fixing capacity and for the transformation of phosphorus into its simplest form for plant use (they participate in P solubilization, mineralization, absorption, and transport processes). Today, the application of microorganisms within sustainable agriculture as biostimulants is a new strategy offered by the agroeconomy, with many advantages over previously known stimulants. Among the advantages offered by using this technique is tolerance to biotic and abiotic stress for the production of excellent quality crops and meeting one of the objectives of the WHO 2030 agenda for food security. Biostimulants are a replacement for agrochemicals that have detrimental effects on the environment and on the health of humans and animals.[40] Despite the advantages offered, there are certain limitations since agriculture and the green strategy that it contemplates are not fully established in the crops of each country, and the regulations have not considered implementing this type of technology. On the other hand, another consortium of microorganisms used are probiotics that act as biostimulants and are applied to the microbiome of crops to improve the development and growth of the plant, in addition to increasing the immune response.[39] The development of endophyte-based bioformulations such as probiotics will modulate different mechanisms involved in the probiotic-plant interaction, such as the acquisition of nutrients, secretion of secondary metabolites, siderophores, and production of phytohormones, as well as an increase in defense mechanisms in response to abiotic and biotic stress.[41] One of the best methods to incorporate microorganisms into the plant system is through bioformulations based on inert carriers (gums, methylcellulose, and starch), which provide stability for the plant and its secondary metabolites. As mentioned above, current agriculture seeks to counteract the effects of phytopathogens from the use of biostimulants.[42]

6.6 FORWARD PERSPECTIVES

Currently, probiotics are used for the development of products that help with the treatment of different human health diseases (digestive, depressive, among others). However, the use of these products is expanding every day since the benefits they can provide is known in other areas of application, such as crops, where, apart from helping against the spread of pathogenic microorganisms that affect these plants, they are attributed the ability to transform organic matter, thus benefiting the development of the plant with some properties mentioned in previous sections. Among the benefits reported so far, some plant growth-stimulating rhizobacteria stand out[20,21] (García-Fraile et al. 2017). Due to the potential that has been found in research on the effects that these microorganisms have, research has been promoted in agricultural areas for products such as biofertilizers.[23] It has also been shown that there are some bacteria that not only offer benefits in terms of production but also improve the nutritional composition, which provides for this field of application: the use of probiotics with dual purposes to combat diseases both in health and in the plant, improve crop development through increased biomass production, shorten production time in food production such as sprouted products, and improve the characteristics of the final products in terms of the contribution of nutrition to consumers.[24,25]

For this reason, it is important to study in depth the uses of probiotics and the different fields of application, whether in the food, pharmaceutical, agricultural, and other industries, as well as the technologies that can be used to promote their application, improve production costs, make them more environmentally friendly, and identify the bioavailability that exists of these naturally and/or isolatedly. This also helps plants to present resistance to extreme conditions, which allows the generation of new trends and thus the use of probiotics from this area to generate agricultural holdings in a way of contributing to new trends in sustainability.

KEYWORDS

- **biostimulation of grains**
- **green technology**
- **plant growth**
- **plant growth-promoting bacteria**
- **plant-associated microbes**

REFERENCES

1. Axelsson, L.; Ahrné, S. Food and Medicine. *Appl. Microb. Syst.* **2000,** *81*, 367–388. https://doi.org/10.1108/eb058860
2. Chandramouli, V.; Kailasapathy, K.; Peiris, P.; Jones, M. An Improved Method of Microencapsulation and its Evaluation to Protect Lactobacillus spp. in Simulated Gastric Conditions. *J. Microbiol. Methods* **2004,** *56*, 27–35. https://doi.org/10.1016/j.mimet.2003.09.002
3. Cruz Casas, D. E.; Cázares Vásquez, M. L.; García Flores, L. A.; Lara Salas, M. A.; Aguilar, C. N.; Rodríguez Herrera, R.; Flores Gallegos, A. C. Probiotics as Functional Foods. In *Advances in Probiotics for Sustainable Food*; Springer Nature, **2021**; pp 121–148. https://doi.org/10.1007/978-981-15-6795-7_6
4. Markowiak, P.; Ślizewska, K. Effects of Probiotics, Prebiotics, and Synbiotics on Human Health. *Nutrients* **2017,** *9*. https://doi.org/10.3390/nu9091021
5. Sanjuán, J.; Nápoles, M. C.; Pérez-Mendoza, D.; Lorite, M. J.; Rodríguez-Navarro, D. N. Microbials for Agriculture: Why Do They Call Them Biostimulants When They Mean Probiotics? *Microorganisms* **2023,** *11*, 1–10. https://doi.org/10.3390/microorganisms11010153
6. Vivek, K.; Mishra, S.; Pradhan, R. C.; Nagarajan, M.; Kumar, P. K.; Singh, S. S.; Manvi, D.; Gowda, N. N. A Comprehensive Review on Microencapsulation of Probiotics: Technology, Carriers and Current Trends. *Appl. Food Res.* **2023,** *3*, 100248. https://doi.org/10.1016/j.afres.2022.100248
7. Antón-Herrero, R.; García-Delgado, C.; Antón-Herrero, G.; Mayans, B.; Delgado-Moreno, L.; Eymar, E. Design of a Hydroponic Test to Evaluate the Biostimulant Potential of New Organic and Organomineral Products. *Sci. Hortic.* **2023,** *310*. https://doi.org/10.1016/j.scienta.2022.111753
8. de Souza Vandenberghe, L. P.; Blandon Garcia, L.; Rodrigues, C. M.; Cândido Camara, M.; Vinícius de Melo Pereira, G.; de Oliveira, J.; Soccol, C. R. Potential Applications of Plant Probiotic Microorganisms in Agriculture and Forestry. *AIMS Microbiol.* **2017,** *3* (3), 629–648. https://doi.org/10.3934/microbiol.2017.3.629
9. Franzoni, G.; Cocetta, G.; Prinsi, B.; Ferrante, A.; Espen, L. Biostimulants on Crops: Their Impact under Abiotic Stress Conditions. *Horticulturae* **2022,** *8* (3). https://doi.org/10.3390/horticulturae8030189
10. Khan, S.; Purohit, A.; Vadsaria, N. Hydroponics: Current and Future State of the Art in Farming. *J. Plant Nutr.* **2020,** *44* (10), 1515–1538. https://doi.org/10.1080/01904167.2020.1860217
11. Lee, S.; Lee, J. Beneficial Bacteria and Fungi in Hydroponic Systems: Types and Characteristics of Hydroponic Food Production Methods. *Sci. Hortic.* **2015,** *195*, 206–215. https://doi.org/10.1016/j.scienta.2015.09.011
12. Menendez, E.; Garcia-Fraile, P. Plant Probiotic Bacteria: Solutions to Feed the World. *AIMS Microbiol.* **2017,** *3* (3), 502–524. https://doi.org/10.3934/MICROBIOL.2017.3.502
13. Mishra, P.; Mishra, J.; Arora, N. K. Plant Growth Promoting Bacteria for Combating Salinity Stress in Plants – Recent Developments and Prospects: A Review. *Microbiol. Res.* **2021,** *252*, 126861. https://doi.org/10.1016/j.micres.2021.126861
14. Pandey, S. S.; Jain, R.; Bhardwaj, P.; Thakur, A.; Kumari, M.; Bhushan, S.; Kumar, S. Plant Probiotics – Endophytes Pivotal to Plant Health. *Microbiol. Res.* **2022,** *263*, 127148. https://doi.org/10.1016/j.micres.2022.127148

15. Raklami, A.; Meddich, A.; Oufdou, K.; Baslam, M. Plants—Microorganisms-Based Bioremediation for Heavy Metal Cleanup: Recent Developments, Phytoremediation Techniques, Regulation Mechanisms, and Molecular Responses. *Int. J. Mol. Sci.* **2022,** *23* (9). https://doi.org/10.3390/ijms23095031
16. Sambo, P.; Nicoletto, C.; Giro, A.; Pii, Y.; Valentinuzzi, F.; Mimmo, T.; Lugli, P.; Orzes, G.; Mazzetto, F.; Astolfi, S.; Terzano, R.; Cesco, S. Hydroponic Solutions for Soilless Production Systems: Issues and Opportunities in a Smart Agriculture Perspective. *Front. Plant Sci.* **2019,** *10.* https://doi.org/10.3389/fpls.2019.00923
17. Sánchez, Ó. J.; Ospina, D. A.; Montoya, S. Compost Supplementation with Nutrients and Microorganisms in Composting Process. *Waste Manag.* **2017,** *69* (26), 136–153. https://doi.org/10.1016/j.wasman.2017.08.012
18. White, J. F.; Chang, X.; Kingsley, K. L.; Zhang, Q.; Chiaranunt, P.; Micci, A.; Velazquez, F.; Elmore, M.; Crane, S.; Li, S.; Lu, J.; Molina, M. C.; González-Benítez, N.; Beltrán-García, M. J.; Kowalski, K. P. Endophytic Bacteria in Grass Crop Growth Promotion and Biostimulation. *Grass Res.* **2021,** *1* (1), 1–9. https://doi.org/10.48130/gr-2021-0005
19. Yakhin, O. I.; Lubyanov, A. A.; Yakhin, I. A.; Brown, P. H. Biostimulants in Plant Science: A Global Perspective. *Front. Plant Sci.* **2017,** *7.* https://doi.org/10.3389/fpls.2016.02049
20. Zia, R.; Malik Nawaz, S.; Muhammad, A.; Muhammad Sohail, A. Sustainable DevelopmentandBiodiversity, In*SoilMicrobiomesforSustainableAgricultureFunctional Annotation*, **2021**; vol *27*. https://doi.org/https://doi.org/10.1007/978-3-030-73507-4_13
21. Chabot, R.; Antoun, H.; Cescas, M. P. Growth Promotion of Maize and Lettuce by Phosphatesolubilizing Rhizobium leguminosarum biovar phaseoli. *Plant Soil* **1996,** *184*, 311–321.
22. Jiménez-Gómez, A.; Menéndez, E.; Flores-Félix, J. D.; García-Fraile, P.; Mateos, P. F.; Rivas, R. Effective Colonization of Spinach Root Surface by Rhizobium. In *Biological Nitrogen Fixation and Beneficial Plant-Microbe Interaction*; González-Andrés, F., James, E., (Eds.); Springer: Cham, **2016**; pp 109–122.
23. Rana, K. L.; Kour, D.; Kaur, T.; Devi, R.; Yadav, A. N.; Yadav, N.; et al. Endophytic Microbes: Biodiversity, Plant Growth-Promoting Mechanisms and Potential Applications for Agricultural Sustainability. *Antonie Van Leeuwenhoek* **2020,** *113*, 1075–1107. https://doi.org/10.1007/s10482-020-01429-y
24. Le, T. A.; Pék, Z.; Takács, S.; Neményi, A.; Daood, H. G.; Helyes, L. The Effect of Plant Growth Promoting Rhizobacteria on the Water-Yield Relationship and Carotenoid Production of Processing Tomatoes. *Hortscience* **2018,** *53*, 816–822. https://doi.org/10.21273/hortsci13048-18
25. Shanta, N.; Schwinghamer, T.; Backer, R.; Allaire, S. E.; Teshler, I.; Vanasse, A.; Whalen, J.; Baril, B.; Lange, S.; MacKay, J.; Zhou, X.; Smith, D. L. Biochar and plant growth promoting rhizobacteria effects on switchgrass (Panicum virgatum cv. Cave-in-rock) for biomass production in southern Québec depend on soil type and location. *Biomass Bioenergy* **2016**, *95*, 167–173. https://doi.org/10.1016/j.bio mbioe.
26. Vizcaíno, R.; Macias-Tomei, C.; Márquez S, J. C.; Morales, A.; Torres, N. Usos clínicos de los probióticos. *Archivos Venezolanos de Puericultura y Pediatría* **2016,** *79* (1), 029–040. Recuperado en 30 de abril de 2023, de http://ve.scielo.org/scielo.php?script=sci_arttext&pid=S0004-06492016000100007&lng=es&tlng=es.
27. Hill, C.; Guarner, F.; Reid, G.; Gibson, G. R.; Merenstein, D. J.; Pot, B.; Morelli, L.; Canani, R. B.; Flint, H. J.; Salminen, S.; et al. Expert Consensus Document. The

International Scientific Association for Probiotics and Prebiotics Consensus Statement on the Scope and Appropriate use of the Term Probiotic. *Nat. Rev. Gastroenterol. Hepatol.* **2014,** *11*, 506–514.
28. Food and Agricultural Organization of the United Nations and World Health Organization: Joint FAO/WHO Working Group Report on Drafting Guidelines for the Evaluation of Probiotics in Food; Food and Agricultural Organization of the United Nations, **2002**.
29. Picard, C.; Baruffa, E.; Bosco, M. Enrichment and Diversity of Plant-Probiotic Microorganisms in the Rhizosphere of Hybrid Maize During Four Growth Cycles. *Soil Biol. Biochem.* **2008,** *40*, 106–115.
30. Vandenberghe, L.; Garcia, L. M. B.; Rodrigues, C.; Camara, M. C.; Pereira, G.; De Oliveira, J.; Soccol, C. R. Potential Applications of Plant Probiotic Microorganisms in Agriculture and Forestry. *AIMS Microbiol.* **2017,** *3*, 629–648.
31. Moreno-Gavíra, A.; Diánez, F.; Sánchez-Montesinos, B.; Santos, M. Paecilomyces variotii as A Plant-Growth Promoter in Horticulture. *Agronomy* **2020,** *10* (4), 597. https://doi.org/10.3390/agronomy10040597.
32. Tang, M. L. Probiotics and Prebiotics: Immunological and Clinical Effects in Allergic Disease. *Nestle Nutr. Workshop Ser. Pediatr. Program.* **2009,** *64*, 219–257.
33. Vael, C.; Desager, K. The Importance of the Development of the Intestinal Microbiota in Infancy. *Curr. Opin. Ped.* **2009,** *21* (6), 794–800.
34. Organización Mundial de la Salud. Estrategia mundial para la alimentación del lactante y el niño pequeño; Ginebra, **2002**. http://www.who.int/nutrition/publications/gs_infant_feeding_text_spa.pdf .
35. Tsuda, K.; Tsuji, G.; Higashiyama, M.; et al. Biological Control of Bacterial Soft Rot in Chinese Cabbage by Lactobacillus plantarum Strain BY Under Field Conditions. *Biol. Control* **2016,** *100*, 63–69.
36. Bhardwaj, D.; Ansari, M. W.; Sahoo, K.; et al. Biofertilizers Function as Key Player in Sustainable Agriculture by Improving Soil Fertility, Plant Tolerance and Crop Productivity. *Microb. Cell Fact.* **2014,** *13*, 1–10.
37. Belorkar, S. A. Biostimulants: Promising Probiotics for Plant Health. In *Biocontrol Agents and Secondary Metabolites*; Woodhead Publishing, **2021**; pp 469–481. ISBN: 9780128229194, https://doi.org/10.1016/B978-0-12-822919-4.00020-X.
38. Shalini, R.; Manoj, K. S.; Anjali, C. S.; Swarnmala, S. Chapter 17 - Microbial Endophytes as Probiotics for the Plant Health: An Overview. In *Microbial Endophytes and Plant Growth,* **2023**; pp 269–281. ISBN: 9780323906203, https://doi.org/10.1016/B978-0-323-90620-3.00011-8.
39. Mojumdar, A.; Tanaya Behera, H.; Das, S.; Ray, L. CHAPTER 13 - Microbe-Based Plant Biostimulants and their Formulations for Growth Promotion and Stress Tolerance in Plants. In *New and Future Developments in Microbial Biotechnology and Bioengineering*, **2022**; pp 213–230. ISBN: 9780323851633, https://doi.org/10.1016/B978-0-323-85163-3.00004-1.
40. Singh, A. L.; Rani, K.; Jena, R.; Kona, P.; Reddy, K.; Gangadhara, K. CHAPTER 8 - Microbes-based bio-stimulants towards sustainable oilseeds production: Nutrient recycling and genetics involved. In *New and Future Developments in Microbial Biotechnology and Bioengineering,* **2022**; pp 111–130. ISBN: 9780323851633, https://doi.org/10.1016/B978-0-323-85163-3.00013-2.

41. Aamir, M.; Rai, K. K.; Zehra, A.; Dubey, M. K.; Kumar, S.; Shukla, V.; Upadhyay, S. 8 - Microbial bioformulation-based plant biostimulants: a plausible approach toward next generation of sustainable agriculture, In *Microbial Endophytes*, **2020**; pp 195–225. ISBN: 9780128196540; https://doi.org/10.1016/B978-0-12-819654-0.00008-9.
42. Mishra, P.; Tiwari, S. K.; Singh, A. K.; Malik, Md. Z. Chapter 8 - Harnessing Endophytic Microbial Diversity for Stress Alleviation: Current Perspectives for Crop Improvement. In *Mitigation of Plant Abiotic Stress by Microorganisms*, **2022**; pp 149–174. ISBN: 9780323905688, https://doi.org/10.1016/B978-0-323-90568-8.00008-0.

CHAPTER 7

Biofuel as Potential Candidates for Sustainable Development

SNEHA SARA VARGHESE[1], VAISAKH. P. H[1], EAPEN PHILIP[1], TRESSIA ALIAS PRINCY PAULOSE[1], and AKHINA H[2]

[1]*Post Graduate and Research Department of Chemistry, Bishop Moore College, Mavelikara, Kerala, India*

[2]*School of Energy Materials, Mahatma Gandhi University, Kottayam, India*

ABSTRACT

Any fuel that comes from biomass, including plant, algae, or animal waste, is referred to as a biofuel. Typically, liquid fuels and blending ingredients made from biomass resources known as feed stocks are referred to as biofuels. The majority of biofuels are employed as motor fuels, but they can also be utilized to generate heat and power. Contrary to fossil fuels like petroleum, coal, and natural gas, biofuel is seen as a source of renewable energy due to the ease with which such feedstock material may be supplied. In light of rising costs of petroleum and growing concern about the role fossil fuels play in global warming, biofuel is frequently promoted as a convenient and environmentally friendly substitute for petroleum as well as other fossil fuels. Four generations of biofuel have been produced as a result of biofuel research and development. By maximizing the carbon cycle on our planet, biofuels contribute to a reduction in the carbon footprint of the transportation and other industries. Many see it as a tool to aid in lowering greenhouse gas emissions and the consumption of fossil fuels, preventing the exhaustion of resources, reducing reliance on unreliable foreign suppliers, promoting rural development, and opening up new markets for agricultural goods.

The Food-Energy-Water Triangle for Sustainable Development: Integrative and Synergic Solutions.
Pedro Aguilar-Zárate, Mónica Chávez-González, Ann Rose Abraham, and A. K. Haghi (Eds.)

7.1 INTRODUCTION

Biofuels have been helpful to mankind for a very long time ever since humans began using firewood to generate heat. Since biofuels have always existed, for example in the form of timber and grass, before humans even realized and utilized fire, they actually have a history that is older than humanity itself.[1] As human civilization progressed, biofuels assumed increasingly significant roles in people's lives. For instance, in the first millennium BC, methane gas, also known as biomethane or biogas, was used to prepare warm bath water. In 1826, ethanol, also known as bioethanol, was used in an internal combustion engine. These applications fundamentally altered and enhanced people's lives everywhere.

Global warming and the growing need for renewable energy are still seen as the two biggest concerns facing society at large. An ongoing source of environmentally friendly, sustainable energy is required for the advancement of the energy sector sustainably. The need for energy will rise along with the economy and population. Currently, coal, natural gas, and crude oil make about 80% of the world's energy consumption. These fossil fuels cannot be renewed quickly; for example, it takes thousands of years for regeneration. They are made from organic ingredients that were created on Earth millions of years ago and cannot be replenished in a short period. Although the recent surge in shale gas relieves the pressure of the fossil fuel scarcity and drives down the price of fossil fuel, the fact that fossil fuels are limited in supply remains unchanged. The fossil regeneration rate or even the fossil finding rate will never meet the consumption rate, according to the most recent data on daily fossil utilization. Depending on the restrictions placed on our usage of fossil fuels, our reserves may survive for 70 or 100 years.

The search for new renewable energy sources before fossil fuels run out is the need of the hour. The attraction of reducing greenhouse gas (GHG) emissions, on the other hand, has been the most hotly debated subject at previous United Nations Climate Change Conferences.

Due to industrialization and expanding population, which heavily rely on fossil fuels for development, transportation, and lifestyle, the globe is currently experiencing energy issues. Exploiting fossil fuels has produced significant greenhouse gas emissions that have contributed to environmental degradation and global warming.[2] Utilizing renewable feedstock, such as lignocellulosic biomass, is one method to minimize greenhouse gas emissions, reduce reliance on fossil fuels, and promote sustainable development. The benefits of developing and using biofuels also include enhancing national energy security, utilizing existing transportation infrastructure, utilizing existing fuel distribution infrastructure, and promoting rural development, in addition to the principal causes outlined above.

The phrase "fossil fuel" was first used in the first half of the nineteenth century to describe fuels like natural gas, coal, and oil that were created from plant and animal remains that were buried beneath the surface of the Earth for thousands of years. A new term, "biofuel," first surfaced and gained popularity in the 1970s. Any fuel, including firewood and petroleum, that is obtained in some way from biomass is referred to as a biofuel. Solid, liquid, or gaseous fuels that are primarily made from biomass are referred to as biofuels or bio-renewable fuels.[3] The usage of renewable biomass energy in the form of biofuels has grown in popularity as a viable alternative to petroleum-based transportation fuels like petrol and diesel. Growing interest has been shown in the large-scale use of biofuels to combat global warming and move away from the increasingly finite, environmentally hazardous, and politically risky petroleum supply.

7.2 GENERAL ASPECTS OF BIOFUEL

The biomass from plants and the processed goods that are burned for energy are referred to as biofuels. Biobased products found in or taken from nature, such as wood, bagasse, or peat, or those created chemically from biomass to create items like charcoal, ethanol, and biogas, may be referred to as biofuels. Biomass used as fuel is another way to describe it. It is a fuel that can also be used to produce power, heat, or energy. The oxygen concentration is the main distinction between petroleum feedstocks and biofuels. Biofuel's chemical characteristics substantially differ from the conventional petroleum products since they contain 10–45% more oxygen than petroleum does. In comparison to fossil fuels, which primarily consist of natural gas, coal, and oil, biofuels typically have a number of advantages. For example, they are renewable, readily available, CO_2/GHG neutral, result in insignificant SOx emissions and lower NOx emissions. They are also environmentally friendly fuels that are capable of being produced locally and are biodegradable.[4] Additionally, biofuels have the following advantages: they use local energy resources; they encourage investment in and development of rural agriculture; they increase supply diversity; they lessen reliance on imports for energy; they improve local livelihoods; etc. As a result, several nations have set a variety of linked goals or policies to advance the biofuels sector. These include papers published in Brazil, China, India, The United States, and the European Union.

Biofuels can be divided into two categories: conventional biofuels and advanced biofuels, depending on the level of commercialization of their underlying technologies. Following is the order of commercialization: commercial > early commercial > demonstration > research. Bioethanol based on sugar and starch, biodiesel based on transesterification, and biomethane derived from anaerobic digestion make up the majority of conventional biofuels. Hydrotreated vegetable oil, one of the advanced biofuels, is in the early phases of commercialization. Lignocellulosic bioethanol, biomass-to-liquid biodiesel, microalgae biodiesel, and biohydrogen, on the other hand, are undergoing intensive research. It should be highlighted that although sophisticated biodiesels are not yet generally accessible, they could do so in the future.

Researchers have categorized biofuels into four classes based on the feedstock sources and generation technologies: first-generation biofuels, second-generation biofuels, third-generation biofuels, and fourth-generation biofuels. The majority of first-generation biofuels are made from food crops like corn, wheat, and soybeans, including biodiesel, bio-alcohols, biogas, and bio-syngas.[5] The first-generation biofuels also include solid biofuels made from agricultural waste, dried dung, and firewood. They are mostly made of vegetable oil, sugar, and starch. The majority of second-generation biofuels are lipid-based and made from insect waste, animal fats, waste vegetable oil, and oleaginous bacteria. These biofuels are mostly made from non-food sources of feedstock primarily lignocellulosic biomass, such as grass, wood, cereal straw, sugarcane bagasse, forest byproducts, specially bred energy crops, and municipal solid waste.[6] The main feedstocks for third-generation biofuels include algae. The combination of genetically altered feedstock with microorganisms that have been genomically prepared results in fourth-generation biofuels. Oil-bearing plants and non-food plants were combined to make 1.5-generation biofuels.[7]

Bioalcohols, vegetable oil, biodiesel, biomethanol, biosyngas, and biogas are the main forms of first-generation biofuels. Bioalcohols, biooil, biodiesel, biodimethylfuran, biomethanol, and biohydrogen are the principal second-generation biofuels. Bioethanol, vegetable oil, biodiesel, biomethanol, and jet fuels are the primary third-generation biofuels. Green aviation fuels, biogasoline, and green diesel make up the majority of fourth-generation biofuels.

For the transportation sector to cover all of its energy needs, biofuel is a cleaner option. Despite the advantages that come with using biofuels, it is important to remember that their manufacturing and final usage have negative effects on the environment, including the eradication of forests, increased soil deterioration, decreased food production, and excessive water use. Due to population growth and demand on natural resources, there is a greater loss of natural habitat. Water resources are under too much pressure due to the need for water for the manufacture of biofuels, which is already under stress and can be seen in many locations. In order to meet the economic needs of nations, it is necessary to develop biofuels on a big scale and study the extraction methods in great depth.

7.3 PRODUCTION TECHNOLOGIES OF BIOFUELS FROM LIGNOCELLULOSIC BIOMASS

By altering the lignocellulosic composition, we can improve the purity and accessibility of cellulose for which biomass has to be exposed to a variety of physical, chemical, physicochemical, and biological processes. Different technologies employed for lignocellulosic pretreatment are specified in Figure 7.1. In general, physical pretreatments make cellulose more accessible, whereas chemical and biological pretreatments can make cellulose more accessible as well as purer. Physical and chemical pretreatment techniques are combined in physicochemical pretreatments. These techniques can be employed independently or in combination to change the accessibility and/or purity of cellulose to increase its reactivity.

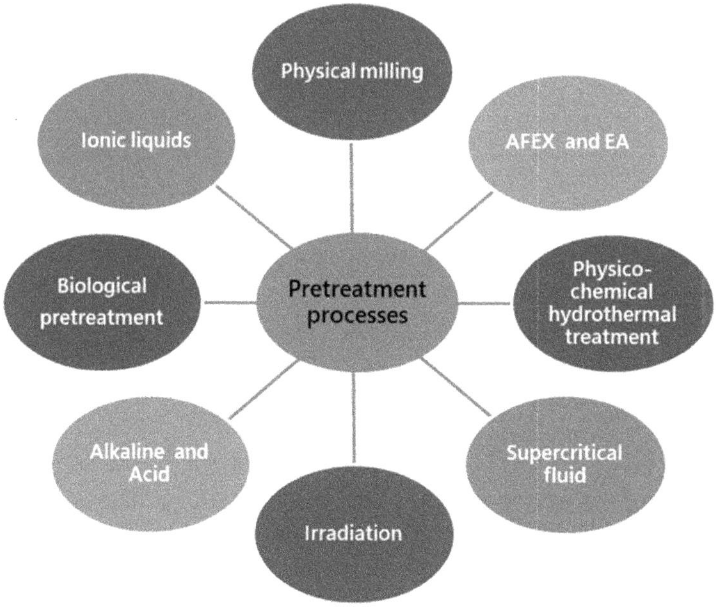

FIGURE 7.1 Pretreatment process.

7.3.1 PHYSICAL MILLING

Since the 1840s, mechanical wood pulping has been employed to produce high-quality cellulose fibers for paper manufacturing, making it perhaps the oldest pretreatment technique. However, the primary function of milling in biorefineries is mainly a reduction in size of particles and crystallinity. In addition, it aims to disrupt the lignocellulose's structure and reduce the level of polymerization. The mechanisms for transferring mass and heat are improved when particle size is reduced because it results in an increase in specific surface area. All of the modifications to lignocellulosic structure brought about by milling make further pretreatments and cellulose conversions more effective. Compression, ball, two rolls, knife, attrition, hammer, colloid, vibro energy, disc, and freeze are required for all types of wet or dry milling treatments. Ball milling is frequently employed in research and significantly increases cellulose reactivity, especially before chemo-catalytic operations with a solid catalyst. Prior to chemo-catalytic methods, dry ball milling is primarily used for cellulose amorphisation. Native cellulose can undergo a thermodynamically more advantageous transition into the cellulose II polymorph during wet ball milling.

7.3.2 IRRADIATION

Gamma ray, microwave, ultrasound irradiation, and electron beam are all types of irradiations that can be used alone or in conjunction with other pretreatment methods, as well as biomass hydrolysis. Electromagnetic radiation can be classified as high-energy, ionizing like gamma and electron beams and low-energy, nonionizing radiations like microwave. Nonionizing radiation only stimulates the rotational and vibrational modes of molecules while ionizing radiation ionizes atoms by removing electrons and causes significant chemical changes. Gamma rays and electron beams can depolymerize and break down lignocellulosic biomass, increasing cellulose's reactivity. When compared to traditional heating from an external source, microwave irradiation effectively generates heat in situ across lignocellulosic substrates, thereby increasing reaction rates.[8] Because of this, additional pretreatment methods and hydrolysis itself usually utilize this irradiation technique. The lignin structure is somewhat disturbed by microwave irradiation, increasing the accessibility of cellulose. Hemicellulose from lignocellulose can be selectively dissolved using microwave-assisted hydrothermal pretreatment without the use of chemicals. By inducing pressure differences in the reaction medium and enhancing mass transfer and mixing processes, ultrasound irradiation increases the accessibility of biomass. By producing radicals, ultrasound can also accelerate delignification.

7.3.3 PHYSICO-CHEMICAL HYDROTHERMAL PRETREATMENT

To enable the breakdown and redistribution of hemicellulose and lignin, hot water or steam is employed in hydrothermal pretreatment. Minerals and other water-soluble metal catalyst inhibitors can also be eliminated via hydrothermal pretreatment. Hemicellulose is broken down during hydrothermal pretreatment, releasing organic acids like acetic acid, which help with subsequent hydrolysis reactions. It is known that lignin can repolymerize, for example, by the production of carbonium ions following the breakage of lignin β-O-4 ether linkages, despite the fact that it can be degraded and solubilized as well. During hydrothermal pretreatment, hydrophobic lignin can also melt, which causes droplet formation and lignin redistribution after cooling.[9] Hemicellulosic and lignin removal and redistribution increase pore volume and enhance cellulose accessibility. Liquid hot water and steam explosion pretreatment are two examples of hydrothermal pretreatment. Up to 60% of the lignocellulose is solubilized by liquid hot water pretreatment at 473–503 K, including 35–60% of the lignin and 90% of the hemicellulose.

7.3.4 AMMONIA FIBER EXPANSION (AFEX) AND EXTRACTIVE AMMONIA (EA)

Similar to steam explosion, ammonia fiber expansion (AFEX), also known as ammonia fiber explosion, uses liquid ammonia rather than steam and does not experience any mass loss throughout its pretreatment. Typically, moist biomas and liquid ammonia are combined for AFEX pretreatment, which is then heated at 313–453 K for 5–60 min before depressurization.[10] During the AFEX procedure, the action of ammonia causes the intermolecular ester bonds between hemicellulose and lignin to be ammonolytic and hydrolytically broken. After depressurization, pores with dimensions ranging from 10 to 1000 nm are produced by migration of solubilized hemicellulose, lignin, and cell wall extracts to the outer cell wall surfaces and corners, improving the accessibility of cellulose. Due to the large amount of water used in ammonia solution, traditional AFEX processing does not significantly modify cellulose crystallinity, unlike liquid anhydrous ammonia. The poor effectiveness of AFEX pretreatment on samples with high lignin contents and the requirement for ammonia recycling are drawbacks. Additionally, many typical solid chemo-catalysts are anticipated to lack adequate pore sizes in the nm range. The rate of enzymatic hydrolysis can be increased up to five times by converting cellulose I to cellulose III using liquid ammonia. In order to simultaneously extract lignin and convert cellulose I to highly digestible cellulose III using liquid ammonia at 393 K, a pretreatment process known as "Extractive Ammonia (EA)" was devised. AFEX has a higher moisture content, which prevents the production of cellulose III. Through EA, almost all polysaccharides can be preserved with only 45% lignin removal. Although AFEX likewise retains the majority of the sugars, after ammonia is evaporated, the dissolved lignin accumulates on the pretreated cell wall. As a result, corn stover that has been EA processed produces greater sugar yield than regular AFEX while requiring 60% less enzyme. Additionally, compared to AFEX, the EA consumes a lot a smaller amount of NH_3. The intense operation pressure of EA is a drawback because it will raise the pretreatment unit's capital cost. It should be noted that adding nitrogen to biomass by ammonia pretreatment may cause further hydrodenitrogenation before burning or NOx to be produced during the combustion of related chemicals. In addition, a chemo-catalytic process' catalysts may become inactive in the presence of nitrogen.

7.3.5 SUPERCRITICAL FLUID

A liquid that can be used above the critical temperature and pressure is known as a supercritical fluid. In addition to having the same density as liquids, supercritical fluid also exhibits gas's diffusibility and transmission qualities. Because of the differences in partition coefficients and component solubilities, supercritical fluid can break through the crystal structure of lignocellulosic materials and streamline the separation procedure. These characteristics significantly increase cellulose digestibility and hold considerable promise for pretreatment. Supercritical CO_2 is widely used as an extraction solvent in industry due to its nontoxicity, recyclability, and minimal impact on the environment. When supercritical CO_2 is utilized in the presence of water or steam, hemicellulose degrades easily. After CO_2 dissolves in water, there is a rise in acidity, which is what causes this. Using additional solvents like ethanol or acetic acid can aid delignification. Wax extraction is another benefit of the supercritical CO_2 pretreatment, which increases the amount of total sugars released following saccharification by 20%.[11] After impregnation, pressure is released during CO_2 mediated pretreatments, rupturing the lignocellulosic structure similarly to a steam or ammonia explosion. It is significantly more efficient to use supercritical CO_2 for the explosion to cause cellulose degradation since it can penetrate the lignocellulose structure much efficiently than subcritical CO_2. Additionally, due to high capital costs for machinery and low pretreatment efficiency, using supercritical CO_2 as a technique for pretreatment is at present unsuitable for industrial applications.

7.3.6 ACID PRETREATMENT

Sulfite pulping was widely used since the 1800s where sulphurous acid and/or its alkali salt combinations are used in this pulping process to cleave and sulfonate lignin to water-soluble lignosulfonates. Because generated lignin carbonium ions are sulfonated during acid sulfite pulping, lignin recondensation and the cleavage of -aryl ether linkages are prevented.

Depending on how severe the procedure is, acid pretreatment can be used to directly produce monomeric sugars from both cellulose and hemicellulose as well as to remove hemicellulose prior to enzymatic hydrolysis. Acid pretreatment of lignocellulose is typically carried out at high and low temperatures, with diluted or concentrated acid solutions. Although HF, HNO_3, H_3PO_4, and SO_2 were widely used to assist steam pretreatment, organic acids such as maleic acid, fumaric acid, oxalic acid, formic acid, acetic acid, and various heteropoly acids have also been employed in the past. HCl and H_2SO_4 are the two most common acids utilized. It has been observed that dicarboxylic acids, such as oxalic acid, hydrolyze -1,4 glycosidic linkages more preferentially than H_2SO_4.[12] Concentrated acid solutions have the ability to dissolve or swell cellulose and can be applied at lower temperatures, which is more advantageous energetically than applying diluted acids at high temperatures. However, soluble acids are difficult to recycle, extremely caustic, and they produce a lot of neutralization waste (particularly at higher concentrations). Formic acid and acetic acid, being concentrated organic acids, can even dissolve lignin. Less strong acid treatments prevent the breakdown of sugar but are less effective at hydrolyzing hemicellulose.

7.3.7 ALKALINE PRETREATMENT

Since the 1940s, kraft pulping has dominated the wood pulping industry. High-quality cellulose fibers for the paper production are made from wood using the Kraft process, which delignifies the wood with a mixture of NaOH and Na_2S. Alkaline pretreatment increases porosity and general cellulose accessibility by removing lignin through the cleavage of lignin ether linkages and solubilizing hemicellulose by the saponification of the intermolecular ester bonds between hemicelluloses and lignin. Alkaline pretreatment chemicals, such as NaOH, $Ca(OH)_2$, and ammonia, can increase delignification when combined with oxidizing agents like O_2 and H_2O_2.[13] Although cellulose is far more robust to basic attack than hemicellulose, it can nevertheless degrade under alkaline circumstances. However, by its depolymerization, alkaline solutions can improve cellulose's accessibility to exoglucanases. With the creation of organic acids like lactic acid and the simultaneous neutralization of the alkaline solution, alkaline cellulose degradation occurs through peeling of reducing end-groups or arbitrary scission of cellulose chains. Higher cellulose conversions are frequently recorded following an alkaline pretreatment relative to an acid because of the significant quantity of delignification that occurs during the alkaline pretreatment.

7.3.8 IONIC LIQUIDS

Ionic liquids (ILs), which differ from molecular solvents by their special chemical composition, structure, organization, and behavior, are salts made up of massive organic cations and inorganic or organic anions. ILs thus provide a special environment for the field of chemistry biocatalysts, separation, material production, and electrochemistry. The most crucial characteristics of typical ILs used to produce furfural are: extremely low volatility, the ability to operate at high temperatures in conventional equipment with minimal emissions, the facilitation of recovering volatile ingredients and products, good dissolving capacity, green nature, chemical and thermal stability, recyclability, nontoxic nature, and nonflammability. It has a tunable nature because their physical characteristics, such as melting point, viscosity, density, solubility, and coordination qualities, can be tailored in accordance with various reactions or processes by choosing the suitable cation and anion. The enhanced catalytic capability and/or the use of energy-efficient processing technologies such as microwave or ultrasound-assisted processing lead to the reduced energy usage. Additionally, the ionic nature of the ILs may contribute to improving the reaction selectivity and catalytic activity as well as stabilizing the reactive catalytic species and/or intermediate. Ionic liquids with high Brönsted acidity exhibit catalytic activity and may behave as multifunctional chemicals, performing multiple functions simultaneously for instance, as solvents, catalysts, or stabilizers for catalysts or intermediates. The aforementioned characteristics provide ILs with the advantage of being effective agents for a variety of biorefinery objectives.

7.3.9 BIOLOGICAL PRETREATMENT

In order to remove hemicellulose, lignin, or antimicrobial compounds from lignocellulose through oxidation, biological pretreatment of lignocellulose may involve treatment with fungi, bacteria, or enzymes.[14] Due to their lignin selectivity, white-rot fungi are the most efficient basidiomycetes for biological pretreatment. Oxidative, lignin-degrading enzymes such as laccases, polyphenol oxidases, and lignin peroxidases work to break down lignin.

7.4 EFFECTS OF PRETREATMENT ON LIGNOCELLULOSIC STRUCTURES

The creation of more effective and affordable pretreatments for overcoming the recalcitrance of the lignocellulosic structure is one of the main objectives to overcome the barriers to the second-generation industry, as this stage is by far the most expensive of the entire endeavor and influences.[15] The following factors must be considered while selecting the method of pretreatment like the kind and yield of the final products, the selection and effectiveness of the downstream procedures, the energy and water demands, the amount of effluents retrieved for treatment, the generation of by-products, the extent of CO_2 emissions, expenses associated with the inclusion of washing, detoxification, solvent or catalyst recovery processes, and the use of special anticorrosive materials, the effectiveness and energy needs of the upstream stages for reducing fiber size,[16] price and yields of the final chemical products resulting from fermentation or from thermochemical or catalytic conversions such as the amount of enzymes added during saccharification or the acquisition of enzymatic complexes, the degree of microorganism's tolerance to high solids concentrations and to the products and byproducts from the transformation of sugar and lignin, the use of enzymatic complexes and microorganisms capable of hydrolyzing and converting glucose and xylose, respectively, on a commercial scale, breaking the cellulose and hemicellulose bonds with the least amount of intervention from pentoses and hexoses degradation reactions, the physical barriers of lignin or of cellulose and hemicellulose oligomers that are still present in the solid fraction of pretreatment, etc. Pretreatments should, in general, make it easier for enzymes to access the carbohydrates in plant cell walls. However, different pretreatment techniques have different effects on the physico-chemical breakdown of lignocellulosic structures, leading to a range of yields and end products. Despite the fact that several pretreatment techniques, like milling, organosolv, and ionic liquids, can greatly increase the digestibility of plant cell walls, their high operational costs severely limit their effectiveness.[17] The most successful and promising pretreatment methods for industrial applications are chemical and physico-chemical pretreatments. In general, chemical pretreatments, particularly alkali and acid pretreatments, can efficiently and cheaply remove lignin and hemicelluloses from lignocellulosic materials. An environmentally benign pretreatment method called steam explosion can solubilize hemicelluloses, disrupt the structure of vegetables, and further enhance the specific surface of lignocellulosic components that are accessible to chemicals or enzymes. In addition, using high pressures or mechanically shrinking the fiber using microwave, irradiation, or other methods will result in less severe operating conditions and less water use, but those combinations will increase the energy needed by the plant.[18] The addition of chemicals, such as NaOH, NH_3, SO_2, CO_2, ethanol, tetrahydrofuran, and H_2O_2, among others, is an additional option for improving the extraction of monomers from hemicellulose and cellulose and dissolving or isolating lignin under mild conditions. However, they raise the costs of recovering the desired chemicals. These techniques can be paired with highly selective biological pretreatments, but they also take a lot of time and precise control of working conditions. In addition, they are being enhanced by the development of microorganisms that can depolymerize the lignocellulosic matrix and convert them into value-added products. As a result, neither the economics nor the yields of carbohydrates or platform products are tied to a particular set of pretreatment process parameters or to the various impacts that these pretreatments have on the lignocellulosic structure, such as a greater degree of lignin, xylose, or acetyl group removal.[19] An establishment that uses thermal energy and electrical power produced by burning lignin might become self-sufficient in utilities and even sell any extra electricity, which would have an impact on the price of their final product in the market. Therefore, strict control of the operational parameters of the pretreatment is essential to reduce costs associated with heating, special materials, and decreasing platform product yields.

7.5 MAJOR BIOFUELS

A form of sustainable energy source called biofuel is produced from microbial, plant, or animal products. Gaseous, liquid, or solid biofuels are all possible. The latter two kinds are the most practical because they are simpler to distribute, provide, and burn cleanly. Examples of biofuels discussed include bioethanol, biodiesel, biobutanol, biohydrogen, and biomethane. A schematic diagram of major biofuels is given in Figure 7.2.

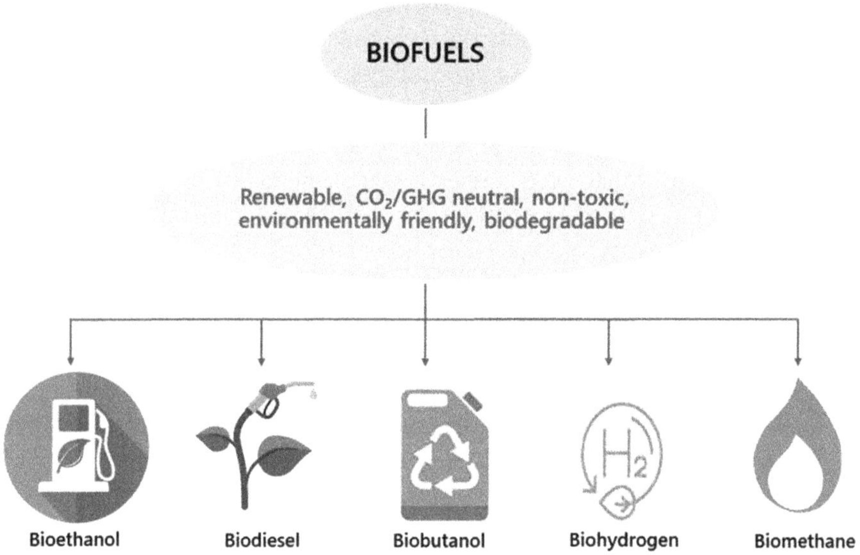

FIGURE 7.2 Biofuels.

7.5.1 BIOETHANOL

Ethanol is one of the most common biofuels currently used and is widely used in the transportation sector, and can be used as a substitute for gasoline or as a starting material for a number of compounds with added value. It is commonly regarded as the most abundant biofuel in the world. Commercial production of the first generation of ethanol uses corn and other starch- and sugar-based resources. This has drawn criticism from all around the world for breaching the ethical considerations raised by the argument over whether to prioritize food or fuel and diverting farmland for the cultivation of food crops primarily to be employed as feedstock for the production of fuel. The manufacture of ethanol from lignocellulosic feedstocks has thus been the subject of extensive investigation.[20] From lignocellulosic biomass, ethanol can be created using two distinct processes, such as thermochemical and biological. The complex lignin molecules must be broken down into simpler intermediates in both conversion processes before being converted to ethanol. By using Fischer--Tropsch catalysis, syngas from the gasification of biomass can be catalytically transformed into fuels including ethanol, butanol, green diesel, and hydrocarbons. Through syngas fermentation, Clostridium species can also be used to make ethanol. Contrarily, in the biological conversion route, cellulases and hemicellulases enzymes are used to hydrolyze the cellulose and hemicellulose present in the biomass into simple sugars through saccharification, which is then followed by bacterial and yeast fermentation to produce ethanol. Second-generation bio-ethanol, which is made from lignocellulose, is regarded as a potential fuel and chemical because of its low toxicity, high volatility, and clean combustion characteristics. Despite the prospective uses for bioethanol, there are still a number of issues that need further research, including the environmental effects of the bioconversion processes as well as its technical and economic viability. The necessity of expensive and complicated biomass pretreatment techniques involving acids, alkalis, enzymes, chemicals, microwaves, ultrasounds, and other physicochemical and mechanical factors is another problem. For a side-by-side comparison of the environmental performance, Mu et al. in 2010,[21] assessed the lifetime of the biological and thermochemical conversion pathways of ethanol production. Utilization of fossil fuels, water use, and GHG emissions were taken into account. According to their findings, the two routes had differing environmental effects despite having comparable ethanol yield and energy efficiency. In the short term, the biological pathway used more fuel and produced more greenhouse gases, although this is expected to change over time. The thermochemical approaches, on the contrary hand, used less freshwater, according to the scientists. According to a more recent lifecycle assessment analysis, the manufacture of second-generation bioethanol from banana waste has the potential to reduce GHG emissions and the depletion of fossil fuels while maintaining a net positive energy balance.[22] Despite the fact that lignocellulosic ethanol production is still in its early stages, some commercial operations, like Razen, Poet-DSM, DuPont, GranBio, and Abengoa, have already made plans to commercialize the manufacture of bioethanol.

7.5.2 BIODIESEL

Long-chain fatty acid alkyl esters are used to create biodiesel, which may be blended with regular diesel or used as a drop-in fuel. In contrast to petrodiesel, which has a flash point of 337 K, biodiesel has a 423 K flash point, making it neither flammable nor explosive. When used as a fuel, biodiesel dramatically minimizes harmful and other pollutants because it is harmless and biodegradable, unlike petroleum-based diesel. Alkyl esters must be taken from a feedstock, such as animal fats or waste vegetable oils, and then transesterified to create fuel in the case of biodiesel fuel manufacturing. The viscosity of the oil is reduced during transesterification. Rudolf Diesel built the first diesel engine in 1885 with the idea of running it on vegetative sources, and biodiesel has a long history as well. This fuel is biodegradable and safe, with an energy density of 33 MJ/L, and because diesel engines naturally have high compression, it can be operated with 20–30% more efficiency than equivalent petrol engines. The main distinction between biodiesel and bio-oil is in the production processes they employ. The production processes and chemical makeup of biodiesel and bio oil are the two main distinctions between them. Transesterification of fats and oils yields biodiesel, while thermochemical cracking of organic molecules containing carbon, hydrogen, and trace amounts of other elements yields bio-oils. Depending on the type of feedstocks and reaction conditions, bio-oils are generally more viscous and acidic than biodiesel.[23] Due to its high water content, bio-oil has poor combustion qualities and is affected by acidity, engine deterioration, polymerization, thermal instability, and excessive carbon emissions when utilized directly. Since fats and oils are transesterified to make biodiesel, the potential for biomass for its production needs to be assessed in terms of the proportion of lipids and oils. By measuring the fatty acid concentration of the precursors, this is ascertained. Its portability, quick availability, renewability, improved combustion efficiency, reduced sulfur and aromatic content, higher cetane number, and higher biodegradability are all benefits of using biodiesel as a diesel fuel.

The primary benefits of biodiesel listed in the scientific literature include the domestic origin, which would lessen a nation's reliance on petroleum imports, its biodegradability, high flash point, and its low cost.[24] The main drawbacks of biodiesel include its greater viscosity, reduced energy content, higher cloud point and pour point, higher NOx emissions, decreased engine speed and power, injector coking, engine compatibility, high cost, and increased engine wear. Blends of biodiesel and fossil diesel have a number of technical drawbacks, including issues with fuel freezing in frigid conditions, a lower energy density, and fuel degradation when stored for extended periods of time. The cost and price of biomass feedstock, which are related to conversion technology, determine how competitive biodiesel is. The relative importance of by-products may vary according to the feedstock used.

With around 95% of the fuel's manufacturing coming from edible oils, biodiesel is thought to be the second most abundant biofuel produced globally behind bioethanol. However, this raises the question of food v/s. fuel, which in turn raises the price of edible feedstocks and has an impact on the usage of arable land and the availability of food.[25] Cooking oil waste and animal fats are viable feedstocks for biodiesel synthesis, which could help to offset these difficulties. These resources, however, are not enough to supply the growing need for sustainable and clean fuels. In order to create biodiesel, renewable feedstocks such lignocellulosic biomass are being investigated. The environmental advantages of producing biodiesel from lignocellulosic feedstock may outweigh its economic value. In the current economic situation, when positive externalities like effects on ecology, employment, climate change, and trade balance are not taken into account in the price mechanism, biodiesel is not cost-competitive with fossil diesel.

The value added to the feedstock, an increase in the number of manufacturing jobs in rural areas, higher income taxes, increased investments in machinery and equipment, an expansion of the manufacturing sector, a higher tax base from plant operations and income taxes, an improvement in the current account balance, and lower health care costs as a result of better air quality and greenhouse gas reduction are some of the economic advantages of a biodiesel industry. The agricultural strategy of subsidising the growing of nonfood crops first makes it easier to produce and use biodiesel. Second, the oil tax does not apply to biodiesel. In 2005, over 89% of the world's biodiesel was produced in the European Union.

In addition to lowering greenhouse gas emissions and a nation's dependency on foreign oil imports, biodiesel also aids domestic agriculture by creating new job prospects and market options for native products. The price of biodiesel fuels fluctuates according to the base stock, geographic region, seasonality of crop production, price of crude oil, and other elements. The high cost of the feedstock plays a significant role in the high cost of biodiesel. As a result, in the current economic climate, biodiesel cannot compete with petroleum-based fuel. The cost of the feedstock derived from biomass and other related expenses determine how competitive biodiesel is. To turn rice bran into biodiesel, Shiu et al. in 2010[26] created an in-situ acid-catalyzed esterification accompanied by a base-catalyzed transesterification. Despite the immense potential of lignocellulosic biomass as a feedstock for the manufacture of biodiesel, their resistant nature needs costly pretreatment techniques before conversion to biodiesel, which affects the process' economic viability. The process's economics could be enhanced by creating low-cost pretreatment techniques, advancing lipid extraction technology, and identifying lipid producers.

7.5.3 BIOBUTANOL

Butanol is a straight-chain alcohol with four carbon atoms and is primarily employed as a significant precursor in the production of polymers, adhesives, plastics, and compounds with added value such as acrylic acid, glycol ethers, etc. The propylene oxo-synthesis method is used to manufacture a large portion of the synthetic butanol that is now on the market from petrochemicals. Using transition metal catalysts, propylene is reacted with CO and H_2 at temperatures ranging from 40 to 200°C under high pressure to produce aldehydes, which are then hydrogenated to produce 1-butanol. About 10–12 billion pounds of butanol are transported through petrochemical processes each year.[27] The propylene market, which is susceptible to fluctuations, is a factor in the creation of synthetic butanol.

Biobutanol, a butanol generated from biomass, is gaining popularity as a cutting-edge biofuel to replace petrol in motor vehicles. In contrast to bioethanol, which needs to be blended up to 85%, butanol can be used as a drop-in fuel in its pure form or blended in a variety of ways. High octane number, low toxicity, fewer risks, and less hygroscopy are a few benefits of biobutanol. When compared to ethanol, biobutanol is known to have better fuel-related characteristics. Biobutanol's fuel characteristics are also comparable to those of petrol. Having an energy content of 29.2 MJ/L, butanol has a slightly lower energy density than petrol and a higher energy density than ethanol. Additionally, butanol has a similar octane rating and air-to-fuel ratio like petrol.

Butanol can be used in the existing transportation pipes and fueling stations without much upgrading because of its comparatively low vapor pressure. Acetone-butanol-ethanol (ABE) fermentation employing solventogenic Clostridium species can produce biobutanol. Acetone, butanol, and ethanol can be produced from a variety of carbon feedstocks using the well-known ABE fermentation process. The choice of feedstock and pretreatment techniques, issues with butanol recovery later, low yield, and butanol's toxicity to microorganisms are a few of the difficulties that limit butanol synthesis on a big scale using ABE fermentation. Numerous strategies have been used to modify butanol-producing organisms in an effort to get beyond these restrictions, including mutagenesis, recombinant DNA technologies, and proteomics.

7.5.4 BIOHYDROGEN

One of the biofuels of the future, biohydrogen, is thought to have the potential to lessen reliance on foreign oil and help reduce greenhouse gas emissions. In the future, hydrogen will play important roles as an environmentally friendly fuel for fuel cells that emit almost zero pollutants and as an intermediary energy carrier for the storage and transportation of renewable energy. Expert organizations from a range of fields are concentrating on the generation and use of biohydrogen while also considering the social consequences of deployment. Production of biohydrogen is seen as a crucial step toward a sustainable worldwide clean energy production and a promising replacement for traditional fossil fuels since, if successful, it could solve the majority of the issues that fossil fuels bring about. Since only clean water is produced when hydrogen is converted to energy, whether by combustion or fuel cells, the main benefit of energy from biohydrogen consists of the elimination of the release of greenhouse gases.[28] Biohydrogen could represent the intended clean outcome of the microbial process with the usage of the right technology.

Because it has the potential to develop into an endless, affordable, and sustainable source of clean energy, the generation of biohydrogen has garnered attention on a global scale. Hydrogen must originate from renewable feedstock in order to be considered renewable. [29] A potential renewable power source and viable replacement for fossil fuels is the creation of biohydrogen. Because biological activities produce hydrogen mostly at room temperature and pressure, they require less energy than chemical or electrochemical processes. Bacteria and microalgae are principally responsible for biohydrogen metabolism, a green energy source of spontaneous bioconversion. It involves a wide range of taxonomically varied species, enzymes, metabolic pathways, and processes within these groupings. The protein hydrogenase or nitrogenase is used in biological processes to produce hydrogen.[30] This enzyme controls how prokaryotes and some eukaryotic species, such as green algae, metabolize hydrogen.

When biological processes produce biohydrogen (H_2) fuel, it is said to be an environmentally conscious energy that can be easily biosynthesized from specific types of biomass. The storage and transportation of gases that are not condensable is one of many difficulties in the H_2 bioproduction process. Therefore, it is possible to generate H_2 through the use of biological byproducts as a potential source via biological processes with great interest by creating a new pathway for the use of renewable and limitless energy sources.[31] Carbohydrates are the primary source of hydrogen, according to numerous investigations of biohydrogen fermentative processes. As a result, wastes and biomass rich in complex carbohydrates seem like good sources of feedstock for the production of biohydrogen. The availability, price, carbohydrate content, and biodegradability of potential substrates for fermentative biohydrogen production must all be taken into consideration. Because they may be easily biodegraded, simple sugars like lactose, sucrose, and glucose are selected as the prototype substrates for the generation of hydrogen. Pure carbohydrate sources, on the other hand, are pricey starting points for the synthesis of hydrogen at scale, which is only commercially feasible when used from renewable and affordable sources. Agricultural crops and their waste byproducts, lignocellulosic materials like wood and wood debris, waste from food processing, aquatic plants, algae, and discharges produced in human habitats are other potential resources in feedstock besides water.[32] Microbial fermentation is the primary method for recovering energy from water-containing biomass, including sewage sludge, agricultural and livestock effluents, and animal excreta. Biohydrogen has been found to have a clean energy carrier property with high energy yields and burning of H_2 fuel, but the main byproduct that is released into the environment is just H_2O molecules.

H_2 fuel has been described as having appealing and sustainable properties since a range of biological processes can be used for its bio-production. H_2 generation by biological channels has been documented in a variety of ways. These methods include photobiological fermentation, anaerobic fermentation, enzymatic or microbial electrolysis, among others. Combining the aforementioned procedures can also result in the creation of hydrogen.[33] The use of bioaugmentation, multiple process integration, and microbial electrolysis processes has been reported in the evaluation of biological routes for efficient and improved hydrogen production; these processes can improve the bioprocess efficiency by focusing on the industrial-level applications.[34]

One of the biofuels of the near future, biohydrogen, is thought to have the potential to both lessen reliance on fossil fuels and reduce the release of GHGs from the transportation and energy sectors. The possibility exists for biomass to contribute significantly to the renewable hydrogen supply. Therefore, by incorporating bioenergy into the bioindustry, nations with strong agricultural economies have the potential to see significant economic growth. Although low-value waste biomass can be used as a feedstock in cost-competitive renewable hydrogen technologies, producing hydrogen from biomass currently costs more than getting it from natural gas. In the future, hydrogen will play a crucial role in energy storage and transportation as well as an alternative fuel for fuel cells that emit almost no emissions. It is evident that the emergence of hydrogen as a source of clean energy will have significant economic repercussions, assuming that scientific and technological obstacles are overcome. However, the function of biohydrogen in an eventual hydrogen economy remains to be seen.

7.5.5 *BIOMETHANE*

A naturally occurring gas known as biomethane is created when dead animal and plant matter, manure, sewage, organic waste, etc., are digested in a process known as anaerobic digestion. It is chemically identical to natural gas, which is also formed from dead plants and animals and is also kept underground. Biogas upgraders, who take the CO_2 out of biogas, turn it into biomethane, a sustainable fuel.[35] Biomethane is made from biogas, which is created from organic elements like sewage, food waste, food waste from distilleries, and agricultural products. Biogas and biomethane are regarded as renewable fuels since the carbon in these substances was recently removed from the atmosphere and is a component of the short-term carbon cycle.

Biomethane is a versatile and convenient fuel that may be used everywhere, natural gas is utilized without requiring any adjustments to machinery made to run on natural gas. There is a ready-made method for distributing biomethane in areas where there is an existing natural gas grid. Biomethane is a renewable energy source, as opposed to methane produced from fossil fuels, which is only seldom available. It is made from organic waste, such as dead animals and plants, sewage, manure, slurry, etc., which will always be available as long as there is life on Earth.

Biomethane may be produced almost anywhere in the world and offers a renewable source of energy. The distribution of the resources, on the other hand, makes it difficult to obtain natural gas and other fossil fuels in most of the world. Additionally, they may be hard to access, like far underneath the ocean floor, making extraction costly and potentially hazardous to both people and the environment. On the other hand, the production of biomethane just calls for the collection of readily accessible organic materials.

Despite producing carbon dioxide and a few other greenhouse gases, biogas plants do not add to the greenhouse effect. Instead, they contribute to a decrease in the emissions of gases like carbon dioxide and others that heat the atmosphere of the Earth. First, the production of biomethane prevents the release of methane, one of the most powerful greenhouse gases, second, it decreases the amount of GHGs released into the atmosphere while organic matter is allowed to decompose naturally, third, it decreases the need for burning fossil fuels, the main contributor to global warming, and finally, it decreases the requirement for firewood and helps preserve the forests, which absorb significant amounts of car emissions.

The natural process of methane synthesis is essentially followed in the making of biomethane by biogas plants. For the anaerobic bacteria that produce gas during digestion, biogas facilities gather various kinds of organic matter and preserve it in airtight tanks. The biogas, which consists mainly of methane and carbon dioxide, is then either extracted to the neighboring petrol storage or stored on the top of the digester tank. Depending on the production techniques and the type of organic matter utilized, this may assist reduce car emissions. Biogas may need to be treated to remove harmful pollutants such as hydrogen sulphide and volatile siloxanes. The majority of contemporary biogas plants also employ purifying techniques that lower the carbon dioxide concentration. A different method of generating biomethane is to collect the gas from landfills. When organic material is compacted and covered by material from above, oxygen cannot reach the organic material and methane is created in landfills. Methane is produced as a result of anaerobic digestion, which can be trapped locally.[36] However, because methane is explosive when it gets exposed to oxygen and since it is challenging to entirely trap it, the creation of biomethane in landfills is hazardous. In addition, a variety of additional gases that either contaminate the air, heighten the greenhouse effect, or both are present in landfill gas.

The creation of biomethane prevents the emission of a significant amount of methane and other hazardous gases into the atmosphere. This is because the synthesis of it prevents organic matter that is degrading from being exposed to the air, which allows methane and other gases to escape into the atmosphere. Additionally, since less fossil fuel is needed when using biomethane, less GHGs are released into the atmosphere. Reducing the demand for firewood also aids in forest preservation, which lowers atmospheric carbon dioxide levels since the trees release pure oxygen while absorbing carbon dioxide. The use of organic waste for the generation of biomethane also enhances rural communities' hygienic conditions and overall standard of life, as well as lowers the intensity of water pollution.

The fact that biomethane does not in any way hurt the environment is another fantastic benefit it has over fossil fuels. Contrary to the extraction of fossil fuels, the generation of biomethane requires hardly any drastic changes to the natural environment and does not significantly increase the risk of contamination or the extinction of biodiversity. Furthermore, as organic waste can be detrimental to the environment in sufficient proportions, it protects the ecosystem against pollution. However, given the fact that biomethane combustion also produces carbon dioxide, the biggest benefit of biomethane over fossil fuels is that it does not increase carbon dioxide emissions. This is because using biomethane essentially simply involves using energy that gets released during the organic matter's natural breakdown.

7.6 CONCLUSION AND FUTURE PERSPECTIVES

While there is a growing need for energy on a worldwide scale, fossil fuels continue to have shortcomings and pose a serious environmental concern. Demand for and the amount of fuel consumed globally is both anticipated to increase quickly, and the usage of fossil fuels has a serious negative influence on the earth's ecology. In order to address the world's energy crisis, sources of renewable energy are essential. Biofuels are a great example of a renewable energy source that may be created utilizing living things and lessen our reliance on fossil fuels.

Global cooperation is crucial for integrating biofuels to our current energy system in order to overcome the energy challenge. Through core mechanistic investigation, cell growth facility development, genetic engineering of specific algal strains, and biofuel production condition optimisation, these collaborative initiatives will support the development of culture methodologies and technology innovation for biofuel production. Although biofuels cannot meet all of the world's demand, they may play a significant role in the transportation sector.

The profitability of biofuels and a number of interlinked variables will determine how much biofuel will be produced globally in the future. Many governments throughout the world are promoting the usage of biofuels. Although the utilization of biofuels as an alternative to fossil fuels has gained global acceptance, there is still room for technological advancement to lower production costs, lower greenhouse gas emissions, and improve viability. The advancement of biofuel as a fossil fuel substitute necessitates global acceptability and the creation of effective biofuel-driven engines. Therefore, reducing climate change and improving environmental quality through efficient biofuel production will contribute to the growth of the global economy. As a result, their early detection and development in the realm of alternative fuels help both industry and the environment.

KEYWORDS

- **biofuel**
- **biomass**
- **bioethanol**
- **biohydrogen**
- **biomethane**

REFERENCES

1. Ruan, R.; et al. Biofuels: Introduction. In *Biofuels: Alternative Feedstocks and Conversion Processes for the Production of Liquid and Gaseous Biofuels*, Second Ed., **2019**; pp 3–43.
2. Nanda, S.; Reddy, S. N.; Mitra, S. K.;Kozinski, J. A. The Progressive Routes for Carbon Capture and Sequestration. *Energy Sci. Eng.* **2016,** *4* (2), 99–122. DOI: 10.1002/ese3.117.
3. Fatih Demirbas, M.; Balat, M.;Balat, H. Biowastes-to-Biofuels. *Energy Convers. Manag.* **2011,** *52* (4), 1815–1828. DOI: 10.1016/j.enconman.2010.10.041.
4. Alaswad, A.; Dassisti, M.; Prescott, T.; Olabi, A. G. Technologies and Developments of Third Generation Biofuel Production. *Renew. Sustain. Energy Rev.* **2015,** *51*, 1446–1460. DOI: 10.1016/j.rser.2015.07.058.
5. Naik, S. N.; Goud, V. V.; Rout, P. K.;Dalai, A. K. Production of First and Second Generation Biofuels: A Comprehensive Review. *Renew. Sustain. Energy Rev.* **2010,** *14* (2), 578–597. DOI: 10.1016/j.rser.2009.10.003.
6. Du, C.; et al. Introduction: An Overview of Biofuels and Production Technologies. An Overview of Biofuels and Production Technologies. In *Handbook of Biofuels Production, Processes and Technology,* Second Ed., **2016**; pp 3–12. DOI: 10.1016/B978-0-08-100455-5.00001-1.
7. Hao, H.; et al. Biofuel for Vehicle use in China: Current Status, Future Potential and Policy Implications. *Renew. Sustain. Energy Rev.* **2017,** *82*, 645–653. DOI: 10.1016/j.rser.2017.09.045.
8. Li, H.; Qu, Y.; Yang, Y.; Chang, S.; Xu, J. Microwave Irradiation - A Green and Efficient Way to Pretreat Biomass. *Bioresour. Technol.* **2016,** *199*, 34–41. DOI: 10.1016/j.biortech.2015.08.099.
9. Li, J.; Henriksson, G.;Gellerstedt, G. Lignin Depolymerization/Repolymerization and its Critical Role for Delignification of Aspen Wood by Steam Explosion. *Bioresour. Technol.* **2007,** *98* (16), 3061–3068. DOI: 10.1016/j.biortech.2006.10.018.
10. Zhang, Y. H. P.; Hong, J.; Ye, X. Chapter 14 Cellulase Assays. In *Business*, 2009; vol *581*, pp 213–231. DOI: 10.1007/978-1-60761-214-8.

11. Kim, K. H.; Hong, J. Supercritical CO2 Pretreatment of Lignocellulose Enhances Enzymatic Cellulose Hydrolysis. *Bioresour. Technol.* **2001,** *77* (2), 139–144. DOI: 10.1016/S0960-8524(00)00147-4.
12. Lu, Y.; Mosier, N. S. Biomimetic Catalysis for Hemicellulose Hydrolysis in Corn Stover. *Biotechnol. Prog.* **2007,** *23* (1), 116–123. DOI: 10.1021/bp060223e.
13. Sun, Y.; Cheng, J. Hydrolysis of Lignocellulosic Materials for Ethanol Production: A Review. *Bioresour. Technol.* **2002,** *83* (1), 1–11. DOI: 10.1016/S0960-8524(01)00212-7.
14. Sindhu, R.; Binod, P.; Pandey, A. Biological Pretreatment of Lignocellulosic Biomass - An Overview. *Bioresour. Technol.* **2016,** *199*, 76–82. DOI: 10.1016/j.biortech.2015.08.030.
15. Behera, S.; Arora, R.; Nandhagopal, N.; Kumar, S. Importance of Chemical Pretreatment for Bioconversion of Lignocellulosic Biomass. *Renew. Sustain. Energy Rev.* **2014,** *36*, 91–106. DOI: 10.1016/j.rser.2014.04.047.
16. Zhuang, X.; et al. Liquid Hot Water Pretreatment of Lignocellulosic Biomass for Bioethanol Production Accompanying with High Valuable Products. *Bioresour. Technol.* **2016,** *199*, 68–75. DOI: 10.1016/j.biortech.2015.08.051.
17. Rasmussen, H.; Sørensen, H. R.; Meyer, A. S. Formation of Degradation Compounds from Lignocellulosic Biomass in the Biorefinery: Sugar Reaction Mechanisms. *Carbohydr. Res.* **2014,** *385*, 45–57. DOI: 10.1016/j.carres.2013.08.029.
18. Liao, Y.; et al. The Role of Pretreatment in the Catalytic Valorization of Cellulose. *Mol. Catal.* **2020,** *487*, 110883. DOI: 10.1016/j.mcat.2020.110883.
19. Seidl, P. R.;Goulart, A. K. Pretreatment Processes for Lignocellulosic Biomass Conversion to Biofuels and Pioproducts. *Curr. Opin. Green Sustain. Chem.* **2016,** *2*, 48–53. DOI: 10.1016/j.cogsc.2016.09.003.
20. Okolie, J. A.; Mukherjee, A.; Nanda, S.; Dalai, A. K.; Kozinski, J. A. Next-Generation Biofuels and Platform Biochemicals from Lignocellulosic Biomass. *Int. J. Energy Res.* **2021,** *45* (10), 14145–14169. DOI: 10.1002/er.6697.
21. Mu, D.; Seager, T.; Rao, P. S.; Zhao, F. Comparative Life Cycle Assessment of Lignocellulosic Ethanol Production: Biochemical Versus Thermochemical Conversion. *Environ. Manag.* **2010,** *46* (4), 565–578. DOI: 10.1007/s00267-010-9494-2.
22. Guerrero, A. B.; Muñoz, E. Life Cycle Assessment of Second Generation Ethanol Derived from Banana Agricultural Waste: Environmental Impacts and Energy Balance. *J. Clean. Prod.* **2018,** *174*, 710–717. DOI: 10.1016/j.jclepro.2017.10.298.
23. Ikura, M.; Stanciulescu, M.; Hogan, E. Emulsification of Pyrolysis Derived Bio-oil in Diesel Fuel. *Biomass Bioenergy* **2003,** *24* (3), 221–232. DOI: 10.1016/S0961-9534(02)00131-9.
24. Knothe, G.; Sharp, C. A.; Ryan, T. W. Exhaust Emissions of Biodiesel, Petrodiesel, Neat Methyl Esters, and Alkanes in a New Technology Engine. *Energy Fuels* **2006,** *20* (1), 403–408. DOI: 10.1021/ef0502711.
25. Rathore, V.; Madras, G. Synthesis of Biodiesel from Edible and Non-edible Oils in Supercritical Alcohols and Enzymatic Synthesis in Supercritical Carbon Dioxide. *Fuel* **2007,** *86* (17–18), 2650–2659. DOI: 10.1016/j.fuel.2007.03.014.
26. Shiu, P. J.; Gunawan, S.; Hsieh, W. H.; Kasim, N. S.; Ju, Y. H. Biodiesel Production from Rice Bran by a Two-Step in-situ Process. *Bioresour. Technol.* **2010,** *101* (3), 984–989. DOI: 10.1016/j.biortech.2009.09.011.

27. MacLean, H. L.; Lave, L. B. Hydrolysis of Oil Palm Empty Fruit Bunch using Dilute Acid. *Prog. Energy Combust. Sci.* **2003,** *29* (1), 1–69. http://linkinghub.elsevier.com/retrieve/pii/S0360128502000321.
28. Claassen, P. A. M.; et al. Utilisation of Biomass for the Supply of Energy Carriers. *Appl. Microbiol. Biotechnol.* **1999,** *52* (6), 741–755. DOI: 10.1007/s002530051586.
29. Show, K. Y.; Lee, D. J.; Zhang, Z. P. Production of Biohydrogen: Current Perspectives and Future Prospects. *Biofuels Altern. Feed. Convers. Process.* **2011,** 467–479. DOI: 10.1016/B978-0-12-385099-7.00021-8.
30. Show, K. Y.; Yan, Y. G.; Lee, D. J. Biohydrogen Production from Algae: Perspectives, Challenges, and Prospects. In *Biofuels from Algae*, Second Ed.; Elsevier B.V., **2018**; pp 325–343.
31. Mirza, S. S.; Qazi, J. I.; Liang, Y.; Chen, S. Growth Characteristics and Photofermentative Biohydrogen Production Potential of Purple Non Sulfur Bacteria from Sugar Cane Bagasse. *Fuel* **2019,** *255*, 115805. DOI: 10.1016/j.fuel.2019.115805.
32. Show, K. Y.; et al. Critical Assessment of Anaerobic Processes for Continuous Biohydrogen Production from Organic Wastewater. *Int. J. Hydrog. Energy* **2010,** *35* (24), 13350–13355. DOI: 10.1016/j.ijhydene.2009.11.110.
33. Kim, S.; Sung, T.; Kim, K. C. Performance and Greenhouse Gas Reduction Analysis of Biogas-Fueled Solid-Oxide Fuel Cells for a Sewage Sludge and Food Waste Treatment Facility. *Energies* **2018,** *11* (3). DOI: 10.3390/en11030600.
34. Srivastava, R. K.; Shetti, N. P.; Reddy, K. R.; Aminabhavi, T. M. Biofuels, Biodiesel and Biohydrogen Production using Bioprocesses: A Review. *Environ. Chem. Lett.* **2020,** *18* (4), 1049–1072. DOI: 10.1007/s10311-020-00999-7.
35. Adnan, A. I.; Ong, M. Y.; Nomanbhay, S.; Chew, K. W.;Show, P. L. Technologies for Biogas Upgrading to Biomethane: A Review. *Bioengineering* **2019,** *6* (4), 1–24. DOI: 10.3390/bioengineering6040092.
36. Xu, N.; et al. Biomethane Production from Lignocellulose: Biomass Recalcitrance and its Impacts on Anaerobic Digestion. *Front. Bioeng. Biotechnol.* **2019,** *7* (AUG), 1–12. DOI: 10.3389/fbioe.2019.00191.

CHAPTER 8

Copper Tin Sulphide: Potential Candidate for Solar Energy Conversion

GEETHU R. and SANTHOSH M. V.

SCMS School of Engineering and Technology, Karukutty, Kerala, India

ABSTRACT

Frequently changing lifestyle entails more and more energy consumption. Depending blindly on conventional energy resources will severely affect future energy needs. Our nature is rich in sustainable energy resources. However, utilizing these energy sources efficiently is still unknown. Researchers are intensively working in this field to develop more and more method to unveil the unknown possibilities and techniques to convert tremendous renewable energy into useful energy. Here in this work, different renewable energy sources are listed and briefly described. Among the renewable energy sources, SUN is inevitable and the energy from Sun called solar energy is the richest among all. A device used to convert solar energy into electricity is solar cells/solar panels. Numerous types of solar cells configurations were developed by researchers around the world. Inorganic solar cells made up of chalcogenide materials are one among them. Recently introduced chalcogenide material in the field of solar cell is Copper Tin Sulfide (CTS). The possibilities and potential of CTS thin film to become an absorber layer in solar cell are evident from the structural, morphological, compositional, and optoelectronic properties of CTS material. Different methods for the deposition of CTS both physical and chemical methods, done and proven by different researchers are discussed here. Changing electrical, optical, and morphological properties of CTS thin films reveals the tunability of the

The Food-Energy-Water Triangle for Sustainable Development: Integrative and Synergic Solutions.
Pedro Aguilar-Zárate, Mónica Chávez-González, Ann Rose Abraham, and A. K. Haghi (Eds.)

material according to the requirement. Various solar cell devices fabricated using CTS are also discussed.

8.1 INTRODUCTION

Most frequently used words, when discussing energy and energy sources are sustainable and renewable energy source. Is it so relevant? Yes, it is because our conventional energy resources are depleting to a larger extend on daily basis. Studies reveal that our non-renewable energy resources will vanish within a few decades. That means we are using the resources that are meant for future generations. The only way to cease this nastiness for future generations is to revise the source of energy that we are using. While changes it should be kept in mind that our current usage should not affect future needs. Such kind of energy sources that are able to meet all the energy requirement of present generation without sacrificing the need of future generation can be called as sustainable energy source. These sources are sustained forever.

Even though research works are immensely going on developing new materials, methods, and technologies for the extraction of energy from renewable energy sources, the use of such sustainable energy for daily needs is still lower. On analyzing the usage of renewable energy, even though an increase in production and usage is reported, utilizing renewable energy is lagging behind the growing energy needs. This trend has to be changed to protect our earth as well as to safeguard our future energy needs.

8.1.1 DIFFERENT TYPES OF SUSTAINABLE ENERGY

Sustainable energy or renewable energy sources that can be used effectively and efficiently for energy extraction are given in Fig. 8.1.

8.1.1.1 SOLAR ENERGY

Radiant light and heat from the Sun can be called as solar energy. This can be harvested using different technologies and converted into useful forms. Solar panels can convert it into electricity and solar thermal heat up or may convert it to electricity. Based on the absorption, utilization, and dispensing of energy, this can be classified as passive and active technologies. Units of photovoltaics, solar power concentrators, etc. come under active solar

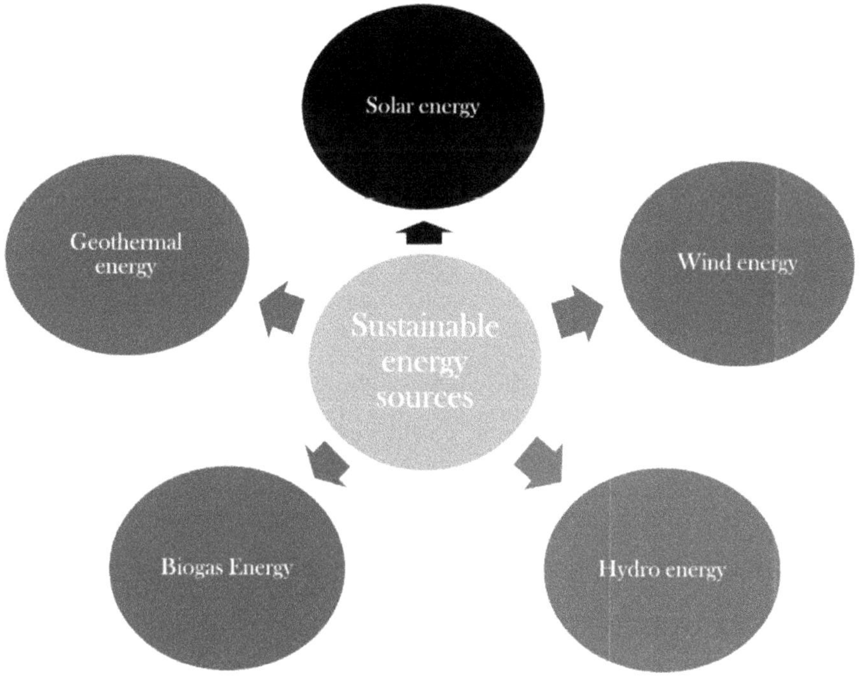

FIGURE 8.1 Different sources of renewable energy.

technology, positioning a building so as to get energy from the Sun and similar comes under passive.,

8.1.1.2 WIND ENERGY

The energy of moving air can be called as Wind energy. Air under movement has kinetic energy which is effectively used to rotate the blades of a huge fan placed at a large height from ground level. This blade will rotate the shaft or the energy generated by rotating the blade is transferred to the shaft and is then generated as electrical energy.

8.1.1.3 HYDRO ENERGY

Similar to the above-mentioned wind energy, here energy from moving water is converted to useful form. Energy from moving water is hydro energy or hydroelectric power. It is one of the oldest renewable methods of energy conversion. Here energy is derived from moving water. In many countries electricity generation is through creating a reservoir by building a dam across the river, thus controlling the flow of water. Water is made to fall from a height, having a high potential energy, and is then allowed to fall down the hills, potential energy is converted to kinetic energy and is made to rotate the blades of a turban which further generates electricity.

8.1.1.4 BIOGAS ENERGY

Energy from waste is termed as biogas. Production of biogas makes use of organic matter like animal waste, waste food, food scraps, etc. This is allowed to degrade in the absence of oxygen by bacteria—anaerobic digestion. Biogas plant—a huge dumping section is required to collect the waste and allow it to degrade without oxygen.

8.1.1.5 GEOTHERMAL ENERGY

Thermal or temperature of the earth is used here to generate electricity or heat. Underground temperature is collected and utilized to generate electricity by making large wells or mines.

8.2 RELEVANCE OF SOLAR ENERGY

The Sun, a source of solar energy is the most dependable form of energy for all the living things on the earth. Various forms of energy from the Sun are utilized by all the living organisms in this world. Heat and light from the Sun are the two energy sources that can be converted to the required form. Solar heaters and solar panels or solar cells are the devices that are used to convert solar energy into useful form. A brief explanation of both is given below.

- Solar thermal energy:
  - Conversion of solar energy into thermal energy.
  - Produce high temperature by collecting and concentrating sunlight—further used for the production of electricity.
  - Reflector and receiver are the two main components of solar thermal energy generation.

- Solar cells or solar panels:
  - Convert light energy to electricity via photovoltaics.
  - Photovoltaic devices consist of p-n junction.
  - Electron–hole are generated at the junction and is then collected and current will flow through the circuit – constitute current.

8.3 MATERIALS USED FOR SOLAR ENERGY CONVERSION

A solar cell device has mainly four layers—transparent conducting oxide (TCO)/bottom contact, p-type layer, n-type layer, and top contact. Basic device structure of an inorganic type solar cell is shown in Fig. 8.2. All the layers have different properties and material should satisfy that specific characteristics to be a part of a solar cell or solar panel. The basic requirements for a material to be a part of the solar cell are tabulated in Table 8.1. Different types of solar cells have different layers and corresponding names and properties. The inorganic solar cell was taken into consideration for developing the table given below since the CTS-based materials are most preferable for such solar cells.

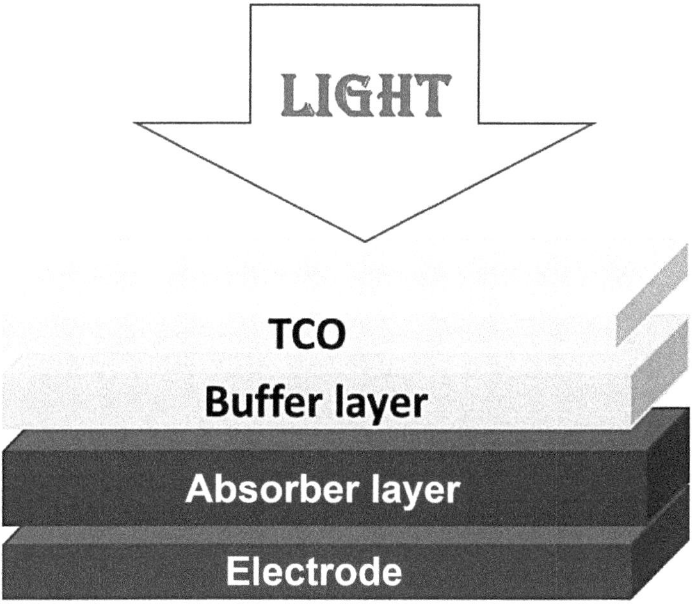

FIGURE 8.2 Basic device structure of an inorganic solar cell.

TABLE 8.1 Explanation of Each Layer, Function of Each Layer, Required Material Property, Commonly Used Materials for Solar Cells are Listed.

Layers of solar cell	Layer function	Material property		Examples/materials used
		Optical	Electrical	
Bottom contact layer – *where light is incident* Transparent conducting Oxide (TCO).	Collect light from the source and passes to the next layer. Collect charge carriers and transfer them to the circuit for the generation of current.	High transmission is required. Band gap greater than 3eV is preferable.	High electrical conductivity is required because charge carrier has to be transferred to the circuit without much loss.	Tin-doped Indium Oxide (ITO) Flourine-doped Tin Oxide (FTO).
n-type layer, Buffer layer.	Collect the light and transmit it to the absorber layer. Forms junction with absorber layer.	Good light transmission.	High conductivity. Thickness must be lower.	Indium Sulfide, Cadmium Sulfide, Zinc Sulfide, Tin Sulfide, Zinc Oxide.
p-type, Absorber layer.	Absorption of light takes place. Charge carrier generation – separation of electrons and holes.	Absorption must be higher. Absorption coefficient high. Transmission must be minimal.	Good conductivity. Higher thickness than other layers.	Copper Indium Sulfide/ Selenide, Copper Zinc Tin Sulfide/ Selenide, Copper Indium Gallium Sulfide/ Selenide, CTS.
Top contact (electrode).	Collect all the charge carriers and transfer them to the circuit.	Highly opaque and reflecting so that no light will transmit through this electrode, instead it will get reflected to the absorber layer.	High conductivity.	Silver, Aluminum, Gold.

A solar cell should contain a p-n junction and two electrodes for energy conversion. p-Layer or layer having majority charge carriers as hole act as the absorber layer and n layer with electron as the majority charge carriers will act as the buffer region. The role of the n-region is to form a junction with the p-region. This junction means there is an energy gap between the p-region and the n-region. It is at this junction and due to this energy gap electron and hole will get separated. The separated charge carriers are then transported toward the other end of the n and p regions where electrodes are connected. Through the electrodes, electrons and holes will be passed to the external circuit resulting in the generation of electric energy.

For developing material suitable for p and n layer, an exhaustive and rigorous study has to be done. Years of intensive research resulted in the development of silicon solar cells, the king of solar cells in this decade. But the face will change in the upcoming years due to the cost of silicon solar cells. The need for replacing silicon with cost-effective material prompted many researchers to develop numerous materials suitable for solar cell application. Hence developed materials were CuInS/Se, CuInGaS/Se, CuZnSnS, InS, ZnO, GaSe, InTe, organic materials like P3HT, PCBM, and other higher order

TABLE 8.2 Optoelectronic, Structural, and Morphological Properties of CTS Thin Film.

	CTS	
Material	• Chalcogenide – I-III-VI • Cost-effective material and deposition technique • Eco-friendly	• Precursors for the deposition of CTS thin films are easily available and low cost
Characteristic properties		
Structural	• Crystalline • Depending on the deposition technique— monoclinic/tetragonal/triclinic structures	• Based on the deposition technique and annealing given to the sample, crystalline structure changes
Optical Properties	• Optical band gap 0.93 eV–1.5 eV • Absorbance 10^4 cm^{-1} • Refractive index 1.8–2.2	• Tunability of optical band gap is an advantage of this material • Higher absorption coefficient—good absorber material
Electrical properties	• Charge carrier concentration $1.27 \times 1021 cm^{-3}$ • Mobility 1.03 cm^2 /Vs • Resistivity ($4.77 \times 10–^{3}$ Ω cm) • P-type	• Low resistivity and p-type—good absorber material

numerous derivatives of thiophene, perovskite materials, etc. Among which first mentioned chalcogenide materials are highly stable in nature. One among the newcomers in the group of chalcogenides is CTS. Research works to date showed that the optoelectronic properties of CTS samples are matching with the required characteristics of a material to be a part of solar cells.

8.4 POSSIBLE DEPOSITION TECHNIQUES FOR THE DEVELOPMENT OF CTS

CTS can be deposited using both physical and chemical routes. Most reported type of technique used for CTS thin film preparation is given in Fig. 8.3. For obtaining pure form CTS, physical deposition is preferable whereas chemical technique is chosen if the need is to develop the material through cost-effective, vacuum-free, and simpler technique. Under physical deposition, commonly used techniques for the deposition of CTS samples are Vacuum Evaporation, Sputtering, Pulsed Laser Deposition (PLD), Electron Beam Deposition, etc. Chemical deposition techniques available and proven for CTS deposition are chemical spray pyrolysis, Chemical Bath Deposition (CBD), Successive ionic layer adsorption and reaction (SILAR), etc.

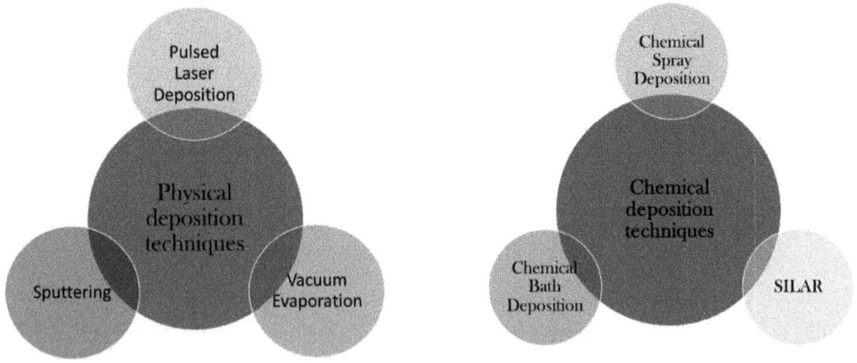

FIGURE 8.3 Physical and chemical deposition techniques for CTS thin film.

8.4.1 CTS DEVELOPED THROUGH PHYSICAL DEPOSITION TECHNIQUES

8.4.1.1 PLD

PLD is a vacuum-assisted deposition technique which makes use of a pulsed laser with high power to vaporize the target material. This vapor is then deposited onto the substrate and kept on the substrate holder. PLD technique requires an ultra-high vacuum for the procedure or should be carried out under the presence of corresponding gases depending on the product required. For example, for obtaining ZnO as the deposited sample the gaseous atmosphere will be oxygen.

PLD deposition required a target material and here it was prepared by hot pressing of copper sulfide (CuS_2) and tin sulfide (SnS_2) powder in equal ratio. A three-stage ultrasonic cleaning process was done for the molybdenum-coated glass and soda lime glass, the substrate for the deposition of CTS. Three-stage process of substrate cleaning was done using (1) acetone, (2) ammonia and ethanol, (3) double distilled water. KrF excimer was the laser source used with a wavelength of excitation of 248 nm having a pulsewidth of 25 ns. The deposition process was done under a pressure of 5×10^{-5} Torr for 30 min. with room temperature as the substrate temperature. Annealing process was done on the deposited samples inside a furnace for 1 h under different temperatures like 200°C, 300°C, and 400°C. Annealing was done in hydrogen sulfide and nitrogen atmosphere. This work was done by Vanalakar et al.

Annealing improved the crystallinity of the sample which was the major observation from XRD data. Morphological studies done on the samples show that CTS sample annealed at 400°C has more crystallized and compact layer in comparison with the samples annealed at other temperatures. FESEM images of the CTS samples at different temperatures are shown in Fig. 8.4.[9] Improvement in sample quality with annealing is clearly evident from SEM image.

Thota et al.[10] prepared pellets by mixing CuS and SnS powder in a 2:1 ratio. This is molded into a pellet by grinding thoroughly followed by pressing. The sintering process was done for the pellet at a higher temperature of 700°C about 12 h. Hence prepared pellet was loaded into the chamber. Four stages of the cleaning procedure were done for the glass substrate before loading. The first step is by ultrasonication using acetone, followed by propanol. This was then cleaned by double deionized water and finally substrate was wiped using nitrogen. PLD was done under a pressure of 6×10^{-6} mbar inside the chamber attained with the help of turbo molecular pump and rotary pump. Laser used to vaporize the target, that is, the above said pellet was Nd-YAG, with 266 nm as the excitation wavelength and 9 ns as pulse width. Here substrate temperature was set at 200°C which was kept at a distance of 4 cm from the source (pellet). The vaporization procedure was well-controlled and systematized in such a manner that the obtained sample was highly uniform. In the presence of Argon, annealing was done as an in situ route at millibarar pressure for 1 h at different temperatures from 350°C in the increasing steps of 50°C up to 550°C.

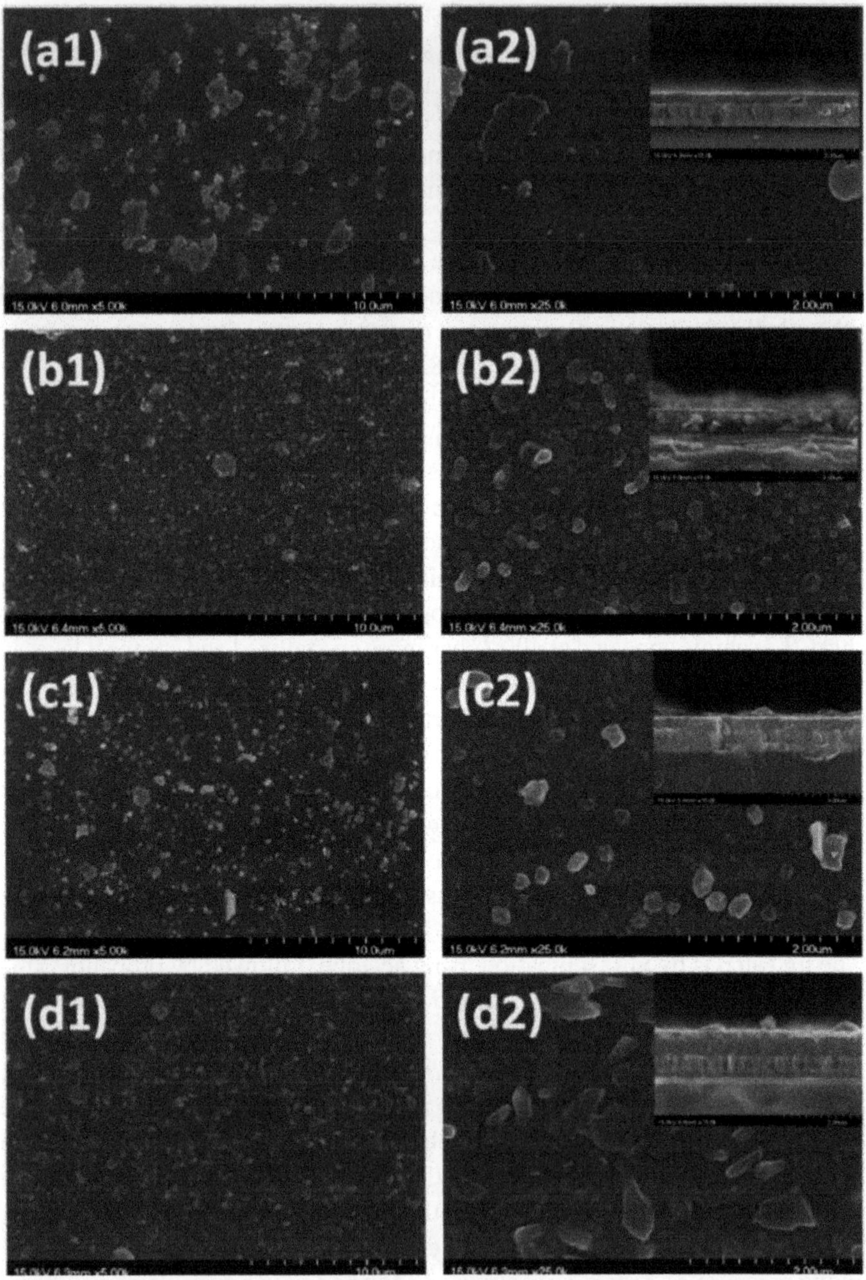

FIGURE 8.4 FESEM images of CTS samples deposited by PLD and annealed at different temperatures, without annealing (a), 200 (b), 300 (c), 400 (d).

Source: Reprinted with permission from Ref. [9] Copyright 2015. Elsevier.

FIGURE 8.5 FESEM images of CTS samples without annealed, annealed at 350°C, 550°C.
Source: Reprinted with permission from Ref. [10] Copyright 2019. Elsevier.

Crystal structure is monoclinic single phase from XRD and Ramana and with a direct band gap of 1.1 eV. FESEM (Fig. 8.5) shows large size pit were formed for the samples deposited without annealing and on annealing, more uniform films were obtained. A similar observation was reported by Vanalakar et al. in 2015[9] (mentioned above in this chapter). From both of these, the importance of annealing the CTS samples is evident.

CTS thin film deposited using different techniques are reported by a number of researchers and a few were tabulated by Lokhande et al.[11] With copyright permission from Elsevier, the tables are included in this chapter as Tables 8.3–8.5.

TABLE 8.3 Data of CTS Thin Film Deposited using PLD Technique.

Method	Fluence (J/cm^2)	Temp	Phase	SEM microstructure	PCE (%)
PLD	1.2	20	Amorphous	Unstructured	-
	0.4	150	Tetragonal	Granular	-
600 nm thick	0.5	250	Cubic+Cu_4SnS_4+SnS	Granular and non-compact	-
PLD	0.25	20	Amorphous	Non uniform	-
750 nm thick		200	Cus+cubic	Rough	-
		300	Cus+cubic	Garin growth	-
		400	Cubic	Compact	0.82

Source: Reprint from Ref. [11] with copyright permission from Elsevier.

8.4.1.2 VACUUM THERMAL EVAPORATION TECHNIQUE

Thermal evaporation or vacuum thermal evaporation is a technique used to deposit thin film under vacuum by resistive heating method. Here precursors must be in their pure solid form or as power. Resistive heating is given to a boat or a basket usually made up of molybdenum. For solid metal, a source boat is used and if the powdered form of the source is used a glass crucible filled with power is inserted into the wire basket. Resistive heating is given to the source through two electrodes. Precursor material heated up and vaporized. This vapor is then deposited on the substrate and kept on the substrate holder.

As for all deposition techniques here also thoroughly cleaned glass substrates were used by Isil et al.[12] Cleaning procedure was done by ultra-sonication of substrate using isopropyl alcohol followed by distilled water for one-forth of an hour and then dried. Precursors of copper (pellet), tin (pellet) are kept separately in the specific boats. Pieces of sulfur were placed in glass tubes. Initially, Sn followed by Cu were deposited on the substrate in such a way that tin form a thin layer of 150 nm thickness and Cu 300 nm thickness. Sulphurization process was done at different quantities of sulfur in order to study the effect of sulfur concentration on the CTS sample.

TABLE 8.4 Data of CTS Thin Films Deposited using Vacuum Evaporation Technique.

Sulfur source	Sulfurization Temp (°C)	Sulfurization Time (min)	Phase	SEM microstructure	PCE (%)
H_2S gas	200	60	Cus,Sns	Square	-
	325		Tetra+mon	Round Shape	-
	500		Tetra+mon	Uneven-Shape	-
S powder	570	30	Monoclinic		0.43
S powder	570	30	Monoclinic	Small grain formed due to nadpoing	4.63

Source: Reprinted with permission from Ref. [11] Copyright 2016. Elsevier.

8.4.1.3 SPUTTERING

One of the best techniques for obtaining highly uniform thin film is sputtering. Initially, gaseous plasma is created and then the ions are accelerated toward the source. These high energy ions erode the source material and it eject neutral particles. These ejected neutral particle in the form of molecules or atomic clusters or may be atoms move in a straight line path and get deposited on the substrate placed in between the traveling path of these particles. CTS sputtering was first reported in 2010 and was done by Fernandes and coworkers.

TABLE 8.5 Data of Sputtered CTS Samples and its Properties Reported by Different Researchers.

Method	Sulfur source	Sulfurization Time (min)	Sulfurization Temp (°C)	E_g(eV)	SEM (microstructure)	Phase	PCE (%)
Rf-mg	H_2S gas	120	350	2.19		Tetragonal	
			375	2.16			
			400	2.03	Nubby garins		
			425	1.77	Liner shape		
Rf	S powder	20 min	400	0.92	Huge agglomeration	Cubic	0
			450	0.96	No agglomeration		0.08
			500	0.98	Large garin		0.16
			550	0.99	compact large		0.28
Dc-mg	S powder	10 min	350	1.35	non-compact in all cases	Tetragonal	
			400	0.96		Cubic	
			520	1.15		Ortho	
Rf-mg	S powder	60 h	400	1.28	Large garins and non-compact	Tetra+ Orth	
Rf	S powder	5	575	1.02		Cubic	0.10
		10	575	1.01			0.17
		20	575	1	Compact		0.20
		30	575	0.98			0.18
mg	Spowder		520	0.92	Compact (Cu:Sn=1.84)	Monoclinic	3.05
					(Cu:Sn=1.45)		1.93
			560	0.92	Voids		
					(Cu:Sn=1.93)	Cu_3SnS_4	1.72
					(Cu:Sn=na)	Cu_3SnS_4	2.37
mg	S powder	120 min	560	1.35	Columnar growth,void free	Cubic	
RF	S powder	70 min	300			CuS	
			400			CTS	
			500			CTS	
		30	400				
		15	520		Compact and void free	CTS Pure	
RF	S powder	15	570		Compact	Monoclinic	0.76

Source: Reprinted with permission from Ref. [11] Copyright 2016. Elsevier.

Kim et al.[14] developed a CTS sample using the radio frequency magnetron sputtering technique. The substrate used was a molybdenum-coated glass substrate. The procedure for Cu coating was preplanned for 35 min and that for Sn coating for 45 min. Rapid thermal annealing was done on the samples in the presence of sulfur atmosphere for 10 min. Better crystalline films were obtained from this method of deposition.

Olgar, 2020[15] solemnly reported a study on sputtered CTS. In the study, the author used DC magnetron sputtering in which metallic layers of Cu, thickness of 175 nm, and Sn, thickness of 230 nm were deposited on a glass/molybdenum substrate. Before the Sulphurization process at 550°C, each samples were taken for soft-annealing treatment in H_2- Ar atmosphere at different temperatures from 200 to 350°C in the steps of 50°C.

8.4.2 CTS DEVELOPED THROUGH CHEMICAL DEPOSITION TECHNIQUES

8.4.2.1 CBD

CBD is one of the simplest method for the deposition of thin film using a precursor in aqueous solution form. Adsorption or deposition of ions onto a substrate (glass/quarts/metal surface, etc.) to form a thin film that is homogeneous in nature. Commonly, CBD technique is used to deposit or to develop thin films of chalcogenide say CuS, CdS, ZnO, CuInS, CuZnS, etc. It is one of the simplest, low cost, easier, non-vacuum technique for the preparation of thin film samples. On the other hand, this technique has some disadvantages like non-uniformity in film surface, possibility of the presence of impurity, etc.–

CTS thin film deposition by stacking and annealing via CBD technique was tried and achieved by Li et al.[20] Tin sulfide coated over molybdenum substrate was then coated with copper to form a molybdenum/tin sulfide/copper. The stacking was done in reverse manner also, to check the difference in the characteristics of the sample, that is, Molybdenum/Copper/Tin Sulfide. To get the proper thickness for SnS sample, double deposition was done using CBD. Devices were fabricated with both the samples and the former one, that is, Mo/SnS/Cu gives the better device performance.

Shelke et al.[21] deposited CTS—Cu_2SnS_3 using CBD technique using stainless steel (conducting) as the substrates. Compared with the other studies here the crystalline structure obtained was triclinic. Similarly for the above said cases, here also relevance of deposition temperature on the sample morphology is explained. But a contrasting result was obtained for optical absorption, a band gap of `~1.7 eV. This is a higher value compared to other report and is better band gap while checking for its application in the field of solar cell's absorber layer.

8.4.2.2 CHEMICAL SPRAY PYROLYSIS

Spray pyrolysis is another simplest method for the deposition of thin films or thick film. This method provides one of the easiest path for doping a sample by varying the doping percentage easily. It is a solution-based technique. Almost all sorts of thin films can be developed using CSP technique provided the precursor is solution. A simple and inexpensive method of preparing thin and thick films, ceramic coatings, and metal and metal oxide powders in large scale is spray paralysis. Reactions happening here are thermally stimulated. Product formation is by nucleation process. As said above thickness of the required film can be easily varied using CSP. A number of reports are there in the literature based on the CTS preparation through Spray Pyrolysis. Below is a report reported by our former group.

Sunny et al.[7] deposited thin films of Cu_2SnS_3 one of the simplest and low cost chemical spray pyrolysis techniques. Here, precursors are prepared separately, and in a particular ratio, it is mixed so as to get the predetermined stoichiometry in the film prepared. Copper chloride is taken as the precursor for copper, the tin precursor is stannous chloride, and for sulfur it is thiourea. Deionized water was the solvent used for dissolving all the chemicals. Compared to copper and tin precursors, thiourea – sulfur precursor, was added in excess because it is highly volatile and evaporates quickly. To maintain the ration sulfur had to be added in excess. The deposition process was held above 300°C.[7]

8.4.2.3 SILAR

SILAR is one of the best suitable, simple, cost-effective, and time-efficient for semiconducting thin film deposition with flexible substrate size. In this method, the substrate is dipped in precursor solutions separately and washed in between using distilled water. Hence this technique can be considered as a cycle process that includes cation adsorption, rinsing, anion adsorption, reaction, rinsing, and repeating the same.[23] Only a few reports are in the literature regarding the CTS developed using SILAR technique and one is explained below.

Sapna et al. in 2019[24] deposited CTS using SILAR technique. Precursors used to conduct the experiments were copper chloride (0.1 M) and tin chloride (0.06 M) solutions. TEA and thioacetamide were used as the cationic precursor and anionic precursor, respectively. Thoroughly cleaned glass substrates were then dipped in precursor solution and allowed to react for a few minutes. The rinsing process was done in between to clear out the un-adsorbed ions. Forty-five cycles of the dipping and rinsing process resulted in CTS thin film with better uniformity and stability.[24]

8.5 APPLICATION OF CTS IN SOLAR CELLS

From the above discussions, it is obvious that wide varieties of techniques under physical method or chemical method are available for the deposition of CTS thin film. Depending on our needs, say if the requirement is to deposit high-quality CTS thin films, physical method is preferable. If the prerequisite is for CTS through simple and cost-effective method, the chemical route will be beneficial. Crystalline nature from structural characterization, optoelectronic properties from electrical and optical properties, morphological nature from the FESEM, etc. obviously revealed the potential of CTS to be a part of photovoltaic devices more specifically inorganic solar cells. Some of the relevant works reported in this field are briefly explained below.

Vanalakar et al. in 2015[9] (deposition technique is mentioned above) developed p-type CTS thin film possessing a band gap of 1.01 eV and hole mobility of 0.51 cm^2/Vs. With this, a thin film of trial work on solar cell with configuration SLG/Mo/CTS/CdS/i-ZnO/AZO/Al was fabricated. This device exhibited a power conversion efficiency of 0.82% and short circuit current density of 11.90 mA/cm^2.[10]

He et al.[25] deposited CTS (Cu_2SnS_3) thin films through the sputtering technique. A detailed analysis of the sputtered samples as well as the influence deposition time of copper was done by checking structural, optoelectronic, morphological, and compositional analysis. Deposition process of Cu in a slow phase resulted in better device efficiency as claimed by the authors. The fabricated solar cell using CTS exhibited an efficiency of 2.39%, open circuit voltage of 208 mV, short circuit current density of 28.92 mA/cm^2, and fill factor of 39.7%.

Kim et al.[26] also utilize the sputtering technique for the deposition of Cu_2SnS_3. Monoclinic crystalline structure with a band gap of 1.03 eV was evident from structural and EQE—external quantum efficiency measurements. Mo/CTS/CdS/i-ZnO/AZO/Al was the configuration chosen to fabricate solar cell. Here, CTS plays the role of absorber layer and CdS as buffer layer. i-ZnO is an intrinsic ZnO layer used to bridge between CdS and AZO. This also adjusts the work function of these two layers and promotes charge carrier transport.[26]

TABLE 8.6 Solar Cell Parameters of Device Fabricated by Kim et al.[26]

Fabricated device structure	Mo/CTS/CdS/i-ZnO/AZO/Al
Open circuit voltage (V_{OC})	147.5 mV
Short circuit current (J_{SC})	28.3 mA/cm^2
Fill factor (FF)	32%
Efficiency (η)	1.35%

Most recent study of CTS sample was reported by Akcay et al. in 2023.[27] They worked on developing CTS sample through magnetron sputtering followed by annealing in sulfur atmosphere. Detailed analysis shows that the prepared sample of Cu_2SnS_3 consists of two crystalline phases, monoclinic, and tetragonal. This work along with He et al.[25] proves that effect of deposition time has a great impact on the deposited sample. Annealing was done on the In_2S_3/Cu_2SnS_3 stack and improvement in device performance was also reported in the paper.[27]

TABLE 8.7 Solar Cell Parameters of Device Fabricated by Akcay et al.[27]

Fabricated device structure	Ni/Al/Ni/AZO/i:ZnO/In2S3/Cu2SnS3/Mo/SLG
Open circuit voltage (V_{OC})	0.265 mV
Short circuit current (J_{SC})	37.90 mA/cm^2
Fill factor (FF)	0.31
Efficiency (η)	3.12%

Kanai and Sugiyama[28] reported the enhancement in device performance on the sulfurization process. Increase in all the device parameters owing to decrease in non-radiative recombination centers, enhanced width of depletion region, lengthening in CTS lifetime, etc. They reported an efficiency of 5.24%.[28]

TABLE 8.8 Tabulated Different Studies on CTS Solar Cells.

Sl. No.	Author name	Device structure	Crystalline structure/ band gap	Solar cell parameters	Year	References
1.	Abdel-Latif	SLG/Mo/CTS/ CdS/ZnO/ AZnO/Ni/Al	Monoclinic/ 0.9 eV	Jsc = 35.59 mA/cm^2 Voc = 159.763 mV FF = 34.13% η = 1.94%	2020	[29]
2	Chierchia	SLG/Mo/CTS/ CdS/ZnO/ ZnO:Al/Al	–	Jsc = 26.2 mA/cm^2 Voc = 243 mV FF = 47.9% η = 3.05%	2016	[30]
3	Zhao	Glass/Mo/CTS/ CdS/i-ZnO/ AZnO/Al		Jsc = 35.00 mA/cm^2 Voc = 230 mV FF = 0.496% η = 3.99%	2020	[31]
4	Hossain	Glass/Mo/CTS/ CdS/ZnO/TCO/ Al	Cubic, Tetragonal, and Orthorhombic		2017	[32]
5	Gincy	TO/CTS/InS/Ag	Tetragonal/1.5 eV	Jsc = 5.9 mA/cm^2 Voc = 0.36 V FF = 40% η = 0.84%	2017	[33]
6	Ghorpade	Glass/Mo/CTS/ CdS/ZnO/Al		Jsc = 20.12 mA/cm^2 Voc = 0.327 V FF = 42% η = 2.77%	2016	[34]
7	Aihara	SLG/Mo/ CTS/CdS/ ZnO:Al/Al	Monoclinic / 0.96–1.00 eV	Jsc = 29 mA/cm^2 Voc = 244 mV FF = 38.5% η = 2.7%	2014	[35]

8.6 CONCLUSION

Necessity of energy is the key parameter for all sorts of about said researchers all around the world. Even though numerous materials are introduced into the world of energy research, only a few come to the mainstream of commercialization and are available to solve the energy crisis. Much more researches have to be done to overcome this. One of the materials in its booming stage is CTS. It was only around one decade ago that this material was known in the renewable energy field of research. Suitable optoelectronic property and tunability in band gap are the major highlights of this material. The above-mentioned studies showed that CTS can have a band gap from 0.9 to 1.75 eV. This range of absorption is best suitable for solar energy conversion. Deposition of CTS as thin film form can be achieved through highly precise vacuum-based physical techniques or through cost-effective, non-vacuum chemical routes. CTS crystal structure can also vary according to the deposition condition mainly via deposition temperature or annealing temperature. CTS, p-type materials were introduced into the solar cells via device configuration glass/Mo/CTS/CdS/ZnO/Al, the most commonly used device structure. Gincy et al.[33] introduced an entirely different device structure with InS as the counterpart buffer layer. But the device performance was very low. The amazing nature of CTS devices is the value of short circuit current density. A possibility for better device performance suggested here is that utilizing this CTS along with organic material may result in better device performance. Hybrid solar cell with high voltage of 1.8 V was reported in Optical Materials in 2022.[36] Incorporating CTS into that hybrid device structure may improve the device in its short circuit current density or introduce organic material into the above-mentioned device structure resulting in an improvement in open circuit voltage in the device. Both will be beneficial and can be tried to obtain better solar cells.

KEYWORDS

- **copper tin sulphide**
- **energy consumption**
- **solar energy conversion**
- **physical and chemical methods**
- **inorganic solar cells**

REFERENCES

1. Solar Energy Perspectives: Executive Summary. *International Energy Agency, 2011, Jan 2012.*
2. Energy. Royal Society of Chemistry, *Apr 2, 2014.*
3. https://www.irena.org/Energy-Transition/Technology/Wind-energy
4. https://education.nationalgeographic.org/resource/hydroelectric-energy/
5. https://www.eia.gov/energyexplained/solar/solar-thermal-power-plants

6. Tiwari, D.; Chaudhuri, T.; Shripathi, T.; Deshpande, U.; Sathe, V. Structural and Optical Properties of Layer-by-Layer Solution Deposited Cu_2SnS_3 Films. J. Mater. Sci. Mater. Electron. **2014,** *25* (9), 3687–3694.
7. Sunny, G.; Kartha, C. S.;Vijayakumar, K. P. In *Tuning Of Opto-Electronic Properties Of Cu2SnS3 Thin Films Through Variation Of Stoichiometry*, Solid State Physics AIP Conference of the Proceedings, **2014**; vol *1591*, pp 1750–1752. DOI: 10.1063/1.4873100
8. Chrisey, D. B.;Hubler, G. K. In *Pulsed Laser Deposition of Thin Films*;, John Wiley & Sons, 1994. ISBN: 0-471-59218-8
9. Vanalakar, S. A.; Agawane, G. L.; Kamble, A. S.; Hong, C. W.; Patil, P. S.; Kim, J. H. Fabrication of Cu2SnS3 Thin Film Solar Cells using Pulsed Laser Deposi-tion Technique. *Solar Energy Mater. Solar Cells* **2015,** *138*, 1–8.
10. Thota, N.; Kumar, M.; Vishwakarma, M.; Mehta, B. R. Structural, Electronic and Optical Properties of Pulsed Laser Deposited Cu2SnS3 Photo Absorber Thin Films: A Combined Experimental and Computational Study. *Thin Solid Films* **2019,** *677*, 62–67.
11. Lokhande, A. C.; Chalapathy, R. B. V.; He, M.; Jo, E.; Gang, M.; Pawar, S. A.; Lokhande, C. D.; Kim, J. K. Development of Cu2SnS3 (CTS) Thin Film Solar Cells by Physical Techniques: A Status Review. *Solar Energy Mater. Solar Cells* **2016,** *153*, 84–107.
12. Songür, I. M.; Yorulmaz, U.; Atay, F.; Akyüz, I. The Effect of Sulphur Amount in Sulphurization Stage on Secondary Phases in Cu_2SnS_3(CTS) Films. *Curr. Appl. Phys.* **2021,** *26*, 64–71.
13. https://www.ajaint.com/what-is-sputtering.html
14. Kim, I. Y.; Lee, J. Y.; Ghorpade, U. V.; Surywanshi, M. P.; Lee, D. S.; Kim, J. H. Influence of Annealing Temperature on the Properties and Solar Cell Performance of Cu_2SnS_3 (CTS) Thin Film Prepared using Sputtering Method. *J. Alloys Compd.* **2016,** *688*, 12–17.
15. Olgar, M. A. Improvement in the Structural and Optical Properties of Cu_2SnS_3 (CTS) Thin Films through Soft-Annealing Treatment. *Superlattices Microstruct.* **2020,** *138*, 106366.
16. Hodes, G. Semiconductor and Ceramic Nanoparticle Films Deposited by Chemical Bath Deposition. *Phys. Chem. Chem. Phys*. **2007,** *9* (18), 2181–2196.DOI: 10.1039/B616684A; ISSN: 1463-9084.
17. Switzer, J. A.; Hodes, G. Electrodeposition and Chemical Bath Deposition of Functional Nanomaterials. *MRS Bull.* **2010,** *35* (10), 743–750. DOI: 10.1557/S0883769400051253; ISSN: 1938-1425.
18. Li, J.; Xue, C.; Wang, Y.; Jiang, G.; Liu, W.; Zhu, C. Cu_2SnS_3 Solar Cells Fabricated by Chemical Bath Deposition–Annealing of SnS/Cu Stacked Layers. *Solar Energy Mater. Solar Cells* **2016,** *144*, 281–288.9
19. Shelke, H. D.; Lokhande, A. C.; Kim, J. H.; Lokhande, C. D. Influence of Deposition Temperature on the Structural, Morphological, Optical and Photoelectrochemical Properties of CBD Deposited Cu_2SnS_3 Thin Films. *J. Alloys Compd.* **2020,** *831*, 154768.
20. Singh, V. K. Thin Film Deposition By Spray Pyrolysis Techniques. *J. Emerg. Technol. Innov. Res.* **2017,** *4* (11).
21. He, M.; Lokhande, A. C. ; Kim, I. Y.; Ghorpade, U. V.; Suryawanshi, M. P.; Kim, J. H. Fabrication of Sputtered Deposited Cu_2SnS_3 (CTS) Thin Film Solar Cell with Power Conversion Efficiency of 2.39%. *J. Alloys Compd.* **2017,** *70*, 901–908.

CHAPTER 9

New Perspectives on Micro-Algal Fuel Cell: An Approach Toward Green Energy

SOOSEN SAMUEL M[1], BINCY VARGHESE[1], BIVITHA[1], and BINOY T THOMAS[2]

[1]*Post Graduate and Research Department of Physics, Catholicate College, Pathanamthitta, Kerala*

[2]*Post Graduate and Research Department of Botany, Catholicate College, Pathanamthitta, Kerala*

ABSTRACT

The most important challenge in front of the future generation will be the utilization of energy along with sustainable development. Algal fuel cells (AFCs) can pave a new milestone in this current scenario. The working of AFCs is similar to that of the microbial fuel cells (MFC) except that the photosynthetic organism, the algae, is used in the cathode chamber where it acts as an oxygen provider and consumes carbon dioxide. Algae can also be used in the anode chamber as a substrate. When compared to other technologies of energy production, in addition to electricity production, AFCs have multiple uses including wastewater treatment, production of value-added products, carbon dioxide removal, and so on. This review paper gives concise information about AFCs. The major factors affecting the AFCs are temperature, pH, electrode, light, carbon dioxide, substrate, and proton exchange membrane (PEM). But the major challenge is the commercial viability. Optimum conditions of these factors along with interdisciplinary work with biotechnology, modeling, nanoparticles, and electrochemistry can bring a breakthrough shortly.

The Food-Energy-Water Triangle for Sustainable Development: Integrative and Synergic Solutions.
Pedro Aguilar-Zárate, Mónica Chávez-González, Ann Rose Abraham, and A. K. Haghi (Eds.)

9.1 INTRODUCTION

9.1.1 INTRODUCTION

Energy is the key factor that drives any country's sustainable development and financial growth.[1] Fossil fuel provides about 86% of total energy production in the world to meet worldwide energy needs but it also contributes to environmental degradation due to the emission of greenhouse gases that lead to global warming.[2] To conquer these problems, renewable energy sources like wind, solar, and other carbon-neutral sources are developed but need more research toward lowering installation and operational costs and improving cycle yield and efficiencies.[2] So, it is necessary to develop alternative sustainable systems without net carbon dioxide emission, with no need for high installation and running expenses but with high energy transformation efficiencies. Novel methods to produce electricity are MFC, and photosynthetic algal microbial fuel cells (PAMFC).

MFC are a novel, promising, cost-effective, environment-friendly technology for sustainable energy production.[3] The idea of generating electric current from bacteria was first reported by Potter in 1911 from living cultures of E. coli and Saccharomyces using platinum electrodes.[4,5] MFC are devices that produce electricity from the anaerobic oxidation of organic matter by electroactive bacteria.[6] MFC consist of a cathode and anode chamber connected through a PEM. The bacterial community is placed along with organic substrates which are placed in an anode chamber to produce electrons during the biological process. The electrons are released and then transferred to the cathode through an external circuit that generates a current. The anode compartment should be maintained under anaerobic conditions because oxygen inhibits electricity generation whereas the cathode compartment is exposed to an oxygen atmosphere.[6] The protons generated at the anode by the anaerobic degradation of organic matter diffuse through the PEM to the cathode.[7] Oxygen is mechanically aerated into the cathode chamber as an electron acceptor. The electrons, protons, and oxygen react to form water.[8] MFC offer many advantages like direct energy production from a substrate, high efficiency at room temperature, do not require any additional energy source, reliable baseload power, low cost of feedstock storage, and low pollution effect.[9] MFC for bioelectricity production are advantageous due to sustainable long-term power generation. However, the performance of MFC is limited by the aeration cost in the cathode chamber.[10]

The generation of electricity from MFC using oxygen has two major drawbacks—the slow rate of oxygen reduction on the surface of graphite or carbon electrodes that prompts a high cathodic overpotential and the development of pH variation from acidic conditions at the anode to alkaline conditions at the cathode, resulting from the gradual diffusion of protons through the PEM in competition with other cations, thereby diminishing the accessibility of protons at the cathode and can harm the anodic biofilm.[7]

The photosynthetic algal microbial fuel cell is a new development of MFC in which microalgae or cyanobacteria are used to produce oxygen through photosynthesis with the uptake of CO_2.[2,8] Different algal species like *Chlorella vulgaris, Chlorella pyrenoidosa, Spirulina, Scenedesmus obliquus, Golenkinia* sp., *Synechococcus* sp., *Chlorococcum* sp., etc. have been used in different systems. Algae play an important role in the aquatic food chain as it provides oxygen and consumes CO_2.[6] All photosynthetic organisms including algae produce oxygen during photosynthesis as well as photolysis of water. This property of algae can be used to make oxygen in the cathode chamber of fuel cells.[11] Algae is rich in carbohydrates, fiber, and protein[12] that could be formed into value-added products like biofuel, feeds, etc.[13] Microalgae can be grown in open ponds, closed containers, or wastewater.[13,14] Microalgae help to remove nutrients, organic compounds, pathogens from wastewater and also reduce the need for external aeration.[15] Microalgae are considered to be a feedstock of third-generation biofuels like bioethanol, biodiesel, biohydrogen, biogas.[16,17] Biofuels are the best replacement for fossil fuels considered to be non-toxic, renewable fuels.[18] Integrating MFC with algal metabolism helps achieve

1. sustainable wastewater treatment with low greenhouse gas emissions and other waste products,
2. direct conversion of chemical energy to electrical energy,
3. use of CO_2 by microalgae,
4. production of value-added products.[13]

This review paper emphasizes the AFC technology, materials, and methods, factors affecting performance, applications, and challenges.

9.1.2 AFC

The increasing demand for renewable energy tends to incorporate phototrophic microorganisms into MFC for generating electricity.[19] The incorporation of photosynthetic organisms into MFC happens in the past years with increasing research in MFC.[19] The oxygen production of photosynthetic organisms is sustained by light and enhances power generation by 42%.[20] The integration can be done in three different ways:

1. Phototrophic microorganisms work as a substrate for donating electrons[19]
2. Photoheterotrophic microorganisms act as a catalyst for half-reaction in the anode[19]
3. Photoautotrophic microorganism gives oxygen as an electron acceptor for the half-reaction in the cathode, and helps to fix carbon dioxide content, nutrient removal, and biomass formation.[10,19,21]

In 2009, Powell et al. designed an AFC in which Saccharomyces cerevisiae in the anode chamber and Chlorella vulgaris in the cathode chamber.[22] The algae act as an electron acceptor by taking CO_2 for photosynthesis and they were able to produce a power density of 2.7 mW/m^2.[9] The development of microalgal techniques to remove the defects of typical MFC by exploiting cathodic reactions leads to economic betterment and nutrient removal capability.[23] The collaborative link between MFC and microalgal reactors is known as AFC.[23]

The main reactions that take place in the device are:

Photosynthesis of organic matter[24]

$$nCO_2 + nH_2O \xrightarrow{hv,\ algae} (CH_2O)_n + nO_2 \quad (9.1)$$

Reduction of oxygen in the cathode[24]

$$nO_2 + 4ne^- + 4nH^+ \rightarrow 2nH_2O \quad (9.2)$$

Oxidation of organic substrate on the anode to generate electrons[24]

$$(CH_2O)_n + nH_2O \rightarrow nCO_2 + 4nH^+ + 4ne^- \quad (9.3)$$

The carbon dioxide formed in the anode chamber is given to the cathode chamber for algal growth through photosynthesis.[24,25] Bioelectricity production from algae is an excellent approach to drawing energy from sunlight in a profitable and conserving manner.[3] This can be implemented by the synchronization of photosynthesis of algae with MFC.[3] The photosynthetic algae can be used either at the cathode to supply oxygen (photo MFC) or at the anode as a substrate for the growth of microorganisms.[3] David et al. developed a photosynthetic algal microbial fuel cell incorporating a photobioreactor that produced a maximum power density of 110 mW/m^2.[26] A study by Das et al. found that quorum sensing molecules extracted from anaerobic sludge increased the productivity of the algae Chlorella sorokiniana, producing power of 24.09 mW/m^2.[27] Chambath et al. showed that the presence of

Chlorella vulgaris in the cathode compartment of a double chamber MFC doubled the maximum power density compared to a mechanically aerated MFC.[8]

9.1.2.1 ALGAE IN THE CATHODE CHAMBER

The utilization of chemicals in the cathode is not a good choice, as it contributes to high costs as well as causes environmental contamination.[3] Air cathodes are used, in which oxygen is used as an electron acceptor but it needs expensive catalysts like platinum.[3] The incorporation of algal photosynthesis with MFC is a reasonable way to overcome all these limitations.[3] Zhang et al. used Chlorella vulgaris as an oxygen provider in the cathode compartment with nutrient ejection and produced a power density of 68 ± 5 mW/m^2.[28] Mechanical aeration in the cathode chamber costs about 50% of the total functioning cost.[2] Therefore, an algal-assisted cathodic chamber is more energetic due to algal photosynthesis and provides oxygen in the cathode compartment.[2] Studies show that algal-assisted cathodes produce 1.3 times high power as compared to non-algal-assisted cathodes.[29] This method enables the use of the surplus biomass from algae as an electron donor in an anode chamber and carbon dioxide from the anode as a carbon source for the growth of algae, which enables electricity production along with carbon dioxide capture.[2] Kakkarla et al. produced a power density of 153 mW/m^2 using Scenedesmus obliquus in the cathode chamber which is 32% greater than the value obtained in the mechanical aerated system.[30] The output generated from an MFC with an algal-assisted cathode is greater than that of an air cathode due to the high dissolved oxygen concentration from photosynthesis.[31]

9.1.2.2 ALGAE IN THE ANODE CHAMBER

The use of algae/algal biomass as substrate in the anode compartment is another efficient way of utilizing the capability of algae and is associated with wastewater treatment and the microorganism in wastewater acts as inoculum.[2,3] Activated sludge, lipid-extracted algae, and pretreated algal biomass are also used in anodic compartments but exhibit low output.[32] The method of using algae as a substrate is analogous to the use of wastewater as a substrate in the anode chamber.[2] A study by Velasquez et al. used two algal species Chlorella vulgaris and Ulva lactuca as substrates in an anode chamber and produced a maximum power density of 980 mW/m^2 associated with *Chlorella vulgaris*.[33] Activated sludge acts as a nutrient source and inoculum with algal biomass is used as substrate.[32] Bacteria such as *Alcaligenes faecalis, Enterococcus gallinarum, Pseudomonas aeruginosa,* and *Shewanella* use algal biomass as nutrients and transfer electrons to the electrodes.[32] Varied concentrations of algal biomass are tested and found that 5 g/L of dry algal biomass produces a maximum power of 1.78 W/m^2.[32] Lipid-extracted algal biomass from the cathode chamber acts as a substrate producing power of 2.7 W/m^3 without any additional carbon supplement.[34] Algal biomass is characterized by a strong cell wall that prevents the utilization by bacteria and this can be solved by pre-treatment techniques like sonication and thermal pre-treatment, but pre-treatment provides low output as compared to non-treated algal biomass.[32] Wang et al. showed that alkaline pretreated algal sludge produced more power density and energy harvesting as compared to raw sludge.[35] Walter et al. used a flow-through technique in which fresh algal biomass is continuously fed into the MFC rather than a pretreated one.[36]

9.1.3 COMPONENTS OF AFC

The major components of AFC are electrodes (anode and cathode), substrate, and PEM.[5]

9.1.3.1 ELECTRODE

The proper selection of electrodes is important in the working of an AFC based on bacterial adherence, electron transfer, and performance efficiency.[37] Electrode designing is a great challenge in the construction of MFC economical manner[3] and also the biofilm developed over electrodes affects the internal resistance and mass travel kinetics.[20] The electrodes should have a high surface area, less resistance, high conductivity, low cost, chemical, and mechanical stability, bio-compatibility, non-corrosive,[3,5,38] porosity, non-fouling property, and be readily available.[39] The resistance can be decreased by increasing the surface area by keeping the volume constant and thus increasing the output of AFC.[37] The high surface area gives more regions for reactions.[37] High electrical conductivity is necessary for electrodes as it makes the flow of electrons with low resistance.[37] The material for electrodes should be durable and must be stable in acidic and alkaline conditions.[37]

Carbon-based electrodes are widely used due to their stability, conductivity, and support of bacterial adherence to form biofilms.[29] Metal electrodes like stainless steel mesh, tin, copper, aluminum, granular graphite, carbon materials, etc. are also used.[3] Granular graphite is less expensive with greater conductivity even if it has a low surface area, granular activated carbon has a high amount of microporosity and catalytic ability.[3] Carbon nanotubes are also used which increases electron transfer ability and electrode surface area[3]

9.1.3.2 ANODE

The anode material influences the biofilm formation and the electron transfer mechanism.[37] Different types of materials are used for the construction of anode which include carbon cloth, carbon paper, carbon foams, reticulated vitrified carbon, graphite rods, graphite granules, graphite brush, conductive polymers, and stainless steel.[37,39] The system with stainless steel mesh as anode and graphite plate as cathode produces more power with the best result of chemical oxygen demand.[6] Metal anodes like stainless steel mesh can be used but copper is toxic to bacteria.[5,40] The graphite plate or rod is less expensive, easy to handle, and has a good surface area, a much larger surface area can be attained by graphite-felt electrodes.[5] A comparative study by Thorne et al. found that the power density of a ceramic anode was 16 times greater than a carbon anode.[41]

9.1.3.3 CATHODE

Most of the materials used as an anode can also be used as the cathode,[39] but a suitable cathode also possesses properties like high mechanical strength, catalytic activity, and high electric and ionic conductivity.[37] For carbon-based electrodes, an extra catalyst is needed due to its low catalytic activity.[37] Different types of cathode materials used are carbon cathode with platinum catalyst and carbon cathode with non-platinum catalyst.[39] Platinum is the commonly used catalyst but it is highly expensive, pH-sensitive, non-sustainable, and leads to the development of non-Pt-based catalysts.[37] Cathode-specific surface area is a major parameter that determines the performance of AFC.[2] Biocathode is advantageous over abiotic cathode because the operation cost can be lowered as a metal catalyst or electron mediators are replaced by algae to produce oxygen through the photosynthetic process, thereby removing the cost of external oxygen supply and also these algae fix carbon dioxide.[42] Biocathodes increase sustainability and microbial metabolism can be used to produce value-added products as well as remove unwanted compounds.[42] Wang et al. showed that the power density of a sediment microbial fuel cell with algae in the cathode compartment produced 21 mW/m^2 which increased to 38 mW/m^2 with the addition of carbon nanotube coating in the cathode, which is 2.4 times higher than the bare cathode.[43]

9.1.3.4 MEMBRANE

Most double-chambered MFC need a membrane to separate the anode and cathode compartments.[5,39,40] Membranes are not required for single-chambered MFC and naturally separated systems like sediment MFC.[40] The membranes are permeable to protons so that the protons developed at the anode can be moved to the cathode.[39] High conductivity and less resistance of protons allow it to move from anode to cathode chamber but do not allow the movement of substrate and oxygen between chambers.[3] The membrane should be designed in a way that it overcomes internal resistance, pH splitting, and oxygen permeability.[3] Various types of membranes are used like ultrafiltration membrane, PEM, Nafion, ultrex, anion exchange membrane, bipolar membrane, and ion exchange membrane.[3,39] Nafion is commonly used because it provides more specific conductivity for protons, increased Coulombic efficiency, and a microorganism lifetime.[3] Instead of Nafion membrane, claywire membrane, ceramic membrane, and natural rubber membrane can also be used.[44] The major drawback of membranes is their high cost and increase in internal resistance, thereby decreasing their power.[39]

9.1.3.5 SUBSTRATE

The most dominant parameter considered in electricity generation is the usage of a substrate.[45] A substrate is a substance placed in an anode compartment and acts as an electron donor, by undergoing oxidation.[5,46] Commonly used substrates are glucose, acetate, wastewater, dairy wastewater, synthetic wastewater, industrial wastewater, and landfill leachate.[47] Acetate is a simple substrate that is used as a carbon source to prompt electroactive bacteria, inert to other microbial processes like fermentation and methane formation at room temperature.[47] Liu et al. showed that the power density with acetate was 506 mW/m^2 up to 66% higher than that produced with butyrate 305 mW/

m^2.[47] Apart from these algal in powder or dry form is also used as substrate in the anode chamber. Khandelwal et al. showed that lipid-extracted algal biomass from *Chlorella vulgaris* as substrate increased the power density to 2.7 W/m^3.[34]

9.1.3.6 CATALYST OR CATHOLYTES

Oxygen is used as an electron acceptor in the cathode chamber due to its high redox potential,[38] abundant availability, sustainability, and the end product formed is water.[5,7] However, a catalyst is required when oxygen is used since the reaction is slow.[5] The commonly used catalyst is platinum but it is expensive, leading to the development of catholytes. The most widely used catholytes are ferricyanide and permanganate.[39] The usage of ferricyanide is not promoted because of environmental safety issues and also decreases energy output.[30] Potassium ferricyanide acts as an electron acceptor in the cathode compartment which improves power generation and oxygen reduction rate.[46] Other noble metals-based catalysts like Ru coated on carbon nanotube, silver nanoparticle coated on carbon paper, and Ti -TiO_2 composites are used.[48]

9.2 TYPES, ANALYSIS, FACTORS

9.2.1 DIFFERENT TYPES OF CONFIGURATION

The configuration of AFC has a crucial role in deciding the power density and performance of a system.[49] Different types of configurations are as follows:

9.2.1.1 SINGLE-CHAMBER AFC

In single-chamber AFC, there is only one chamber, where bacterial and algal co-cultures are grown together. Here membranes are absent. A biofilm of microalgae is formed on the anode and usually, some air cathodes are configured by them. Algae simultaneously consume CO_2 emitted by both autotrophic and heterotrophic organisms. Single-chamber AFCs are worthwhile and easy to handle.[49] The system developed by Nishio et al. by making use of Chlamydomonas Reinhardtii, Lactobacillus, and Geobacter produces electricity of 0.078 W/m^2 is an example of Single-chamber AFC.[50]

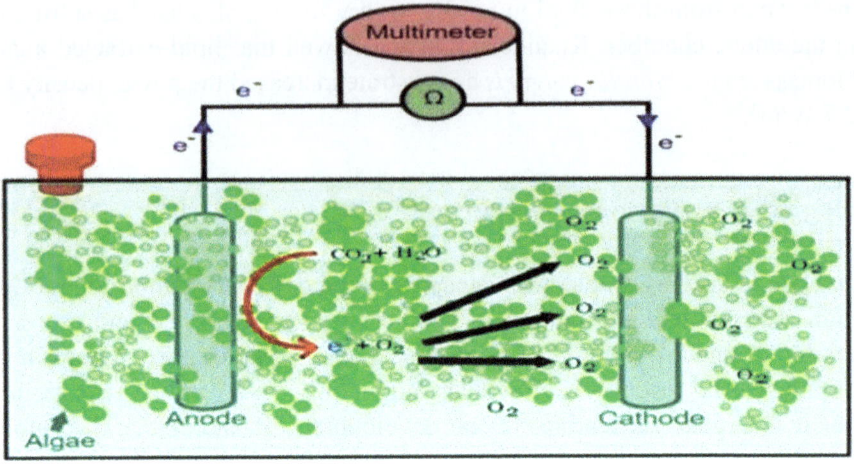

FIGURE 9.1 Single-chambered AFC.

9.2.1.2 DUAL-CHAMBERED AFC

Microalgae and bacteria are grown in two chambers separated by a membrane. The high internal resistance of membranes is a demerit of this method.[49] The system developed by Lakaniemi et al. using *Chlorella vulgaris* and *Dunaliella tertiolecta* with graphite plate electrodes produced an output of 0.015 W/m^2 is an example of dual Chamber AFC.[50]

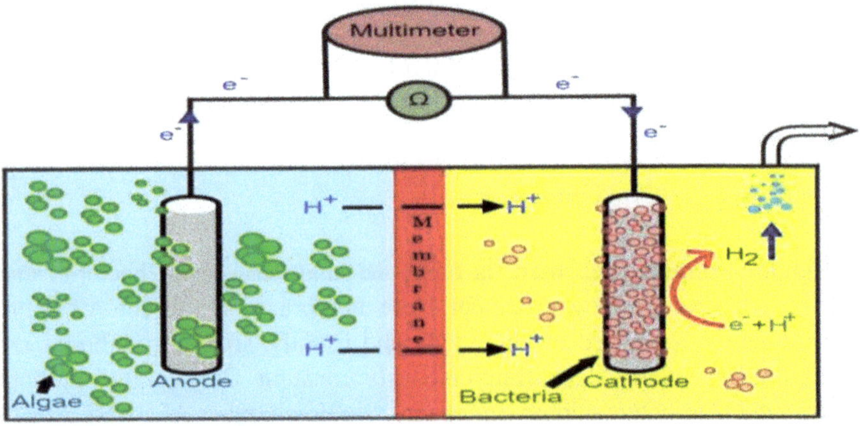

FIGURE 9.2 Dual-chambered AFC.

9.2.1.3 THREE-CHAMBERED AFC

A third chamber, having saltwater, is present in addition to the chambers of anode and cathode. Stress to production power is given by this middle chamber. The movement of cations toward the cathode and anion toward the anode results in partial desalination in the third chamber.[49]

9.2.1.4 ALGAL-BASED MICROBIAL CARBON CAPTURE CELLS [MCC]

In Algal-based Microbial Carbon Capture Cells, algae present in the cathode chamber causes carbon sequestration via carbon dioxide fed from the anode chamber. Algal-based MCC is better than MFC in many ways including improved voltage outputs, emission of CO_2 from both anode and cathode compartments are minimal, and concentration in the cathode compartment is high. Scaling up of MCC can be done by increasing the specific surface area of the cathode and spacing between electrodes. The cost of external aeration can be evaded in algal-based MFC because the voltage produced in MCC with external aeration and those without are comparable, 630 ± 30 mV and 610 ± 50 mV, respectively.[2]

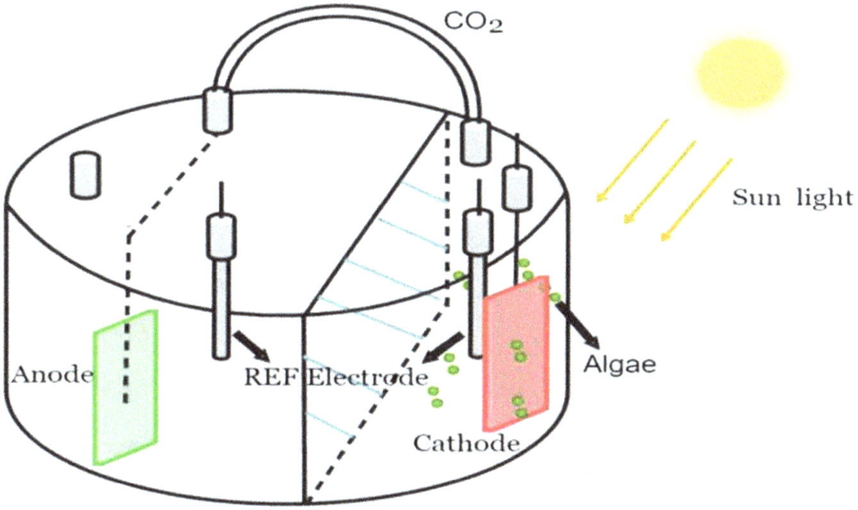

FIGURE 9.3 Schematic representation of MCC with CO_2 exchange from the anode compartment.

9.2.1.5 SEDIMENT AFC

In sediment AFC, the anode is placed in the sediment and the cathode compartment having microalgae is put on top of the sediment. CO_2 produced by anodic bacteria will be consumed by algae in the cathode chamber. By utilizing the oxygen thus produced, power is generated.[49,50] In sediment MFCs for obtaining better electricity production, carbon nanoparticles, and carbon nanotubes are used as sediment.[2]

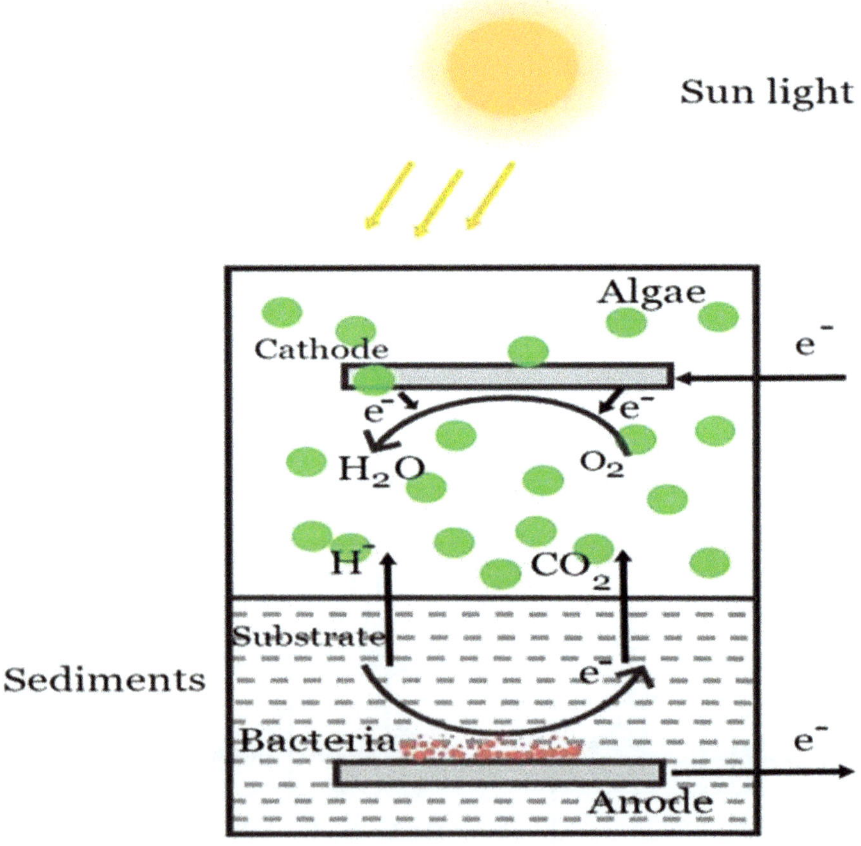

FIGURE 9.4 Schematic representation of sediment AFC.

Some other ways of incorporating algae in MFCs are as follows:

i. Anode-catalyzed microalgae MFCs

Electron production and transfer of electrons to the anode in the anode compartment can be done without electromediators by making use of photosynthetic bacteria or microalgae.[50] Under optimal conditions, Chang et al. constructed an MFC of the peak power density of 6030 mW/m^2, by utilizing Chlorella pyrenoidosa as an electron donor in the anode compartment.[50]

ii. Algae as substrate supplier in dark MFCs anodic end

By incorporating live algae and dry algal biomass as feedstock in the anode compartment we can reduce pollution and cost. Algal biomass is rich in proteins, carbohydrates, and lipids.[50] Lakaniemi et al. reported a peak power density of 5.3mW/m^2 with Dunaliella tertiolecta and a thrice higher peak density of 15 mW/m^2 with Chlorella vulgaris at the cathode.[51]

iii. `Integrated photo-bioelectrochemical systems

As the name suggests, the incorporation of MFCs with an algal-based bioreactor results in integrated photo-bio electrochemical systems.[50] These systems can be used in the production of algal biomass and electricity.[50] Xiao et al. constructed such a system with a peak power density of 2.2W/m^3 associated with a COD of 92%, and removal of phosphate and ammonium nitrogen at 82% and 98%, respectively.[50]

9.2.2 *ANALYSIS TECHNIQUES*

The cell voltage is measured at regular intervals using a digital multimeter or a data acquisition system. Various instruments used are Keithley 2000 multimeter,[52–54] Keithley instruments 2700,[55,56] PCI-6280,[57] fluke multimeter,[22] METRAVI-45,[34] 16- channel ADC-24 Picolog recorder data logger,[58] and so on.

Using Ohm's law current can be calculated[44]

$$I = V_{emf}/R_{ext}$$

where I – current

V_{emf} – measured closed circuit voltage

R_{ext} – external resistance

Power can be calculated using the equation[39]

$$P = I \times V$$

A polarization curve is used to characterize current as a function of voltage.[39] By varying resistances on the circuit different voltages are obtained and are plotted. The internal resistance of MFC can be obtained by the polarization slope method.[39]

Power density = $(I \times V)$/ anodic volume[59]

Or power density $= \frac{V^2}{A \times R}$, where A can be anode/cathode surface area[8]

Or power density= power/anode surface area[60]

The maximum power density value is obtained from the power curve plot of current density as a function of power density.[39] Coulombic and energy efficiency can be calculated to know the efficiency of the system.

Coulombic efficiency is defined as the fraction of electrons recovered as current versus that in the starting organic matter.[39]

$$C_E = \frac{Coulombs\,recovered}{total\,Coulombs\,in\,substrate}$$

$$C_E = \frac{M_s \int_0^t I\,dt}{F b_{es} V_{An} \Delta C}$$

where ΔC – substrate concentration changes over the batch cycle

F – Faraday's constant

M_s –molecular weight of the substrate

V_{An} – the volume of liquid in the anode compartment

Energy efficiency is defined as the energy recovered in the system compared to the energy content of the starting material; expressed as[39]

$$\eta_{MFC} = \frac{\int_0^t E_{MFC} I dt}{\Delta H n_s}$$

ΔH –the heat of combustion

n_s – amount (mol) of substrate added.

Other instruments like lux meter, oximeter, COD photometer, DO meter, pH meter, the photometer are also used. Using modeling, we can predict the performance of devices under different conditions.[1,61]

9.2.3 FACTORS AFFECTING THE PERFORMANCE OF THE AFC

The different factors that affect the performance of an algal MFC are electrochemical reactions that occur between anode and cathode, microbes in the anode chamber, the porosity of PEM, supply of oxygen, fuel type, the concentration of substrate, operating factors like pH, temperature.[62]

9.2.3.1 TEMPERATURE

The temperature changes the conductivity of the substrate and the activity of microbes used.[6] The high operating temperature is advantageous for AFCs which increases current density and Coulombic efficiency and decreases internal resistance. An optimum temperature for the working of fuel cells is found to be 35°C, but it differs from microbes used in fuel cells.[3] The internal resistance decreases with an increase in temperature as the ionic conductivity increases with temperature.[6] A comparative study by Saba et al. found that a maximum power density of 193.8 mW/m^3 occurs at 30°C for single-chambered as compared to 174 mW/m^3 at 35°C in batch mode.[3] The dissolved oxygen solubility also lowers as temperature increases.[29] Thus, the operation of a fuel cell depends on temperature and it cannot be operated at low temperature as the reaction inside the chamber takes place at a slow pace.[62]

9.2.3.2 PH

The anode chamber favors low concentration and neutral pH whereas the cathode chamber can be of the same value of pH that is pH of 7 or a little higher.[3] Any variation in concentration of oxygen level, electron, or proton leads to pH gradient and reduces the output. The introduction of buffers like carbonate, phosphate, borax, saline catholyte, etc. helps to maintain pH.[3]

9.2.3.3 ELECTRODE

An MFC configuration using *Chlorella vulgaris* with stainless steel mesh as anode and graphite plate as cathode yields a power of 9 mW/m^2.[6] Polyaniline (PANI) graphite nanosheet MFCs have higher electricity than the normal cathode. This is because PANI has the catalytic ability of oxygen at room temperature.[46] Potassium ferricyanide is one of the electron acceptors used in the cathode chamber which increases power production and oxygen reduction rate.[46] The incorporation of a rotating electrode results in higher power density [49 mW/m^2] compared to an ordinary cathode [29 mW/m^2] by increasing oxygen utilization in the chamber.[46] Carbon nanotubes can be used as an electrode that promotes electron transfer through their high conductivity and large surface area.[43] The surface area of the electrode should be large enough so that power produced within the chamber can be used to run other components that are connected to the fuel cell.[62] Partially submerged electrodes [50%] can retrieve oxygen from the air to obtain a high working voltage of 297 ± 26 mV, 99 mW/m^2.[31] AFC anode coated with conductive polymers increases the voltage rate and light amplitude response.[57] Covered anodic chambers have high voltage, power density, Coulombic efficiency as compared to exposed anodic chambers.[63]

9.2.3.4 LIGHT

There are few studies related to the influence of light on the performance of MFC. Autoclave or irradiating algal cells with microwave radiation for 20 min enhances electricity production.[19] Wang et al. reported that output voltage and power improve as illumination increases from 5000 to 10,000 lux and after that, it decreases.[23] The power density obtained at 10,000 lux is 466.9 mW/m^3 and at 15,000 lux is 337.7 mW/m^3 using *Chlorella vulgaris*.[23] Wu et al. found 1.5 times high power generation when light intensity increased from 1500 lux to 3500 lux.[29] The light intensity between 5000 and 6000 lux is suitable for Chlorella photosynthesis and improved power production.[64] Thus, intermediate illumination promotes a more efficient and stable operation of AFC.[10] Power density from photosynthetic microalgae depends on light and its intensity because it affects chlorophyll production, photosynthesis, and the stomatal opening of algal cells.[65] Xing et al. found that irradiation of fuel cells to incandescent light improves power density by 8–10% for glucose substrate and 34% for acetate as compared with fuel cells under a dark atmosphere. Other studies investigated the effect of light intensity and its types on the growth of *Chlorella vulgaris*.[49] They analyzed that fluorescent light is more effective in the growth of algae of 0.029 g dry cell weight with a light intensity of 9 W/m^2.[49] Nannochloropsis sp. shows a maximum growth rate of 0.64 d^{-1} for blue, 0.51 d^{-1} for red, 0.54 d^{-1} for green, 0.58 d^{-1} for white light.[49] A study by Lan et al. about the effect of the light source and the intensity in a photosynthetic microbial fuel cell using *Chlamydomonas reinhardtii* for the different intensities of blue and red light found that red LED light provides more power density 12. 95 m W/m^2 than blue light.[9] More the light intensity, the better the working of AFC.

9.2.3.5 *CARBON DIOXIDE*

Optimum concentration is necessary for the stable growth of microalgae.[66] An increase in CO_2 concentration improves the amount of biomass and lipid in microalgae.[67] The cells form polyunsaturated fatty acids under high concentrations of CO_2.[49] Studies show that the optimum carbon range for the cultivation of *Chlorella vulgaris* is 5–20% CO_2 level.[66] But in certain cases, a high concentration of CO_2 hurts the growth of algae because the dissolved CO_2 decreases the pH of electrolyte and it is difficult for the algal to overcome this pH.[49]

9.2.3.6 *SUBSTRATE*

Power density depends on the complexity of the substrate, as the substrate is more complex, the power generated is less.[42] Campo et al. found that acetate (506 mW/m^2) has more power as compared to swine wastewater (261 mW/m^2) and domestic wastewater (146 mW/m^2) using *Chlorella vulgaris*. Studies show that easily digestible substrates such as glucose, starch, fatty acid are more favorable rather than complex substrates due to their easy degradation path.[3] The power density was ten times larger than the power produced by complex substrates such as dairy wastewater.[58] A higher concentration of glucose decreases the bacterial culture by hindering proteinaceous enzymes.[6] The mixture of substrates produces more power and current.[62]

9.2.3.7 *PEM*

The ion transfer can be improved by decreasing the thickness of the PEM and increasing the surface-to-volume ratio of electrodes.[68] Membrane fouling, high COD, low pH can affect the performance of PEM.[29] Saar et al. developed biophotovoltaic devices with a membrane-less system by using the laminar flow to separate anolyte, catholyte, and genetically modified algae to increase the power density.[69] PEM supplanted by biofilm made from the algal species *Spirulina platensis* increases the energy output.[46] The absence of a PEM permits the oxygen to flow from cathode to anode and leads to low Coulombic efficiency.[7]

9.3 APPLICATIONS, CHALLENGES

9.3.1 *APPLICATIONS OF AFC*

9.3.1.1 *ELECTRICITY GENERATION*

The primary focus of the majority of AFC-related work is electricity production. Commault et al. recorded that the maximum power density obtained was doubled to 34.2 mW/m^2 by including algae (Chlorella vulgaris) at the cathode.[7] Factors affecting output voltage and power density are discussed in the previous section. The results obtained in various AFCs for electricity generation are tabulated.

9.3.1.2 *WASTEWATER TREATMENT*

Wang et al. developed the first immobilized AFC for the removal of nutrients together with the recovery of bioenergy from domestic wastewater. They recorded removal efficiency of up to 95.9% of NH_4+-N, 93.2% COD, 82.7% of PO_4^{3-} -P, 95.1% of TN, and a peak power density of 466.9 mW/m^3.[23] Under optimal conditions (5000 lux and 25°C), He et al. recorded 92.1% COD removal efficiency and a peak power density of 2572.8 mW/m^3 for AFC constructed by using an immobilized Chlorella vulgaris in a cathode chamber.[70] Cathodic algal biofilm MFC equipped with a bioactive oxygen consuming unit [AB-OCU-MFC] proposed by Elmaadawy et al. had a better yield of algal biomass and removal of COD, NH_4^+-N, TN than algal biofilm MFC alone and control reactors.[71] Yang et al. reported that ABMFC had better performance when compared to AB or MFC alone. They obtained TN, COD, and TP removal of 96.01%, 80.2%, and 91.5% respectively in ABMFC.[56] Mixed indigenous microalgae (MIMA) in the temperature range of 20–30°C showed more than 70% of ammonia and 75% of phosphorus removal.[72]

9.3.1.3 *CO_2 REMOVAL*

The autotrophic growth of microalgae makes use of CO_2 produced in the anodic compartment in light to produce oxygen for the cathodic reaction.[2] Assimilation of CO_2 by algae in the cathode compartment and CO_2 released in the anode compartment affects the removal of CO_2 in MFCs.[2] Assimilation of CO_2 by algae in the cathode compartment and CO_2 released in the anode compartment affects the removal of CO_2 in MFC.[2] pH evaluation can be considered an indicator of CO_2 capturing efficiency.[72]

9.3.1.4 VALUE-ADDED PRODUCTS AND ALGAL BIOMASS

Through various steps, we can obtain value-added products from algal-based MFCs.[2] Harvesting of algal biomass present in the cathode chamber [e.g., formation of biofilm] is the first step.[2] Low-price harvesting techniques like cell immobilization, bio-flocculation, membrane separation, or biofilm formation can be used to separate algal biomass from the cathode chamber.[2] Value-added chemicals are recovered from algae in the next step. Retreatment of the obtained algal biomass may be done in this process.[2] Pigments like β -carotene, chlorophyll, lutein, canthaxanthin, zeaxanthin can be extracted from algal biomass.[2] The lipid-extracted biomass or biomass can be used as animal feed. Biogas can also be produced from algal biomass.[2] Algal polysaccharides obtained from algal biomass can be used for producing synthetic plastic.[29] The growth of algal biomass and thereby value-added products can be increased by stimulating the growth of microalgae under stress conditions like the stress of nutrients [nitrogen] and light.[29] In the production of antioxidants, nutraceuticals, and pharmaceuticals, we can utilize lutein, secondary carotenoids, and β-carotene.[29] Biomass can be utilized for the production of biodiesel, bioethanol, and biohydrogen.[3]

9.3.1.5 ALGAL MXC APPLICATIONS

MXCs are bioelectrochemical systems based on the same principle applied in MFC. Microbial electrolysis cells [MECs] and microbial solar cells [MSCs] can be constructed from MXC. Several problems like bioremediation, desalination, and wastewater treatment can be solved using this technology. Oxygen is required for the removal of organic matter and the role of electron donors is served by algae.[62]

9.3.2 CHALLENGES

Commercial viability is one of the main challenges in AFC.[2] When compared to usual systems for wastewater treatment, MFCs need a large area for effective utilization of light which increases the cost.[2] Usage of algal biomass directly in MFC is complicated by pre-treatment techniques.[2] Infrastructure cost is high. pH membrane gradient adversely affects output voltage and power.[49] The commonly used membranes are of high cost.[46] Microorganisms can be affected if lipid-extracted algae are used as substrates in the anode.[2] Production of oxygen and bioelectricity is adversely affected by the change in light intensity throughout the year.[2] High light intensity may result in photo-inhibition.[29] Mohan et al. observed low activity in summer when compared to winter. At low temperatures reactions inside MFCs occur at a slow rate.[62]

Replacement of cathode by biocathode affects biofilm formation and hence is challenging effective bioelectricity generation.[2] The usage of algae as substrate in anode lowers the output.[29] The presence of oxygenic species at the anode lowers performance.[73] Power generation is also affected by a slow oxygen reduction rate.[49] Another challenge will be the control of the consumption and production rate of O_2, CO_2, and biomass generated in full-scale operation.[25] Algal growth and energy production should be effectively balanced.[46] Finding an optimal solution for the growth of algae is another barrier.[46] MFCs suffer power overshoot at high currents.[25] The affordable ceiling value of anodic biofilm limits the power output. Usually, the ceiling value is of the order of a few hundred mW/m^2.[25] Using a combination of substrates is more efficient than a single substrate.[62] Yang et al. noted that in algal biofilm-assisted MFC, anode reactions are adversely affected by algal sedimentation in the anode chamber.[56] Continuous operation in practical application and algal harvest for biofuels are limited by algal suspensions.[56] Value-added products obtained have to be purified and are still a problem.[46] Particularly power output reduces on increasing the scale.[59] Even though current density was increased three times in a parallel MFC,[74] the electricity production in stacked cells was lower than expected, but energy harvested from algae was higher.[54] Electrode and membrane biofouling is accelerated by using algae biomass as fuel.[54]

9.4 CONCLUSION

An AFC is an emerging technology with multiple advantages. Shortly, AFCs would be a prominent solution to the demand for renewable energies. It is found that algae can assist MFC as a substrate or catalyst in an anode chamber or it can act as an electron acceptor as well as a CO_2 fixer in the cathode chamber. AFCs have wide-ranging applications including wastewater treatment, CO_2 fixation, production of value-added products, and so on. Different studies cited that temperature, pH, electrode, light, CO_2, substrate, and PEM are the factors affecting MFC's performance. Under optimal conditions, incorporating biotechnology, modeling, electrochemistry, and nanoparticles in AFCs would be a promising solution for the future.

9.4.1 *FUTURE PERSPECTIVES*

Researchers are trying to find a new sustainable approach for the production of energy sources with fewer environmental problems. In the case of photosynthetic MFC, although no large-scale technology has been developed, only a few lab-scale systems with satisfactory results have now been developed. The lipid production in algal MFC can be increased by selecting suitable algal strains or bioengineered algae strains and thus lead to maximum power production. The incorporation of nanotechnology into AFC technology will help to enhance the performance of the system. Many researchers replaced conventional carbon electrodes with single-walled or multi-walled carbon nanotubes, and other metal nanoparticles as catalysts thereby achieving stable power density. The efficient use of nanoparticles will help to attain high-performance energy systems. In the future, it is possible to use PMFC technology for large-scale applications like wastewater treatment, energy conversion, and nutrient removal.

KEYWORDS

- **algae**
- **AFC**
- **bioelectricity**
- **photosynthetic microbial fuel cell**
- **sustainable**

REFERENCES

1. Masadeh, M. A.; Kuruvinashetti, K.; Shahparnia, M.; Pillay, P.; Packirisamy, M. Electrochemical Modeling and Equivalent Circuit Representation of a Microphotosynthetic Power Cell. *IEEE Trans. Ind. Electron.* **2017,** *64* (2), 1561–1571. https://doi.org/10.1109/TIE.2016.2618755
2. Arun, S.; Sinharoy, A.; Pakshirajan, K.; Lens, P. N. L. Algae Based Microbial Fuel Cells for Wastewater Treatment and Recovery of Value-Added Products. *Renew.*

Sustain. Energy Rev. **2020,** *132* (December 2019), 110041. https://doi.org/10.1016/j.rser.2020.110041
3. Shukla, M.; Kumar, S. Algal Growth in Photosynthetic Algal Microbial Fuel Cell and its Subsequent Utilization for Biofuels. *Renew. Sustain. Energy Rev.* **2018,** *82* (October 2016), 402–414. https://doi.org/10.1016/j.rser.2017.09.067
4. Pant, D.; Van Bogaert, G.; Diels, L.; Vanbroekhoven, K. A Review of the Substrates used in Microbial Fuel Cells (MFCs) for Sustainable Energy Production. *Bioresour. Technol.* **2010,** *101* (6), 1533–1543. https://doi.org/10.1016/j.biortech.2009.10.017
5. Tekle, Y.; Demeke, A. Review on Microbial Fuel Cell. *Int. J. Eng. Technol.* **2015,** *8* (November), 424–427.
6. Abdulwahab, Y. D.; Mohammed, A. K.; Abbas, T. R. Improving the Performance of Constructed Wetland Microbial Fuel Cell (CW-MFC) for Wastewater Treatment and Electricity Generation. *Baghdad Sci. J.* **2021,** *18* (1), 7–17. https://doi.org/10.21123/bsj.2021.18.1.0007
7. Commault, A. S.; Laczka, O.; Siboni, N.; Tamburic, B.; Crosswell, J. R.; Seymour, J. R.; Ralph, P. J. Electricity and Biomass Production in a Bacteria-Chlorella based Microbial Fuel Cell Treating Wastewater. *J. Power Sources* **2017,** *356,* 299–309. https://doi.org/10.1016/j.jpowsour.2017.03.097
8. Yahampath Arachchige Don, C. D. Y.; Babel, S. Comparing the Performance of Microbial Fuel Cell with Mechanical Aeration and Photosynthetic Aeration in the Cathode Chamber. *Int. J. Hydrogen Energy* **2021,** *46* (31), 16751–16761. https://doi.org/10.1016/j.ijhydene.2020.09.068
9. Baicha, Z.; Salar-García, M. J.; Ortiz-Martínez, V. M.; Hernández-Fernández, F. J.; de los Ríos, A. P.; Labjar, N.; Lotfi, E.; Elmahi, M. A Critical Review on Microalgae as an Alternative Source for Bioenergy Production: A Promising Low Cost Substrate for Microbial Fuel Cells. *Fuel Process. Technol.* **2016,** *154,* 104–116. https://doi.org/10.1016/j.fuproc.2016.08.017
10. Gajda, I.; Greenman, J.; Melhuish, C.; Ieropoulos, I. Self-sustainable Electricity Production from Algae Grown in a Microbial Fuel Cell System. *Biomass and Bioenergy* **2015,** *82,* 87–93. https://doi.org/10.1016/j.biombioe.2015.05.017
11. Altaher, Q.; Alshekhmoham, K.; Naser, F.; MakanI, H.; Salunke, A. Construction of Algal Fuel Cell for Production of Electricity Using Scenedesmus Spp. *Int. J. Adv. Res.* **2017,** *5* (6), 668–676. https://doi.org/10.21474/ijar01/4464
12. Hou, Z. Q.; Lou, M. Y.; Yang, Y. T.; Zhou, J. C.; Liu, L. C.; Cai, J. J. Algae-based Carbons: Design, Preparation and Recent Advances in Their use in Energy Storage, Catalysis and Adsorption. *Xinxing Tan Cailiao/New Carbon Mater.* **2021,** *36* (2), 278–303. https://doi.org/10.1016/S1872-5805(21)60023-18
13. Bolognesi, S.; Cecconet, D.; Callegari, A.; Capodaglio, A. G. Combined Microalgal Photobioreactor/Microbial Fuel Cell System: Performance Analysis Under Different Process Conditions. *Environ. Res.* **2021,** *192* (September 2020), 110263. https://doi.org/10.1016/j.envres.2020.110263
14. Alami, A. H.; Alasad, S.; Ali, M.; Alshamsi, M. Investigating Algae for CO2 Capture and Accumulation and Simultaneous Production of Biomass for Biodiesel Production. *Sci. Total Environ.* **2021,** *759,* 143529. https://doi.org/10.1016/j.scitotenv.2020.143529
15. Muñoz, R.; Guieysse, B. Algal-bacterial Processes for the Treatment of Hazardous Contaminants: A Review. *Water Res.* **2006,** *40* (15), 2799–2815. https://doi.org/10.1016/j.watres.2006.06.011

16. Anto, S.; Mukherjee, S. S.; Muthappa, R.; Mathimani, T.; Deviram, G.; Kumar, S. S.; Verma, T. N.; Pugazhendhi, A. Algae as Green Energy Reserve: Technological Outlook on Biofuel Production. *Chemosphere* **2020,** *242*. https://doi.org/10.1016/j.chemosphere.2019.125079
17. Chen, H.; Li, T.; Wang, Q. Ten Years of Algal Biofuel and Bioproducts: Gains and Pains. *Planta* **2019,** *249* (1), 195–219. https://doi.org/10.1007/s00425-018-3066-8
18. Chi, N. T. L.; Anto, S.; Ahamed, T. S.; Kumar, S. S.; Shanmugam, S.; Samuel, M. S.; Mathimani, T.; Brindhadevi, K.; Pugazhendhi, A. A Review on Biochar Production Techniques and Biochar Based Catalyst for Biofuel Production From Algae. *Fuel* **2021,** *287* (May), 119411. https://doi.org/10.1016/j.fuel.2020.119411
19. Xiao, L.; He, Z. Applications and Perspectives of Phototrophic Microorganisms for Electricity Generation from Organic Compounds in Microbial Fuel Cells. *Renew. Sustain. Energy Rev.* **2014,** *37,* 550–559. https://doi.org/10.1016/j.rser.2014.05.066
20. Gajda, I.; Greenman, J.; Melhuish, C.; Ieropoulos, I. Photosynthetic Cathodes for Microbial Fuel Cells. *Int. J. Hydrogen Energy* **2013,** *38* (26), 11559–11564. https://doi.org/10.1016/j.ijhydene.2013.02.111
21. Liu, T.; Rao, L.; Yuan, Y.; Zhuang, L. Bioelectricity Generation in a Microbial Fuel Cell with a Self-sustainable Photocathode. *Sci. World J.* **2015,** *2015*. https://doi.org/10.1155/2015/864568
22. Powell, E. E.; Evitts, R. W.; Hill, G. A.; Bolster, J. C. A Microbial Fuel Cell with a Photosynthetic Microalgae Cathodic Half Cell Coupled to a Yeast Anodic Half Cell. *Energy Sources, Part A: Recovery, Util. Environ. Eff.* **2011,** *33* (5), 440–448. https://doi.org/10.1080/15567030903096931
23. Wang, Y.; Lin, Z.; Su, X.; Zhao, P.; Zhou, J.; He, Q.; Ai, H. Cost-effective Domestic Wastewater Treatment and Bioenergy Recovery in an Immobilized Microalgal-Based Photoautotrophic Microbial Fuel Cell (PMFC). *Chem. Eng. J.* **2019,** *372* (April), 956–965. https://doi.org/10.1016/j.cej.2019.05.004
24. Zhang, Y.; Zhao, Y.; Zhou, M. A photosynthetic Algal Microbial Fuel Cell for Treating Swine Wastewater. *Environ. Sci. Pollut. Res.* **2019,** *26* (6), 6182–6190. https://doi.org/10.1007/s11356-018-3960-4
25. Lee, D. J.; Chang, J. S.; Lai, J. Y. Microalgae-Microbial Fuel Cell: A Mini Review. *Bioresour. Technol.* **2015,** *198,* 891–895. https://doi.org/10.1016/j.biortech.2015.09.061
26. Strik, D. P. B. T. B.; Terlouw, H.; Hamelers, H. V. M.; Buisman, C. J. N. Renewable Sustainable Biocatalyzed Electricity Production in a Photosynthetic Algal Microbial Fuel Cell (PAMFC). *Appl. Microbiol. Biotechnol.* **2008,** *81* (4), 659–668. https://doi.org/10.1007/s00253-008-1679-8
27. Das, S.; Das, S.; Ghangrekar, M. M. Quorum-sensing Mediated Signals: A Promising Multi-Functional Modulators for Separately Enhancing Algal Yield and Power Generation in Microbial Fuel Cell. *Bioresour. Technol.* **2019,** *294* (September), 122138. https://doi.org/10.1016/j.biortech.2019.122138
28. Zhang, Y.; Noori, J. S.; Angelidaki, I. Simultaneous Organic Carbon, Nutrients Removal and Energy Production in a Photomicrobial Fuel Cell (PFC). *Energy Environ. Sci.* **2011,** *4* (10), 4340–4346. https://doi.org/10.1039/c1ee02089g
29. Saba, B.; Christy, A. D.; Yu, Z.; Co, A. C. Sustainable Power Generation from Bacterio-algal Microbial Fuel Cells (MFCs): An overview. *Renew. Sustain. Energy Rev.* **2017,** *73* (May 2016), 75–84. https://doi.org/10.1016/j.rser.2017.01.115

30. Kakarla, R.; Min, B. Photoautotrophic microalgae Scenedesmus Obliquus Attached on a Cathode as Oxygen Producers for Microbial Fuel Cell (MFC) Operation. *Int. J. Hydrogen Energy* **2014,** *39* (19), 10275–10283. https://doi.org/10.1016/j.ijhydene.2014.04.158
31. Ling, J.; Xu, Y.; Lu, C.; Lai, W.; Xie, G.; Zheng, L.; Talawar, M. P.; Du, Q.; Li, G. Enhancing Stability of Microalgae Biocathode by a Partially Submerged Carbon Cloth Electrode for Bioenergy Production from Wastewater. *Energies* **2019,** *12* (17). https://doi.org/10.3390/en12173229
32. Rashid, N.; Cui, Y. F.; Muhammad, S. U. R.; Han, J. I. Enhanced Electricity Generation by Using Algae Biomass and Activated Sludge in Microbial Fuel Cell. *Sci. Total Environ.* **2013,** *456–457,* 91–94. https://doi.org/10.1016/j.scitotenv.2013.03.067
33. Velasquez-Orta, S. B.; Curtis, T. P.; Logan, B. E. Energy from Algae Using Microbial Fuel Cells. *Biotechnol. Bioeng.* **2009,** *103* (6), 1068–1076. https://doi.org/10.1002/bit.22346
34. Khandelwal, A.; Vijay, A.; Dixit, A.; Chhabra, M. Microbial Fuel Cell Powered By Lipid Extracted Algae: A Promising System for Algal Lipids and Power Generation. *Bioresour. Technol.* **2018,** *247,* 520–527. https://doi.org/10.1016/j.biortech.2017.09.119
35. Wang, H.; Lu, L.; Cui, F.; Liu, D.; Zhao, Z.; Xu, Y. Simultaneous Bioelectrochemical Degradation of Algae Sludge and Energy Recovery in Microbial Fuel Cells. *RSC Adv.* **2012,** *2* (18), 7228–7234. https://doi.org/10.1039/c2ra20631e
36. Walter, X. A.; Greenman, J.; Taylor, B.; Ieropoulos, I. A. Microbial Fuel Cells Continuously Fuelled by Untreated Fresh Algal Biomass. *Algal Res.* **2015,** *11,* 103–107. https://doi.org/10.1016/j.algal.2015.06.003
37. Mustakeem. Electrode materials for microbial fuel cells: Nanomaterial approach. *Mater. Renew. Sustain. Energy* **2015,** *4* (4), 1–11. https://doi.org/10.1007/s40243-015-0063-8
38. Schneider, K.; Thorne, R. J.; Cameron, P. J. An Investigation of Anode and Cathode Materials in Photomicrobial Fuel Cells. *Philos. Trans. A Math. Phys. Eng. Sci.* **2016,** *374* (2061). https://doi.org/10.1098/rsta.2015.0080
39. Logan, B. E. *Microbial Fuel Cell.* John Wiley & sons Publications, 2008.
40. Logan, B. E.; Hamelers, B.; Rozendal, R.; Schröder, U.; Keller, J.; Freguia, S.; Aelterman, P.; Verstraete, W.; Rabaey, K. Microbial Fuel Cells: Methodology and Technology. *Environ. Sci. Technol.* **2006,** *40* (17), 5181–5192. https://doi.org/10.1021/es0605016
41. Thorne, R.; Hu, H.; Schneider, K.; Bombelli, P.; Fisher, A.; Peter, L. M.; Dent, A.; Cameron, P. J. Porous Ceramic Anode Materials for Photo-Microbial Fuel Cells. *J. Mater. Chem.* **2011,** *21* (44), 18055–18060. https://doi.org/10.1039/c1jm13058g
42. Gonzalez del Campo, A.; Perez, J. F.; Cañizares, P.; Rodrigo, M. A.; Fernandez, F. J.; Lobato, J. Study of a Photosynthetic MFC for Energy Recovery from Synthetic Industrial Fruit Juice Wastewater. *Int. J. Hydrogen Energy* **2014,** *39* (36), 21828–21836. https://doi.org/10.1016/j.ijhydene.2014.07.055
43. Wang, D.; Bin, Song, T. S.; Guo, T.; Zeng, Q.; Xie, J. Electricity Generation from Sediment Microbial Fuel Cells with Algae-assisted Cathodes. *Int. J. Hydrogen Energy* **2014,** *39* (25), 13224–13230. https://doi.org/10.1016/j.ijhydene.2014.06.141
44. Xu, C.; Poon, K.; Choi, M. M. F.; Wang, R. Using Live Algae at the Anode of a Microbial Fuel Cell to Generate Electricity. *Environ. Sci. Pollut. Res.* **2015,** *22* (20), 15621–15635. https://doi.org/10.1007/s11356-015-4744-8
45. Polontalo, N. F.; Joelyna, F. A.; Filardli, A. M. I.; Hadiyanto, H.; Zakaria, Z. A. Microalgae Microbial Fuel Cell (MMFC) using Chlorella Vulgaris and "Batik"

Wastewater as Bioelectricity. *E3S Web Conf.* **2021,** *226,* 1–8. https://doi.org/10.1051/e3sconf/202122600032

46. Peng, K. T.; Wang, X.; Peng, G.; Yu, L.; Li, H. Y. Recent Progress in Solar-Induced Direct Biomass-to-Electricity Hybrid Fuel Cell Using Microalgae as Feedstocks. *Front. Bioeng. Biotechnol.* **2021,** *9* (March), 1–7. https://doi.org/10.3389/fbioe.2021.638971
47. Rosenbaum, M.; He, Z.; Angenent, L. T. Light energy to bioelectricity: Photosynthetic microbial fuel cells. *Curr. Opin. Biotechnol.* **2010,** *21* (3), 259–264. https://doi.org/10.1016/j.copbio.2010.03.010
48. Singh, S.; Verma, N. Fabrication of Ni Nanoparticles-dispersed Carbon Micro-nanofibers as the Electrodes of a Microbial Fuel Cell for Bio-energy Production. *Int. J. Hydrogen Energy* **2015,** *40* (2), 1145–1153. https://doi.org/10.1016/j.ijhydene.2014.11.073
49. Sivakumar, P.; Ilango, K.; Praveena, N.; Sircar, A.; Balasubramanian, R.; Sakthisaravanan, A.; Kannan, R. Algal Fuel Cell. *Microalgal Biotechnol.* **2018.** https://doi.org/10.5772/intechopen.74285
50. Saratale, R. G.; Kuppam, C.; Mudhoo, A.; Saratale, G. D.; Periyasamy, S.; Zhen, G.; Koók, L.; Bakonyi, P.; Nemestóthy, N.; Kumar, G. Bioelectrochemical Systems using Microalgae – A Concise Research Update. *Chemosphere* **2017,** *177,* 35–43. https://doi.org/10.1016/j.chemosphere.2017.02.132
51. Lakaniemi, A. M.; Tuovinen, O. H.; Puhakka, J. A. Production of Electricity and Butanol from Microalgal Biomass in Microbial Fuel Cells. *Bioenergy Res.* **2012,** *5* (2), 481–491. https://doi.org/10.1007/s12155-012-9186-2
52. González Del Campo, A.; Cañizares, P.; Rodrigo, M. A.; Fernández, F. J.; Lobato, J. Microbial Fuel Cell with an Algae-assisted Cathode: A Preliminary Assessment. *J. Power Sources* **2013,** *242,* 638–645. https://doi.org/10.1016/j.jpowsour.2013.05.110
53. Lobato, J.; González del Campo, A.; Fernández, F. J.; Cañizares, P.; Rodrigo, M. A. Lagooning Microbial Fuel Cells: A First Approach by Coupling Electricity-Producing Microorganisms and Algae. *Appl. Energy* **2013,** *110,* 220–226. https://doi.org/10.1016/j.apenergy.2013.04.010
54. Asensio, Y.; Fernandez-Marchante, C.; Villaseñor, J.; Lobato, J.; Cañizares, P.; Rodrigo, M. A. Algal Biomass as Fuel for Stacked-MFCs for Profitable, Sustainable and Carbon Neutral Bioenergy Generation. *J. Chem. Technol. Biotechnol.* **2018,** *93* (1), 287–293. https://doi.org/10.1002/jctb.5354
55. Kondaveeti, S.; Choi, K. S.; Kakarla, R.; Min, B. Microalgae Scenedesmus Obliquus as Renewable Biomass Feedstock for Electricity Generation in Microbial Fuel Cells (MFCs). *Front. Environ. Sci. Eng.* **2014,** *8* (5), 784–791. https://doi.org/10.1007/s11783-013-0590-4
56. Yang, Z.; Pei, H.; Hou, Q.; Jiang, L.; Zhang, L.; Nie, C. Algal Biofilm-Assisted Microbial Fuel Cell to Enhance Domestic Wastewater Treatment: Nutrient, Organics Removal and Bioenergy Production. *Chem. Eng. J.* **2018,** *332,* 277–285. https://doi.org/10.1016/j.cej.2017.09.096
57. Yongjin, Z.; Pisciotta, J.; Billmyre, R. B.; Baskakov, I. V. Photosynthetic Microbial Fuel Cells with Positive Light Response. *Biotechnol. Bioeng.* **2009,** *104* (5), 939–946. https://doi.org/10.1002/bit.22466
58. Salar-García, M. J.; Gajda, I.; Ortiz-Martínez, V. M.; Greenman, J.; Hanczyc, M. M.; de los Ríos, A. P.; Ieropoulos, I. A. Microalgae as Substrate in Low Cost Terracotta-Based Microbial Fuel Cells: Novel Application of the Catholyte Produced. *Bioresour. Technol.* **2016,** *209,* 380–385. https://doi.org/10.1016/j.biortech.2016.02.083

59. Khandelwal, A.; Chhabra, M.; Yadav, P. Performance Evaluation of Algae Assisted Microbial Fuel Cell Under Outdoor Conditions. *Bioresour. Technol.* **2020,** *310* (February), 123418. https://doi.org/10.1016/j.biortech.2020.123418
60. Naina Mohamed, S.; Ajit Hiraman, P.; Muthukumar, K.; Jayabalan, T. Bioelectricity Production from Kitchen Wastewater Using Microbial Fuel Cell with Photosynthetic Algal Cathode. *Bioresour. Technol.* **2020,** *295,* 122226. https://doi.org/10.1016/j.biortech.2019.122226
61. Tanneru, H. K.; P, R. S. M.; Ramanan, A. V.; Shahparnia, M.; Packirisamy, M.; Pillay, P.; Williamson, S.; Juneau, P.; Rengaswamy, R. Integrated Bio-electrochemical Model for a Micro Photosynthetic Power Cell [Online]. October **2016.** https://doi.org/10.1142/S2339547816500114
62. Enamala, M. K.; Dixit, R.; Tangellapally, A.; Singh, M.; Dinakarrao, S. M. P.; Chavali, M.; Pamanji, S. R.; Ashokkumar, V.; Kadier, A.; Chandrasekhar, K. Photosynthetic Microorganisms (Algae) Mediated Bioelectricity Generation in Microbial Fuel Cell: Concise Review. *Environ. Technol. Innov.* **2020,** *19,* 100959. https://doi.org/10.1016/j.eti.2020.100959
63. Juang, D. F.; Lee, C. H.; Hsueh, S. C. Comparison of Electrogenic Capabilities of Microbial Fuel cell with Different Light Power on Algae Grown Cathode. *Bioresour. Technol.* **2012,** *123,* 23–29. https://doi.org/10.1016/j.biortech.2012.07.041
64. Bazdar, E.; Roshandel, R.; Yaghmaei, S.; Mardanpour, M. M. The Effect of Different Light Intensities and Light/dark Regimes on the Performance of Photosynthetic Microalgae Microbial Fuel Cell. *Bioresour. Technol.* **2018,** *261* (April), 350–360. https://doi.org/10.1016/j.biortech.2018.04.026
65. Lan, J. C. W.; Raman, K.; Huang, C. M.; Chang, C. M. The Impact of Monochromatic Blue and red LED Light Upon Performance of Photo Microbial Fuel Cells (PMFCs) Using Chlamydomonas reinhardtii Transformation F5 as Biocatalyst. *Biochem. Eng. J.* **2013,** *78,* 39–43. https://doi.org/10.1016/j.bej.2013.02.007
66. Cui, Y.; Rashid, N.; Hu, N.; Rehman, M. S. U.; Han, J. I. Electricity Generation and Microalgae Cultivation in Microbial Fuel Cell Using Microalgae-Enriched Anode and Bio-cathode. *Energy Convers. Manag.* **2014,** *79,* 674–680. https://doi.org/10.1016/j.enconman.2013.12.032
67. Singh, S. P.; Singh, P. Effect of CO2 Concentration on Algal growth: A Review. *Renew. Sustain. Energy Rev.* **2014,** *38,* 172–179. https://doi.org/10.1016/j.rser.2014.05.043
68. Shahparnia, M.; Packirisamy, M.; Juneau, P.; Zazubovich, V. Micro Photosynthetic Power Cell for Power Generation From Photosynthesis of Algae. *Technology* **2015,** *03* (02n03), 119–126. https://doi.org/10.1142/s2339547815400099
69. Saar, K. L.; Bombelli, P.; Lea-Smith, D. J.; Call, T.; Aro, E. M.; Müller, T.; Howe, C. J.; Knowles, T. P. J. Enhancing Power Density of Biophotovoltaics by Decoupling Storage and Power Delivery. *Nat. Energy* **2018,** *3* (1), 75–81. https://doi.org/10.1038/s41560-017-0073-0
70. He, H.; Zhou, M.; Yang, J.; Hu, Y.; Zhao, Y. Simultaneous Wastewater Treatment, Electricity Generation and Biomass Production by an Immobilized Photosynthetic Algal Microbial Fuel Cell. *Bioproc. Biosyst. Eng.* **2014,** *37* (5), 873–880. https://doi.org/10.1007/s00449-013-1058-4
71. Elmaadawy, K.; Hu, J.; Guo, S.; Hou, H.; Xu, J.; Wang, D.; Liang, T.; Yang, J.; liang, S.; Xiao, K.; Liu, B. Enhanced Treatment of Landfill Leachate with Cathodic Algal

Biofilm and Oxygen-consuming unit in a Hybrid Microbial Fuel Cell System. *Bioresour. Technol.* **2020,** *310* (January 2021). https://doi.org/10.1016/j.biortech.2020.123420
72. Mohamad, S.; Fares, A.; Judd, S.; Bhosale, R.; Kumar, A.; Gosh, U.; Khreisheh, M. Advanced Wastewater Treatment Using Microalgae: Effect of Temperature on Removal of Nutrients and Organic Carbon. *IOP Conf. Ser.: Earth Environ. Sci.* **2017,** *67* (1). https://doi.org/10.1088/1755-1315/67/1/012032
73. Greenman, J.; Gajda, I.; Ieropoulos, I. Microbial Fuel Cells (MFC) and Microalgae; Photo Microbial Fuel Cell (PMFC) as Complete Recycling Machines. *Sustain. Energy Fuels* **2019,** *3* (10), 2546–2560. https://doi.org/10.1039/c9se00354a
74. Jafary, T.; Rahimnejad, M.; Ghoreyshi, A. A.; Najafpour, G.; Hghparast, F.; Daud, W. R. W. Assessment of Bioelectricity Production in Microbial Fuel Cells Through Series and Parallel Connections. *Energy Convers. Manag.* **2013,** *75,* 256–262. https://doi.org/10.1016/j.enconman.2013.06.032
75. Hou, Q.; Cheng, J.; Nie, C.; Pei, H.; Jiang, L.; Zhang, L.; Yang, Z. Features of Golenkinia sp. and Microbial Fuel Cells used for the Treatment of Anaerobically Digested Effluent from Kitchen Waste at different Dilutions. *Bioresour. Technol.* **2017,** *240,* 130–136. https://doi.org/10.1016/j.biortech.2017.02.092
76. Colombo, A.; Marzorati, S.; Lucchini, G.; Cristiani, P.; Pant, D.; Schievano, A. Assisting Cultivation of Photosynthetic Microorganisms by Microbial Fuel Cells to Enhance Nutrients Recovery from Wastewater. *Bioresour. Technol.* **2017,** *237,* 240–248. https://doi.org/10.1016/j.biortech.2017.03.038
77. Ma, J.; Wang, Z.; Zhang, J.; Waite, T. D.; Wu, Z. Cost-effective Chlorella Biomass Production From Dilute Wastewater Using a Novel Photosynthetic Microbial Fuel Cell (PMFC). *Water Res.* **2017,** *108,* 356–364. https://doi.org/10.1016/j.watres.2016.11.016
78. Gouveia, L.; Neves, C.; Sebastião, D.; Nobre, B. P.; Matos, C. T. Effect of Light on the Production of Bioelectricity and Added-value Microalgae Biomass in a Photosynthetic Alga Microbial Fuel Cell. *Bioresour. Technol.* **2014,** *154,* 171–177. https://doi.org/10.1016/j.biortech.2013.12.049
79. Pei, H.; Yang, Z.; Nie, C.; Hou, Q.; Zhang, L.; Wang, Y.; Zhang, S. Using a Tubular Photosynthetic Microbial Fuel Cell to Treat Anaerobically Digested Effluent From Kitchen Waste: Mechanisms of Organics and Ammonium Removal. *Bioresour. Technol.* **2018,** *256* (November 2017), 11–16. https://doi.org/10.1016/j.biortech.2018.01.144
80. Nguyen, H. T. H.; Kakarla, R.; Min, B. Algae Cathode Microbial fuel Cells for Electricity Generation and Nutrient Removal from Landfill Leachate Wastewater. *Int. J. Hydrogen Energy* **2017,** *42* (49), 29433–29442. https://doi.org/10.1016/j.ijhydene.2017.10.011
81. Uggetti, E.; Puigagut, J. Photosynthetic Membrane-less Microbial Fuel Cells to Enhance Microalgal Biomass Concentration. *Bioresour. Technol.* **2016,** *218,* 1016–1020. https://doi.org/10.1016/j.biortech.2016.07.062
82. Bejjanki, D.; Muthukumar, K.; Radhakrishnan, T. K.; Alagarsamy, A.; Pugazhendhi, A.; Naina Mohamed, S. Simultaneous Bioelectricity Generation and Water Desalination using Oscillatoria sp. as Biocatalyst in Photosynthetic Microbial Desalination Cell. *Sci. Total Environ.* **2021,** *754,* 142215. https://doi.org/10.1016/j.scitotenv.2020.142215
83. Gupta, S.; Nayak, A.; Roy, C.; Yadav, A. K. An algal Assisted Constructed Wetland-Microbial Fuel Cell Integrated with sand Filter for Efficient Wastewater Treatment and Electricity Production. *Chemosphere* **2021,** *263,* 128132. https://doi.org/10.1016/j.chemosphere.2020.128132

84. Wu, X. Y.; Song, T. S.; Zhu, X. J.; Wei, P.; Zhou, C. C. Construction and Operation of Microbial Fuel Cell with Chlorella Vulgaris Biocathode for Electricity Generation. *Appl. Biochem. Biotechnol.* **2013,** *171* (8), 2082–2092. https://doi.org/10.1007/s12010-013-0476-8
85. Ng, F. L.; Phang, S. M.; Periasamy, V.; Yunus, K.; Fisher, A. C. Enhancement of Power Output by Using Alginate Immobilized Algae in Biophotovoltaic Devices. *Sci. Rep.* **2017,** *7* (1), 1–8. https://doi.org/10.1038/s41598-017-16530-y
86. Kakarla, R.; Min, B. Evaluation of Microbial Fuel Cell Operation using Algae as an Oxygen Supplier: Carbon Paper Cathode vs. Carbon Brush Cathode. *Bioproc. Biosyst. Eng.* **2014,** *37* (12), 2453–2461. https://doi.org/10.1007/s00449-014-1223-4
87. Zhou, M.; He, H.; Jin, T.; Wang, H. Power generation Enhancement in Novel Microbial Carbon Capture Cells with Immobilized Chlorella Vulgaris. *J. Power Sources* **2012,** *214,* 216–219. https://doi.org/10.1016/j.jpowsour.2012.04.043
88. Wang, X.; Feng, Y.; Liu, J.; Lee, H.; Li, C.; Li, N.; Ren, N. Sequestration of CO2 Discharged from Anode by Algal Cathode in Microbial Carbon Capture Cells (MCCs). *Biosen. Bioelectron.* **2010,** *25* (12), 2639–2643. https://doi.org/10.1016/j.bios.2010.04.036
89. Hou, Q.; Nie, C.; Pei, H.; Hu, W.; Jiang, L.; Yang, Z. The Effect of Algae Species on the Bioelectricity and Biodiesel Generation Through Open-air Cathode Microbial Fuel Cell with Kitchen Waste Anaerobically Digested Effluent as Substrate. *Bioresour. Technol.* **2016,** *218,* 902–908. https://doi.org/10.1016/j.biortech.2016.07.035
90. Manchanda, T.; Tyagi, R.; Nalla, V. K.; Chahar, S.; Sharma, D. K. Power Generation by Algal Microbial Fuel Cell Along with Simultaneous Treatment of Sugar Industry Wastewater. *J. Bioprocess. Biotech.* **2018,** *08* (03). https://doi.org/10.4172/2155-9821.1000323
91. Yadav, A. K.; Nayak, S. K.; Acharya, B. C.; Mishra, B. K. Algal-assisted Microbial Fuel Cell for Wastewater Treatment and Bioelectricity Generation. *Energy Sources, Part A: Recovery, Util. Environ. Eff.* **2015,** *37* (2), 127–133. https://doi.org/10.1080/15567036.2011.576422
92. Angioni, S.; Millia, L.; Mustarelli, P.; Doria, E.; Temporiti, M. E.; Mannucci, B.; Corana, F.; Quartarone, E. Photosynthetic Microbial Fuel Cell with Polybenzimidazole Membrane: Synergy Between bacteria and Algae for Wastewater Removal and Biorefinery. *Heliyon* **2018,** *4* (3), e00560. https://doi.org/10.1016/j.heliyon.2018.e00560
93. Lin, C. C.; Wei, C. H.; Chen, C. I.; Shieh, C. J.; Liu, Y. C. Characteristics of the Photosynthesis Microbial Fuel Cell with a Spirulina Platensis Biofilm. *Bioresour. Technol.* **2013,** *135,* 640–643. https://doi.org/10.1016/j.biortech.2012.09.138

CHAPTER 10

A Comprehensive Overview of Magnetic Energy Storage Applications of Barium Titanate Nanoparticles: A Review

DHANYA RAJ[1] and MANJUSHA M.V[2]

[1]*Research Scholar, Department of Physics, The Cochin College, Kochi, Ernakulam, Kerala, India*

[2]*Assistant Professor, Department of Physics, The Cochin College, Kochi, Ernakulum, Kerala, India*

ABSTRACT

The growing global demand for renewable energy is driving scientists from all over the world to study valuable materials with good energy storage ability. The dielectric and ferroelectric materials used to fabricate energy storage capacitors must have high saturation polarization, high dielectric constant, strong fracture resistance, low loss, excellent temperature stability, and high fatigue limit. It is very difficult in practice to realize all these qualities in a single device. Lead-based components are also being researched around the world due to their better energy storage. High-value ceramic capacitors and ultrasonic transducers, sensors, and actuators have long been manufactured from many lead zirconate titanates (PZT). The major disadvantage of PZT is that it contains lead, which is dangerous and is now illegal in many countries around the world. Barium titanate ($BaTiO_3$) is an excellent choice when looking for a lead-free alternative to PZT, because the use of PZT in commercial portable electronics applications is very limited due to many environmental concerns. Various $BaTiO_3$ materials have been investigated in the past to achieve high performance and strength. $BaTiO_3$ is a versatile

The Food-Energy-Water Triangle for Sustainable Development: Integrative and Synergic Solutions.
Pedro Aguilar-Zárate, Mónica Chávez-González, Ann Rose Abraham, and A. K. Haghi (Eds.)

material with a wide and strong dielectric constant that can be used in many electric current applications. The properties of $BaTiO_3$ can be improved by doping with suitable elements such as Mn, Co, Ni, Fe, $LiFe_5O_8$, and Sm, and using various synthesis methods, including sol–gel method, hydrothermal method, chemical alloying method, coprecipitation method, and polymer precursor method. The main objective of this study is the main technology and theoretical models for the improvement of energy density and energy efficiency of various $BaTiO_3$-based materials and to provide an analysis of the commercial use of single and multiphase mixtures of $BaTiO_3$.

10.1 INTRODUCTION

The world population is increasing day by day and therefore the needs of all living things including those to which we belong, are increasing. Due to the increase in the world population and economic growth, the demand for energy in various electronics and automotive industries is increasing rapidly. The International Energy Agency reported that the world's electricity needs have tripled in the last 40 years. Environmental problems such as climate change, air pollution, and the destruction of energy sources necessitate the use of green energy. Although fossil fuels account for more than 60% of the world's energy consumption, human society now relies on many environmentally friendly and sustainable energy sources such as hydro, wind, solar, tidal, geothermal, and biomass. The availability, use, and conservation of energy are paramount in the 21st century because it is the only way we can save energy for future generations.

Initially, materials such as PZT were used as electrical materials. PZT is an important material for energy storage devices such as high-efficiency electronics, ultrasonic transducers, sensors, and actuators. However, because it contains a lot of lead, it is dangerous for the health of organisms. It causes serious damage to the environment, suggesting lead-free, clean, renewable, and energy storage technologies. Extensive research is required to develop electronic products using lead-free materials. The best option is to convert this energy into another form, usually electricity. Reliable and efficient energy solutions are fast energy combined with high energy efficiency, and progress in this important area helps in the efficient use of energy by this resource. The importance of energy saving for different types of energy storage is provided by energy storage systems especially superconducting magnetic energy storage systems (SMESS), supercapacitors, intermediate and gas flow lamps, flywheels, fuel cells, and dielectric/ferroelectric capacitors. When it comes time to create a lead-free energy storage device, magnetic storage devices are an excellent choice because of much research. A lot of research is still being done in this area today. Creating an efficient and environment-friendly energy storage system is the basis of all these research studies. Magnets can be used to store energy in a magnetic generator.

Magnetic electronics are playing an increasing role in everyday life as they are more efficient and cheaper. They serve many purposes, such as providing electricity during power outages, driving electric vehicles, storing energy, etc. Electromagnetic devices are becoming more common and play an important role in energy production. They can be used to store energy for current and future use. This has many applications and can store energy for later use. In electric vehicles, magnetic energy storage devices are often used to store energy that can be used to power the vehicle. Magnetic nanoparticles (NPs) have attracted much attention in recent years due to their controlled and optimized properties and their potential applications in high-speed data, energy storage, chemical applications, and biological applications. Apart from minor electronic applications, NPs are now of interest in many applications due to their different physical properties. The case of magnetic NPs is of particular interest because of the similarity between magnetic properties and nonmagnetic NPs. Therefore, single-domain ferromagnetic (FM) and superparamagnetic (SPM) NPs are produced, which are two important aspects of magnetic behavior in NPs. The surface-to-volume ratio of these nanoscale magnets sets them apart from larger materials, have received much attention in recent years due to their potential applications in high-speed data, energy storage, and chemical applications. The surface-to-volume ratio of these nanoscale magnets sets them apart from larger materials.

The main properties of magnetic NPs are controlled and optimized by properties such as blocking temperature (T_B), saturation magnetization (M_s), permanent residual magnetization (M_r), coercive force (H_C), and relaxation time (t_N and t_B). When everything is combined in the direction of the field, it interacts with its neighbors and with the field at a precise moment. M_s is the magnetization followed by all magnetic moments. In any case, the sample retains magnetization even after the external magnetic field is reduced to zero. M_r is the magnetization that remains after the magnetic field (H) is removed and saturation occurs. The coercivity, or H_C, is the amount of H required to bring the net magnetization to zero when the field is reversed (opposite the residual direction). The T_B is the temperature of the thermal energy over the energy anisotropy of the NPs.

One of the most successful materials used in electronic applications is barium titanate ($BaTiO_3$). Due to the unique properties of this material, a lot of research has been done on it for over 60 years. $BaTiO_3$ is frequently used in the manufacture of electronic devices such as capacitors and multilayer capacitors, as it exhibits ferroelectric properties above room temperature with high dielectric constant and low loss properties. $BaTiO_3$'s color and properties change when mixed with other elements, in its pure form it appears as a white powder. Piezoelectric, thermistor, and semiconductor devices are the main applications of doped $BaTiO_3$. $BaTiO_3$ differs from other perovskite ferroelectric materials because it is biologically dense and capable of transporting chemicals. Recently, many articles have been published claiming that the intermediate or doping compound can improve the strength and durability of $BaTiO_3$-based materials. In this review, the magnetic properties of various materials with $BaTiO_3$ will be discussed.

10.2 MAGNETIC ENERGY STORAGE APPLICATIONS OF BaTiO3 NPS

The classification of magnetic NPs is diamagnetic, paramagnetic, FM, ferrimagnetic and anti-FM. Ferroelectricity refers to the properties of certain materials that allow their electrical polarization to be changed by the introduction of external electricity. Multiferroic composites with magnetoelectric properties have various ferroic-level parameters such as ferroelectric, FM or piezoelectric and ferroelastic phases. Rochelle salt (sodium tartrate tetrahydrate $NaKC_4H_4O_6$ $4H_2O$), a crystal, was the first material to exhibit ferroelectricity and was discovered in 1921.[1] For a material to be classified as a ferroelectric material, it must meet two conditions: (1) Polarization must be present and (2) polarization must be shown to be reoriented. A field is a polarized region with one polarization direction and the crystal symmetries affect how fields interact. One of the most important properties of ferroelectric materials is the polarization property *P* vs electric field *E*, which is called the hysteresis loop property. Figure 10.1 shows the hysteresis loop for a ferroelectric device. $BaTiO_3$anate and Rochelle salt are examples of ferroelectric materials. These products are made of crystals, and the building blocks of crystals are tiny electrons. That is, the areas between positive and negative charges are slightly different in each cell.[2]

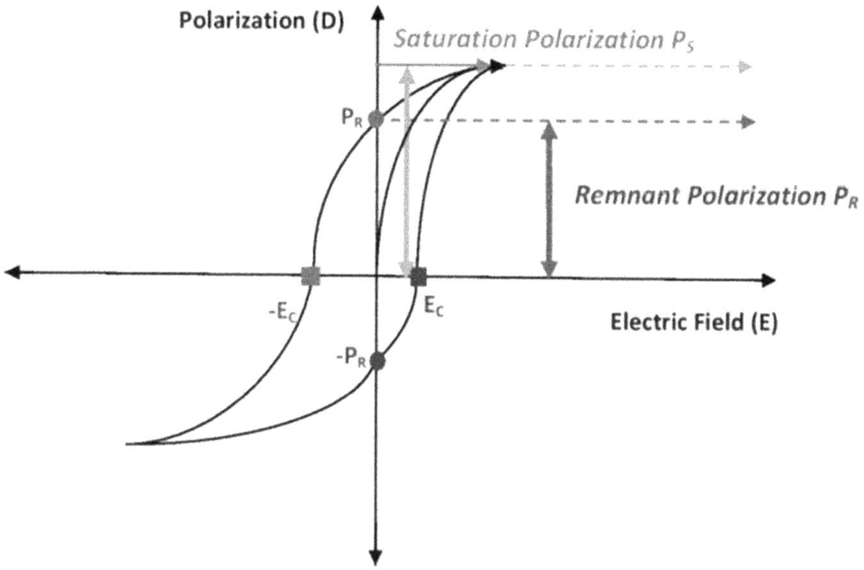

FIGURE 10.1 Typical polarization vs electric field (*P*-*E*) hysteresis loop of ferroelectrics.

The best magnetic materials will generate a magnetic field strong enough to maintain a strong magnetic moment even after exposure to a magnet. High M_r is the average of the hysteresis loop. However, it must be difficult to demagnetize to maintain stability over time. This translates into high H_C unfortunately, high remnant magnetization materials are generally magnetically "soft" (H_C less than 1000 Oe), while low remnant magnetization materials are generally magnetically "hard" H_C greater than 1000 Oe). Until recently, choosing the best magnetic material required to maintain these two advantages, are the choice of high magnetic moment and high H_C.[3] Kolhatkar et al. have shown that the magnetic properties of NPs depend on their size, surface-to-volume ratio, shape, phase, composition, and shell–core structure.[4]

$BaTiO_3$, is a perovskite material with general formula ABO_3 and it can be doped in both A and B sites.[5] Ba^{2+} (barium) has coordination number 12 and Ti^{4+} (titanium) has coordination number 6. Figure 10.2 shows the perovskite structure of $BaTiO_3$ in which barium (Ba), titanium (Ti), and oxygen (O) atoms are arranged in various configurations. Ti atoms are in the center, oxygen atoms are in the center of the face, and Ba atoms are in the corner.[6] The first material used for the manufacturing of multilayer capacitors, dielectric ceramic capacitors, etc. is $BaTiO_3$. It is used in this application because of its high dielectric constant and low dielectric loss. Purity, density, grain size, temperature, and frequency are the properties that will affect the value of the dielectric constant.[7]

Ferroelectric $BaTiO_3$'s Curie temperature is 120°C, its dielectric constant is up to 1500 at high temperatures, and the dielectric loss is very small. The structure of $BaTiO_3$ can change with temperature. They exist in 5 different phases, from the order of increasing temperature, they are listed as follows: rhombohedral, orthorhombic, tetragonal, cubic, and hexagonal. The original cubic unit cell is stable in the temperature range of 120–1460°C. The hexagonal structure is stable at this temperature.[8] When the temperature is lower than the Curie point, $BaTiO_3$ will change the crystallization. The first transition occurs at about 120°C and is a ferroelectric transition between cubic, paraelectric, and ferroelectric phases of a tetragonal structure. It turns into an orthorhombic structural phase at 5°C, and into a triangular low-temperature phase at −90°C.[9]

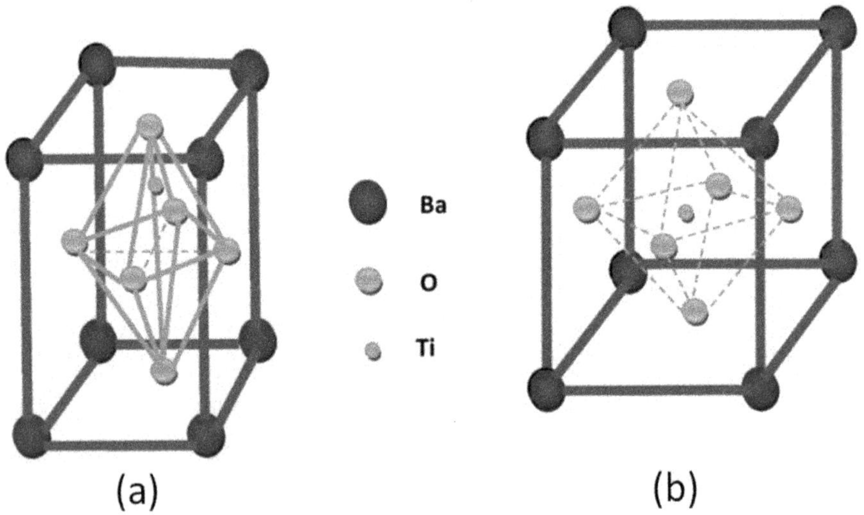

FIGURE 10.2 Schematic illustration of the perovskite structure of $BaTiO_3$. (a) Tetragonal lattice (above Curie temperature, <120°C). (b) Cubic lattice (above Curie temperature, >120°C).

In certain cases, weak ferromagnetism at room temperature has been found for NPs of various nonmagnetic materials. The ferromagnetism found in nanostructured materials is generally thought to be the result of defects found on the surface of NPs. Oxygen vacancies (V_o) in $BaTiO_3$ can create magnetic moments between nearby cations. According to Pan et al., the lack of oxygen causes $BaTiO_3$ NPs to form Ti^{3+}-V_0 and Ti^{3+}-O pairs. Spin couplings result in small FM clusters at reasonable distances due to carrier-induced FM states.[10] $BaTiO_3$ NPs are created using a variety of techniques. There are sol–gel method, reaction phase, hydrothermal method, microwave hydrothermal method, coprecipitation method, polymer precursor method, biological method, mechanochemical method, Pechini method, oxalate coprecipitation method, citric acid salt method, solvothermal method, and emulsification combustion method. Among them, sol–gel method, hydrothermal method, and solid-state reaction are the three most important technologies. These methods are used commercially because they are inexpensive and work in large quantities. For the preparation of NP of size less than 100 nm, sol–gel method and hydrothermal method are most preferred. Jain et al. suggested that the bottom-up sol–gel process can produce small grains with high microstructural properties. The sol–gel process is one of the best methods for producing $BaTiO_3$-based nanomaterials because it produces materials with excellent tightly packed particle shapes. Another method that has the special potential to crystallize chemicals at high temperatures is the hydrothermal method of nanomaterial production. It can be clearly stated that sol–gel and hydrothermal processes can be used to produce very fine powders since they produce fine powders of phase-pure and homogeneous quality.[11]

Emelianov et al. compared the magnetic properties of bulk and nanostructured $BaTiO_3$, and they found that the diamagnetic response in nanostructured $BaTiO_3$ was higher than that of bulk materials. It is caused by many different lattice defects, some of which lead to ferromagnetism and others to diamagnetic fields. When heated in the same way as the bulk sample, the self-magnetization of the nanostructured $BaTiO_3$ increases, but the diamagnetic field decreases. This is because working in an H_2 environment increases the amount of oxygen in the surrounding air. They worked on nanostructured $BaTiO_3$ with simultaneous FM, diamagnetic, and ferroelectric properties.[12] Ramakanth et al. used sol–gel method to prepare pure $BaTiO_3$ NPs, they measured the magnetic field (*M-H*) and found that samples with higher charge carriers had higher magnetic moments and lower band gap values.[13]

$BaTiO_3$ is doped with Co, Fe, Mn, Ni, and other elements and can be used in electronics, magnetism, piezoelectricity, and organics. To realize ferromagnetism in $BaTiO_3$, Nakayama et al. recommended Cr, Mn, Fe, and Co is as required doping elements. There has been interest in the deliberate doping of transition metals in recent nonmagnetic perovskite materials to create versatile materials with applications in fields such as spintronics, data storage, and multistate memory.[14] Using a simple one-step sol–gel precipitation method, Liu et al. prepared undoped and codoped $BaTiO_3$ samples. $BaTiO_3$ particles were formed in the cubic phase and magnetic studies showed that Co ions in codoped $BaTiO_3$ exist in different paramagnetic regions without FM exchange.[15] Sm-doped $BaTiO_3$ NPs were produced by Es-saddik et al. The average particle size obtained by the sol–gel method is 33.2 nm. They use a vibrating sample magnetometer (VSM) to measure the magnetism and magnetic properties of objects. When they did this, they found that the magnetism showed a decrease between 20 K and 75 K with temperature compared to temperature for the *H*. This anomaly will be seen especially when the *H* increases. At low temperatures, hysteresis loops indicate the coupling of the FM and anti-FM phases. They show that at high temperatures, a diamagnetic phase coexists with the FM phase, causing the dominant FM phase below 50 K to become less dominant. In addition, the thermal behavior of the entropy change is related to the *H*.[16]

G. R. Gajula et al. synthesized pure and LF doped $BaTiO_3$. From X-ray diffraction analysis, they found that tetragonal $BaTiO_3$ was formed but the center of the peak did not change as the percentage of LF content increased, indicating that the pattern is ferroelectric and there is a difference between ferroelectric and FM components. When the percentage of LF content in BT increases, the magnetic transition temperature and Neel temperature of the composite also increase, indicating magnetic behavior in the composite. They concluded that the piezoelectric properties and magnetoelectric sensitivity of BT ceramics were improved by the combination of BT and LF. Therefore, this material will find application in actuators, storage, and magnetoelectric sensor applications.[17]

The causes and properties of the FM or SPM behavior of Co-doped $BaTiO_3$ ceramics with doping up to 10% mole were carefully analyzed by HT Langhammer et al. The mixed oxide powder process is used to produce hexagonal $BaTiO_3$ ceramic powder with the desired composition. Partial reconversion to 3C-$BaTiO_3$ after annealing in a reducing environment was previously observed in the hexagonal Mn-doped $BaTiO_3$ study. Metallic precipitation of additives also occurred when $BaTiO_3$ samples containing additional 3D dopants Sc, V, Cr, Cu, and Zn were heated in a highly reduced atmosphere. In addition, Fe- and Ni-doped $BaTiO_3$ samples show distinct magnetic hysteresis and net magnetization saturation behavior in the FM condition.[18] Nickel is a FM material with good oxidation resistance at room temperature. Therefore, Ni-doped $BaTiO_3$ systems will exhibit a combination of ferroelectricity and magnetism. Ni-doped $BaTiO_3$ NPs prepared by K. Das et al. Conventional solid-state reaction techniques are used to create ceramic structures. In the pure $BaTiO_3$ sample, the tetragonal phase is present, and the tetragonal phase changes to the hexagonal phase when the Ni concentration increases. There is the formation of octahedral stacking of TiO_6 in the hexagonal phase. Thus, the hexagonal $BaTiO_3$ creates two Ti domains compared to the tetragonal $BaTiO_3$, which has only one Ti domain. V_0 is formed when dopants are accepted to replace Ti. The number of hexagonal levels increases as more and more octahedrons change from the corner-sharing to the face-sharing structure.[19]

Z. Liu et al. has been shown that the application of a thin layer of Al_2O_3 or B_2O_3–SiO_2 to ceramic materials increases the breakdown strength of $BaTiO_3$ ($BaTiO_3$) ceramics. Therefore, the energy storage density of dielectric ceramics increases a lot. Based on the behavior of the structure and the dielectric material, they investigated the process of improving the material. The surface coating of Al_2O_3 and B_2O_3–SiO_2 on $BaTiO_3$ particles results in an increased energy storage density. The increase in scattering energy brought about by changes in particle size is responsible for the increase in energy storage density. This study shows that the surface modification of ceramic materials is a way to produce dielectric ceramics with reliable energy.[20] K. Samuel et al. studied ceramics made of Fe-doped $BaTiO_3$ and sintered at 1200°C by a solid-state reaction. The X-ray diffraction pattern of the sample sintered at room temperature shows that it is a single-phase structure. Using BDS at different temperatures, the properties of different samples are described. According to the microstructure investigation using field emission scanning electron microscopy (FE-SEM), they showed that by measuring the different temperature potential, the frequency will increase as the capacity decreases and the final grain growth depends on the duration of the sintering heat.[21]

Fe ions replace Ti ions in BT, which change the crystal structure and lower the ferroelectric Curie temperature and dielectric constant, while at the same time having a significant effect on the magnetic properties of the material. In addition, the band gap energy of $BaTiO_3$ is reduced and its optical properties are affected by Fe doping. Both A-site and B-site Fe-doped BT exhibit FM behavior at room temperature. In contrast to Fe substitution at site A, which leads to M_s, Fe substitution at site B was found to increase the coercive field. Devendra Singh et al. did the synthesis of pure Fe-doped tetragonal $BaTiO_3$ using the solid-state reaction method. The dielectric constant of BT is not significantly larger and has lower dielectric loss at low frequencies. The dielectric constant decreases with frequency at all frequencies from 101 to 106 Hz. The decrease in permittivity with frequency may be due to the resistance of the electric dipoles to the action of the applied field. Compared to Fe-doped materials, the dielectric constant increases slightly with temperature. As the Fe concentration rises above 100°C, the dielectric constant increases.[22]

10.3 CONCLUSION

In this study, the use of $BaTiO_3$ in magnetic electronic devices and some of its effects on ferroelectric, dielectric, and piezoelectric properties were investigated. The first ferroelectric perovskite discovered was $BaTiO_3$. A series of three-phase transitions connects their ferroelectric properties. Thus, quality $BaTiO_3$ can be used with new smart products. For the production of $BaTiO_3$ powder, solid-state reaction, biological synthesis, and wet processes such as sol–gel, hydrothermal coprecipitation, and polymer precursor processes are often used. $BaTiO_3$ is a material with strong ferroelectric and dielectric properties, low dielectric loss, and high dielectric constant. By mixing $BaTiO_3$ with positive elements, we can use its properties to develop energy storage materials. We include research on $BaTiO_3$ with various dopants including cobalt, nickel, iron, LF, and samarium. All these additives improve the dielectric, ferroelectric, and magnetic properties of $BaTiO_3$, making it an ideal material for energy storage. According to the information obtained, $BaTiO_3$ is an excellent energy storage material for optoelectronic devices, transducers, ultrasonic devices, ceramics, and other electronic devices. The materials of the future require materials with high temperature, improved piezoelectric and dielectric properties, and fracture toughness.

KEYWORDS

- **barium titanate**
- **magnetic energy storage**
- **ferroelectric**
- **hysteresis loop**
- **dopants**
- **dielectric properties**

REFERENCES

1. Vijatovic, M. M.; Bobic, J. D.; Stojanovic, B. D. History and Challenges of Barium Titanate: Part I. *Sci. Sinter.* **2008,** *40* (2), 155–165, doi: 10.2298/SOS0802155V.
2. Ferroelectricity | physics | Britannica. 2023. https://www.britannica.com/science/ferroelectricity (accessed May 03, 2023).
3. Frey, N. A.; Peng, S.; Cheng, K.; Sun, S. Magnetic Nanoparticles: Synthesis, Functionalization, and Applications in Bioimaging and Magnetic Energy Storage. *Chem. Soc. Rev.* **2009,** *38* (9), 2532, doi: 10.1039/b815548h.
4. Kolhatkar, A.; Jamison, A.; Litvinov, D.; Willson, R.; Lee, T. Tuning the Magnetic Properties of Nanoparticles. *Int. J. Mol. Sci.* **2013,** *14* (8), 15977–16009, doi: 10.3390/ijms140815977.
5. Rahman, M. A. Understanding of Doping Sites and Versatile Applications of Heteroatom Modified $BaTiO_3$ Ceramic. *J. Asian Ceram. Soc.* **2023,** 1–10, doi: 10.1080/21870764.2023.2203635.
6. Tewatia, K.; Sharma, A.; Sharma, M.; Kumar, A. Factors Affecting Morphological and Electrical Properties of Barium Titanate: A Brief Review. *Mater. Today Proc.* **2021,** *44,* 4548–4556, doi: 10.1016/j.matpr.2020.10.813.
7. Vijatovic, M. M.; Bobic, J. D.; Stojanovic, B. D. History and Challenges of Barium Titanate: Part II. *Sci. Sinter.* **2008,** *40* (3), 235–244, doi: 10.2298/SOS0803235V.
8. Cho, W.-S. Structural Evolution and Characterization of BaTiO3 Nanoparticles Synthesized from Polymeric Precursor. *J. Phys. Chem. Solids* **1998,** *59* (5), 659–666, doi: 10.1016/S0022-3697(97)00227-8.
9. Koleżyński, A.; Tkacz-Śmiech, K. From the Molecular Picture to the Band Structure of Cubic and Tetragonal Barium Titanate. *Ferroelectrics* **2005,** *314* (1), 123–134, doi: 10.1080/00150190590926300.
10. Phan, T.-L.; Zhang, P.; Yang, D. S.; Thanh, T. D.; Tuan, D. A.; Yu, S. C. Origin of Ferromagnetism in $BaTiO_3$ Nanoparticles Prepared by Mechanical Milling. *J. Appl. Phys.* **2013,** *113* (17), 17E305, doi: 10.1063/1.4799473.
11. Jain, A.; Wang, Y. G.; Shi, L. N. Recent Developments in BaTiO3 based Lead-free Materials for Energy Storage Applications. *J. Alloys Compd.* **2022,** *928,* 167066, doi: 10.1016/j.jallcom.2022.167066.
12. Emelianov, N. A.; Korotkov, L. N.; Al Mandalawi, W. M.; Kashirin, M. A.; Roldan Lopez, J. A.; Korotkova, T. N. In *Ferromagnetism in Barium Titanate Nanoparticles. Effect of Annealing in Hydrogen Atmosphere on Magnetic Properties*, 2018 IEEE ISAF-FMA-AMF-AMEC-PFM Joint Conference (IFAAP), Hiroshima: IEEE, May **2018**, pp. 1–3. doi: 10.1109/ISAF.2018.8463349.
13. Ramakanth, S.; Hamad, S.; Venugopal Rao, S.; James Raju, K. C. Magnetic and Nonlinear Optical Properties of $BaTiO_3$ Nanoparticles. *AIP Adv.* **2015,** *5* (5), 057139, doi: 10.1063/1.4921480.
14. Hiroyuki Nakayama, H. N.; Hiroshi Katayama-Yoshida, H. K.-Y. Theoretical Prediction of Magnetic Properties of $Ba(Ti_{1-x}M_x)O_3$ (M=Sc,V,Cr,Mn,Fe,Co,Ni,Cu). *Jpn. J. Appl. Phys.* **2001,** *40* (12B), L1355, doi: 10.1143/JJAP.40.L1355.
15. Liu, H.; Cao, B.; O'Connor, C. J. Structural and Magnetic Properties of Single-Crystalline Co-doped Barium Titanate Nanoparticles. *J. Magn. Magn. Mater.* **2010,** *322* (7), 790–793, doi: 10.1016/j.jmmm.2009.11.004.

16. Es-saddik, F.; Limame, K.; Sayouri, S.; Hamzah, M.; Khenfouch, M.; Srinivasu, V. V. Magnetic Properties of Sm-doped Barium Titanate (Ba0.99Sm0.01TiO3) Prepared by Sol–Gel Route. *J. Mater. Sci. Mater. Electron.* **2020,** *31* (22), 19910–19919, doi: 10.1007/s10854-020-04513-1.
17. Gajula, G. R.; Buddiga, L. R.; Ch, A. K.; Chidambara Kumar, K. N.; Nethala, G. P.; Dasari, M. Studies on Structural, Dielectric, Conductivity, Magnetic and Magneto-electric Properties of Barium Titanate Doped with Lithium Ferrite. *Phys. B Condens. Matter* **2018,** *543,* 38–45, doi: 10.1016/j.physb.2018.05.015.
18. Langhammer, H. T.; et al. Ferromagnetic Properties of Barium Titanate Ceramics Doped With Cobalt, Iron, and Nickel. *J. Mater. Sci.* **2016,** *51* (23), 10429–10441, doi: 10.1007/s10853-016-0263-3.
19. Das, S. K.; Mishra, R. N.; Roul, B. K. Magnetic and Ferroelectric Properties of Ni Doped BaTiO3. *Solid State Commun.* **2014,** *191,* 19–24, doi: 10.1016/j.ssc.2014.04.001.
20. Liu, Z.; Zhang, Z.; Zhang, F.; Li, Y. Enhanced Energy Storage Properties of Barium Titanate Ceramics Made From Surface Modified Particles. *J. Adv. Dielectr.* **2013,** *03* (03), 1350023, doi: 10.1142/S2010135X13500239.
21. Samuvel, K.; Ramachandran, K. Structure, Electrical and Magnetic Property Investigations on Fe-doped Hexagonal BaTiO3. *Optik* **2016,** *127* (4), 1781–1786, doi: 10.1016/j.ijleo.2015.10.240.
22. Singh, D.; Dixit, A.; Dobal, P. S. Ferroelectricity and Ferromagnetism in Fe-doped Barium Titanate Ceramics. *Ferroelectrics* **2021,** *573* (1), 63–75, doi: 10.1080/00150193.2021.1890464.

CHAPTER 11

Era of Managing and Reviving Water Footprint

REKHA MIRLE, AGNESS CHACHA, VEDASHREE N, PRIYANKA GHOW, and KEERTHANA M

School of Sciences, Jain University, Bengaluru, Karnataka, India

ABSTRACT

In today's developing world, nothing seems impossible to achieve except eliminating human ignorance with respect to sustainability. As a result, we are water. In faced with numerous obstacles to accomplishing simple tasks such as saving this chapter we will be focusing on the ways that could be implemented in getting the extra mile in preserving water and water resources. Reusing and recycling water helps to minimize the utilization of water resources and reinforce the exploitation of other substitute resources. With sustainable management of water, we will be able to diminish the risk of global disease burden, improve health, and increase the overall productivity of society. Moreover, this review provides a brief insight into various risk factors involved in conserving our water resources for upcoming generations, various ways to overcome them which leads to the management of water in a sustainable manner.

11.1 INTRODUCTION

Water is a vital part of sustainable development and a crucial factor affecting socioeconomic development in fields such as human health, economy, environment, agriculture, and overall population. It is important in reducing the global burden of diseases, improving health, and supporting

The Food-Energy-Water Triangle for Sustainable Development: Integrative and Synergic Solutions. Pedro Aguilar-Zárate, Mónica Chávez-González, Ann Rose Abraham, and A. K. Haghi (Eds.)

biodiversity. It also serves us a crucial link between climatic change, human society, and the environment.[1] The Earth' surface is covered with water about 70%, among which 3–4% constitutes freshwater and 1.2% is drinkable. Water sustainability is the process of utilizing water to fulfill our needs and striking a balance with the needs of the upcoming generations. The demand for freshwater is escalating, and scientists feel that a shortage of water in the future is inevitable.[2] Water footprint refers to the measure of the amount of freshwater utilized to produce goods and services. This water footprint also gives a rough sum-up of the total amount of water consumed in production activities, providing services at the individual level, business level and globally too. Water is a sustainable resource that needs protection to meet current and future needs, to do so water sustainability and management applications are necessary since they encircle all the policies, strategies, and activities that are made to manage and protect water resources.[2] Increases in population, agriculture, climatic change, and industrialization affect the amount of water used, thus leading to water scarcity. Better water management has its social and economic benefits helping to protect our environment; thus, water sustainability should be a compulsory practice worldwide.

11.2 DIFFERENT METHODS OF WATER CONSERVATION

We cannot produce water artificially, but we can remain reliant on the available water resources on the earth. Management of water resources involves managing water flows and water storage. Different methods have been observed to provide positive results in conserving water.

A few methods are discussed in the following subsections.

11.2.1 RAINWATER HARVESTING

The method of obtaining and storing rainwater is called rainwater harvesting. Rainwater is usually collected from rooftops through various methods. Rainwater harvesting is not only the simplest among all the methods but also the oldest and the most common method of conserving water.[3] Rainwater harvesting dates back to 1700 BC and has been traditionally practiced in regions of India and Pakistan. It was invented independently in different parts of the world later. It is the easiest, most reliable and least polluted source.[4] Rainwater harvesting installations are designed based on the infrastructure of the buildings to serve different purposes in schools, hospitals, household activities, and other civil services.[5] Rainwater from rooftops is suitable for domestic needs such as drinking, irrigation, and livestock management. The collected water can also be used to recharge the groundwater table and store water for a long duration.[6] With the growing number of freshwater-flooded forests, it is possible to harvest rainwater efficiently from water-logged land. Rainwater can also be harvested from solar power panels. The water collected from the solar panels is free of contamination with less salt content produced by easy filtration and the addition of disinfectant.[7–9] Other innovations are rooftop catchment, rain saucer, and free-standing rainwater collection.

Rainwater harvesting is a boon for developing countries that are suffering from water scarcity and helps improve their economy. It also provides water during dry seasons; it is cost effective and environmentally friendly. The water obtained has negligible contamination and is saline free. It also reduces the demand for groundwater thus sustaining its levels.

FIGURE 11.1 Rain water harvesting model.

11.2.2 GROUNDWATER RECHARGE

This refers to a hydrological process that involves the movement of surface water into the ground. This water is then referred to as groundwater. Groundwater recharge is made possible due to the presence of aquifers and its movement occurs in a region below the plant roots. Movement of water from the water table to a region of porous rocks filled with water (saturated zone) is also referred to as groundwater recharge.[10] Groundwater recharge can either occur by natural means or by human activities. Naturally, the process of recharging groundwater is aided by natural water sources such as surface water (rivers; lakes) to a great extent, or rain and in a few cases snow. The process of recharge can be interfered with by man activities such as agriculture and road constructions that result in the loss of the fertile soil layer which reduces penetration of water into the ground. The process helps in the leaching of excess salts from roots into other soil layers and also the presence of trees in an area helps in minimizing water-runoff.[11] Even during rainy seasons, waterlogged areas help in the process of groundwater recharge.[12] Climatic change, urbanization, wastelands due to agricultural processes, constructions, and pavement development are some factors that affect groundwater recharge.

11.2.3 SEWAGE WATER TREATMENT

It is the process wherein wastewater from households, industries, and business centers is treated for the removal of contaminants for reuse purposes. Mostly the treated water is used in the agricultural sector for irrigation or even used in construction sites.

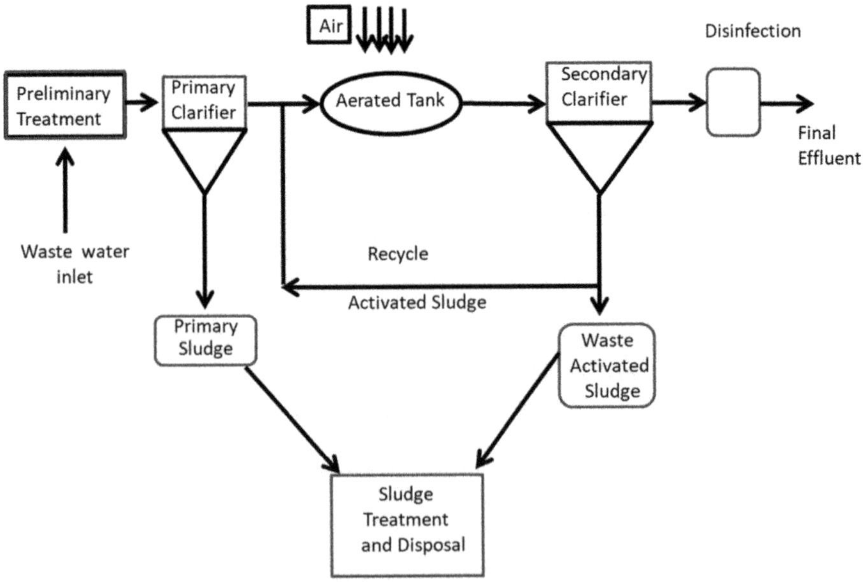

FIGURE 11.2 Flowchart of sewage water treatment process.

Sewage treatment of water mainly involves the following four processes:

i. Pretreatment: removal of large solids particles that can cause blockage.
ii. Primary treatment: elimination of suspended solids and organic matter.
iii. Secondary treatment: Use of microorganisms to eliminate the organic matter causing pollution.
iv. Tertiary treatment: Nutrient removal and disinfectant.

Wastewater treatment also known as water reclamation, serves the main purpose of reducing environmental pollution and distributing water to areas with water scarcity. It can also be used for drinking purpose and will serve as a solvent for industrial purposes.

11.2.4 DESALINATION

Desalination is the elimination of dissolved salts from seawater, forms a main source for saline water.[13] Desalinated water can be used for various purposes such as irrigation, industrial activities, domestic use, etc.[14] The main goal of this process is to provide a cost-effective solution for the availability of water

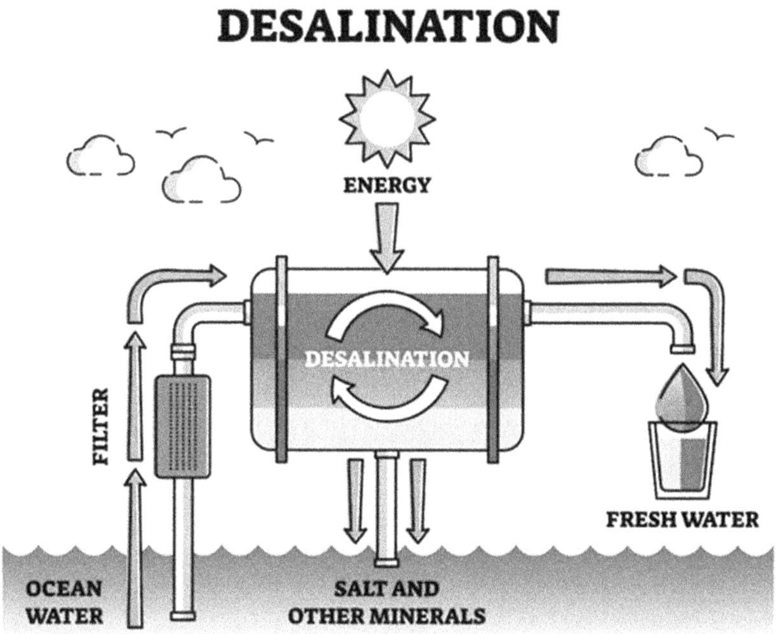

FIGURE 11.3 Desalination of water.

Image Credit: VectorMine/Shutterstock.com

sources for human consumption.[15] Desalinating seawater requires a huge amount of energy and is generally more expensive. Although it is expensive, desalination is a reliable technique that is not on the verge of depletion, unlike other water-sustainable resources.[16,17] Desalination technology uses high amount of energy that can be derived from fossil fuels thermally or mechanically through evaporation process.[18]

11.2.5 AQUIFER STORAGE AND RECOVERY

Aquifer storage is an intended process of storing water in aquifers built underground to serve the need for water during its scarcity. This process is considered beneficial as it restores water and is used when required. Injection and extraction are done through wells that are constructed using materials such as permeable rocks, and porous materials such as unconsolidated sand that allow the passage of water to the ground. As water stays in the aquifer for a long duration of time, it is necessary to treat it with disinfectants such as chlorine before injecting it into the wells, to prevent the growth of pathogens in water. A major drawback of this method is the formation of disinfectant spinoffs, the presence of soluble precipitates of carbonates, and disposal of harmful metals from chemical industries into the aquifers. These pose a threat to the individual to consume water.

11.2.6 CONJUNCTIVE USE

This is a unified way of consuming water resources such as surface water and groundwater for their efficiency, conservation of water for long-term usage and preventing various water sources from unfavorable consequences. Ideally, it optimizes the use of both surface and groundwater during the wet and dry seasons. Conjunctive administration programs are implemented in various parts of the world as a shield to control the deterioration of groundwater levels and dearth. This program can be accomplished either locally or universally.

The advantages of conjunctive management are listed as in the following:

- It enhances the availability of water to local areas.
- It reduces the overflow of groundwater.
- It reduces the chances of water logging.
- It helps to achieve our socioenvironmental needs.

11.3 IMPACTS OF SUSTAINABLE MANAGEMENT OF WATER ON SOCIETY

The overall aim of water sustainability is to ensure that the world no longer suffers from water scarcity in years to come, thus different methods are implemented in many countries, especially developing countries. Water acts as an important factor in different sectors of our life; being able to manage the use of water in a sensible manner helps to overcome other threats that the world might face in the future such as food shortage, sustenance of animal and plant life on earth, etc. The impacts of water sustainability on society are discussed below briefly.

Agriculture: Agriculture is the cornerstone of any country. Apart from water, the world depends on agriculture to survive. Agriculture is the largest consumer of water. A significant amount of water is wasted and used to support agriculture. It raises the importance of utilizing methods that will reduce the amount of water consumed daily in agriculture. Methods such as the use of drought-resistant crops, capturing and storing rainwater in wells so that they can be used for the process of irrigation and also some methods decide the growth of crop depending on the moisture content of the soil during dry seasons.[20]

Drip irrigation: It is one of the successful methods that have reduced the utilization of water in agriculture. It prevents maximum water loss and evaporation as it directly delivers water to the crop roots at the required amount. Being the largest consumer of water, sustainable management methods have supported the agricultural sector by offering different techniques in modifications of crops to cater to various situations.[20]

Industrial sector: Industrial waste is one of the major pollutants induces the risk of environmental pollution. Hence, treating wastewater plays a significant role in minimizing the waste of water throughout the process, as it can be recycled and reused and hence reduces the loss of water. Moreover, sewage treatment helps to recover nutrients that can be potential raw material for other processes too.[21,22]

Domestic use: Water sustainability and management methods such as rainwater harvesting have been seen to provide a positive outcome as a source of water. Water harvested from rain can be stored and later be used to serve household purposes such as drinking, cooking, gardening, and other chores. Due to its easy accessibility and reliability, improves water availability for domestic use even during dry seasons. For areas that are faced with water shortage aquifer storage has been seen to be a big help to provide individuals with water for domestic use.[23]

Ecosystem: There are different ecosystems on Earth, and they all are dependent on one crucial compound called water. In an ecosystem, water is essential for the growth and development of plants and other species, acts as a breeding ground for some organisms, and provides habitat for aquatic organisms such as the marine ecosystem.[24] Availability of water leads to biodiversity which increases the richness of species in a particular region. Due to its distinct characteristics, water is a unique solvent, it is impossible for the existence of another substance that can substitute water; thus, water sustainability is important for the ecosystem.[25]

11.4 CONCLUSION

Water is an essential resource that supports life on Earth. Due to the population explosion, the world is facing major challenges regarding water safety, control of water pollution, and providing a healthy environment to the ecosystem. Climate change and water diversions threaten the environment and ecosystem; however, the presence of traditional methods such as rainwater harvesting helps to minimize the effect to a smaller extent.[26] Provision should be made to implement the policies to achieve sustainable water management goals. The predictable increased variability, availability and distribution of freshwater resources demands political commitment in supporting and advancing technology for the collection and analysis of hydrological data. More up-to-date information will enable policy makers to make better decisions regarding water resource management leading to sustainability.[26]

KEYWORDS

- **water**
- **water footprint**
- **conservation**
- **sustainability**
- **management**
- **society**
- **risk factors**

REFERENCES

1. Un.Org/water for life decade/water _and _sustainable _development .schtml.
2. Watconman.org/introduction/zibeline international publishing.
3. Gosmartbricks.com/ rain _ water _ harvesting _ methods _ in _ individual.
4. Patil Abhijit, S.; Patil Akshay, A.; Patil Somnath, P.; Saundade Sushant, D.; Chugule Kusum, A.; Katavare Shital, B.; Panhalkar Sachin, S. Analysis of the Agriculture Drought Severity and Spatial Extent Using Vegetation Health Index (VHI) in Manganga Watershed of Maharashtra. **2021**.
5. Network, R. W. S. Rural Water Supply Network Self-supply Site [Online] **2017**. http://www.ruralwater-supply.net/en/self-supply (accessed Mar 19, 2017).
6. Kinkade-Levario, H. *Design for Water: Rainwater Harvesting, Stormwater Catchment, and Alternate Water Reuse*; New Society Publishers, **2007**.
7. Dhanraj, S.; Gopi, R.; Loganathan, G. B. In *An Efficiency Study On Water Extraction From Air Using Thermophoresis Method*, IOP Conference Series: Materials Science and Engineering, Vol. 574, No. 1, p. 012003; IOP Publishing, **2019**.
8. Hanley, S. Inverted Umbrella Brings Clean Water & Clean Power to India. *Clean Technica* **2017**.
9. Pham, D. New Rooftop Solar Hydro Panels Harvest Drinking Water and Energy at the Same Time [Online] **2017** (accessed Jan 7, 2020).
10. Freeze, R. A.; Cherry, J. A. *1979. Groundwater*; Printice Hall lnc: Englewood Cliffs, New Jersey, **1994**.
11. Szatko, A.; Tang, Z.; Tian, S.; Rodie, S. Gray or Green Infrastructure Approaches: Regulating Stormwater Management in Land use Development. *Land Use* **2012,** *13*.
12. Major Floods Recharge Aquifers. University of New South Wales Science [Online]. January 24, 2011. (accessed Oct 31, 2012).
13. Dictionary, A. H. *The American Heritage Science Dictionary*; **2005**.
14. Panagopoulos, A.; Haralambous, K. J.; Loizidou, M. Desalination Brine Disposal Methods and Treatment Technologies-A Review. *Sci. Total Environ.* **2019,** *693,* 133545.
15. Fischetti, M. Fresh from the Sea. *Sci. Am.* **2007,** *297* (3), 118–119.
16. Ebrahimi, A.; Najafpour, G. D.; Kebria, D. Y. Performance of Microbial Desalination Cell for Salt Removal and Energy Generation Using Different Catholyte Solutions. *Desalination* **2018,** *432,* 1–9.
17. Buell, P. D.; Anderson, E. N.; de Pablo Moya, M.; Oskenbay, M. Food by Country. In *Crossroads of Cuisine*; Brill, **2020**; pp 204–286.
18. Cohen, Y.; Ed., Advances in Water Desalination Technologies. *World Sci.* **2021**, *17*.
19. Shamsuddin, M. K. N.; Sulaiman, W. N. A.; Suratman, S.; Zakaria, M. P.; Samuding, K. Conjunctive use of Surface Water and Groundwater via the Bank Infiltration Method. *Arab. J. Geosci.* **2014,** *7,* 3731–3753.
20. Obaideen, K.; Yousef, B. A.; AlMallahi, M. N.; Tan, Y. C.; Mahmoud, M.; Jaber, H.; Ramadan, M. An Overview of Smart Irrigation Systems Using IoT. *Energy Nexus* **2022,** 100124.
21. Weerasekara, P. The United Nations World Water Development Report 2017 Wastewater. *Future Food: J. Food, Agric. Soc.* **2017,** *5* (2), 80–81.
22. Grizzetti, B.; Liquete, C.; Antunes, P.; Carvalho, L.; Geamănă, N.; Giucă, R.; Woods, H. Ecosystem Services for Water Policy: Insights Across Europe. *Environ. Sci. Policy* **2016,** *66,* 179–190.

23. Clasen, T. Household Water Treatment and Safe Storage to Prevent Diarrheal Disease in Developing Countries. *Curr. Environ. Health Rep.* **2015,** *2* (1), 69–74.
24. Postel, S.; Richter, B. *Rivers for Life: Managing Water for People and Nature*; Island Press, **2012**.
25. Boelee, E. *Ecosystems for Water and Food Security, Nairobi: United Nations Environment Programme*; International Water Management Institute: Colombo, **2011**.
26. Cosgrove, C. E.; Cosgrove, W. J. The United Nations World Water Development Report–N 4–The Dynamics of Global Water Futures: Driving Forces 2011–2050 (Vol. 2). UNESCO, **2012**.

CHAPTER 12

Water Quality Index: Composite Measure of Quality of Water

PARVATHI JAYASANKAR[1], SATHEESH KUMAR K V[2], VEDAVALLI V[3], AGAMPODI SANDALI INUPAMA DE ZOYSA[3], and MEGHANA U[3]

[1]*Department of Chemistry & Biochemistry, JAIN (Deemed-to-be) University, India*

[2]*Postgraduate and research department of chemistry, Govt. College, Kasaragod, Kerala, India*

[3]*JAIN. Deemed- to-Be University, India*

ABSTRACT

Water is the prime natural resource that is facing deterioration globally. The measurement of the quality of this vital natural resource is important. Scientifically it is recommended that there should be a need for development of parameters for water quality assessment. Water Quality Index (WQI) is a summarized method of collating and indexing data of various water quality parameters, for reporting to the public. The various parameters of the water samples are analyzed and assigned an average weightage (wi) for each parameter. Further, a quality rating scale is assigned for each parameter taking into account national/international standards. At the end, WQI is expressed as an index number which is a representation of all the tested parameters that express the overall quality of water.

This chapter gives an idea about the advantages of using WQI and the various methods of calculation of WQI. The application and context of usage of these methods are discussed to give a comparative awareness. In conclusion, the significance of WQI in assessing the quality of water, different methods of calculation of WQI, and their limitations are discussed.

The Food-Energy-Water Triangle for Sustainable Development: Integrative and Synergic Solutions.
Pedro Aguilar-Zárate, Mónica Chávez-González, Ann Rose Abraham, and A. K. Haghi (Eds.)

12.1 INTRODUCTION

Water is the elixir of human life and water quality is a significant criterion in meeting the demands of a population with the supply of water resources available in the area. The availability of fresh water with adequate quality is of great significance for a civilized society. Also, agriculture demands the right quality of water for irrigation which is essential for the production of healthy crops.[1]

It is imperative to distribute and consume good quality water and maintain the hygiene of the water resources which is necessary for the sustainable growth of a civilized society. Both surface water and ground water act as the major water source in various parts of the globe and India. However, the municipal water supply must be monitored continuously for quality assurance of water supplies in various sectors viz industrial, agriculture, domestic uses and for ecosystem health and hygiene. According to Horton, the quality of water should be monitored in various aspects-considering physical, chemical, and biological parameters.[2] In developing countries, about 1.8 million people, including many children, die every year due to water-borne diseases.[3]

12.1.1 SIGNIFICANCE OF WQI IN WATER QUALITY TESTING

Water quality depends on various physical, chemical, and biological parameters. However, it is expensive and tedious for researchers to test, report, and compare various parameters to convey a comprehensive analysis of quality of water.[4] WQI analysis delivers the various test results to a simple expression in order to give an effective interpretation of the data, allowing public to access the water quality data.[5]

WQI analysis gives the idea about different contaminants present in the water source for various types of uses viz agriculture usage, industrial usage, domestic usage, and usage for maintaining aquatic life. WQI also gives an idea of the change in the quality of a water body over a period of time. Many of the water quality analyses using WQI reveal the fact that water quality improved in many of the water bodies during post-COVID sessions due to minimal anthropogenic activities. However, conventional methods of water testing will not yield results as the water body is not always static with respect to quality. Also, the nature of pollution in a water source varies across the water body with varied use of water across the different sections of water body. In the above conditions, the value of a relative index of water quality is handy in determining the relative quality of water across the water body. The index gives a quick view to the public and policy holders of a determination of the state of quality vis a vis different area of the water body or changes in the water quality over a period of time.

However, WQI should be used appropriately depending on the end use. The choice of parameters (ranging from 5 to 18) is generally used to arrive at a composite index of WQI. Also, weightages of different parameters have to be carefully chosen for the end objective in mind. Certain confined water bodies viz ponds are not real candidates where WQI can be used effectively to determine the quality of water.[6]

12.1.2 WATER QUALITY PARAMETERS AND SELECTION OF PARAMETERS FOR TESTING

To determine the composite water quality, chemical, physical, and biological conditions of the water need to be tested and compared. The type of pf parameters usually taken for these tests are the following:

a. Physical Indicators including temperature, electrical conductivity, taste, Total Suspended Solids (TSS), turbidity of water, Odor, and Total Dissolved Solids (TDS).
b. Chemical Indicators including acidity, alkalinity, Biological Oxygen Demand (BOD), Chemical Oxygen Demand (COD), Dissolved Oxygen (DO), Hardness, contaminants including phosphates, pesticides or nitrates, presence of surfactants and heavy metals.
c. Biological indicators including various types of bacteria, viruses, worms, and other aquatic animals.

To arrive at the WQI of a system, more than 5–15 parameters/indicators are used. The intent of the use of WQI viz; human consumption, agriculture or aquatic life, and the availability of data across the water body over a period of time are the prerequisites for the choice of parameters. The parameters are given relative importance based on the intent of the usage of the index. National and international standards are used to determine the weighted averages of the indicators in computation on the index (WQI).

Water quality standards are numeric values that describe water quality and fall within the following three categories:

a. Assigned uses of the water body as related to water supply—aquatic life, agriculture, or recreation.
b. Water quality criteria and general statements that describe good water quality and specific numerical concentrations for various parameters.
c. Policy and criteria designed to maintain and protect the existing water use for each water body.

Some of the BIS standards used for various parameters are given in Table 12.1. Based on the standards, the whole information is converted into a single value giving information about that particular water body at a specific duration.[7]

TABLE 12.1 Standards Chosen to Assess WQI.

Parameters	BIS standards (India)
Temperature	25–35°C
Acidity (mg/L)	30
Alkalinity (mg/L)	200
Conductivity (mg/L)	400
pH	6.5
TDS (mg/L)	500
TSS (mg/L)	1
COD (mg/L)	250
Total hardness (mg/L)	200
Chloride (mg/L)	250

The variation in concentration of any parameter (below and above the permissible limit) will contribute to the overall increase in the index value; therefore, the index is called as Integrated Water Quality Index (IWQi).[8]

There are different ways to interpret WQI. WQI value of 100 means that the water is pure, while the low number of zero means that the water is heavily polluted and contaminated, index can compare sites and track changes at one site over time.[9] The higher values of the WQI indicate better quality water, while the lower values show water of poor quality.[10] WQI values show an ascending scale, in case of increased with the degree of pollution (water pollution indices). However, indices have decreasing scale in less polluted water.[11] However, Table 12.2 gives a comparative view of the quality of water based on the final WQi values.[12]

TABLE 12.2 Water Quality Based on WQI Values.

WQi value	Quality assessment
91–100	Excellent
71–90	Good
51–70	Medium
26–50	Bad
0–25	Very bad

12.2 HISTORIC USE OF WQI AND METHODOLOGY OF COMPUTATION

Historically, WQI has been used in various forms and shapes since 1965. Horton developed the first WQI model in the 1960s in the US which based on 10 water quality parameters was considered to assess the quality of waterbodies.[13] The Environmental Department of Oregon and the Canadian Council of Ministers came up with different versions of WQI in 1970 and 1990, respectively. The Weighted Arithmetic Water Quality Index Method (WQI) is a better-suited method for determining the WQI as it gives a composite WQI giving weighted importance of the said parameters. Thus, numerous methods of calculation of WQI prevail in different parts of the world depending on their needs and areas of interest. There is no stable method suggested by experts for a worldwide usage as no method gives 100% objectivity and accuracy.[14]

To understand the different methods, four representative indexes and computation techniques are elaborated as representative samples.

a) National Sanitation Foundation: In 1970, National Sanitation Foundation (NSF) came up with a WQI calculation. NSF WQI is primarily used to compare water quality across multiple water bodies. In this index, nine water quality parameters are selected as the basis of the computation of NSF WQi. This is based on survey conducted among 142 water quality scientists.[2] The nine quality parameters chosen by the scientists are TDS (total of dissolved solids), pH, turbidity, phosphates, nitrates, BOD (biochemical consumption of oxygen), Coliforms, OD (dissolved oxygen) ,and temperature. Depending on the contribution, each parameter is given a weightage as per Table 12.3.[12]

TABLE 12.3 Weightage of Parameters for NSF WQI Calculation.[11]

Parameter	Weightage
TDS	0.07
Turbidity	0.08
OD	0.17
Coliform	0.16
BOD, pH	0.11
Temperature, nitrates, phosphates	0.1

NSF WQI can be calculated according to the following formula

$$NSF\ WQi = \sum_{i=1}^{n} WiQi$$

where W_i is the weightage of each parameter and Q_i is the sub-index quality of the *i*th parameter obtained from the conversion curve. Once NSFS WQi is calculated, ranges are determined to assess the quality as good, Bad, Excellent, etc. as per Table 12.4.

TABLE 12.4 Water Quality Assessment as Per NSF WQI Calculation.[12]

NSF WQi value	Quality assessment
90–100	Very good
70–90	Good
50–70	Medium
25–50	Bad
0–25	Very Bad

b) Oregon Water Quality Index (OWQi): It is a derivation of NSF WQi where a different method of calculation is done to arrive at the index. There are only eight quality parameters used for calculation of this WQI and the are temperature, dissolved oxygen, biochemical oxygen demand, pH, ammonia, nitrate nitrogen, total phosphorus, total solids, and fecal coliform. The main purpose of this index is to provide a simple method of determining the quality index for recreational purposes. The key difference is that weightages are not assigned and the arithmetic average is used as a basis for this calculation.

$$OWQI = \sqrt{\frac{n}{\sum_{i=1}^{n} \frac{1}{SIi^2}}}$$

where n is the number of parameters (total eight in number) and S_i is the sub-index of the *I*th parameter. The calculus of computing sub-index is explained in the paper.[12] The obtained value of the indexes is again graded on quality on ranges from Excellent to Very poor quality of water (Table 12.5).

TABLE 12.5 Water Quality Assessment as Per O WQI Calculation.[11]

O WQI value	Quality of water
90–100	Excellent
85–89	Good
70–84	Fair
60–70	Poor
0–59	Very Poor

Canadian Council of Ministers of the Environment Index (CCME–WQi): The first version of this index came in 2010 and is based on a formula agreed by British Columbia Ministry of Defence.[12] The following parameters are tested viz temperature, conductivity, color, turbidity, dissolved oxygen (OD), pH, alkalinity, Ca, Na, Mg, K, SO4, Cl-, F-, Dissolved Organic Carbon, P, Nitrates, SiO2, Al, As, Ba, Be, Cd, Co, Cr, Cu, Fe, Hg, Li, Mn, Mo, Ni, Pb, Se, Sr, V, Zn.[13] Essentially three elements are analyzed and studied (a) F1 (% of values which does not meet the objectives at least once during the period—*How many*), (b) F2 (percentage of individual tests which did not meet the goals at all—*How often*) and (c) F3 (is the factor which represents the amount through which the test values did not respect their guideline values—*How much*) are calculated.[12]

$$\text{F1 (Scope)} = (\text{No. of failed variables / Total no. of variables}) * 100$$

$$\text{F2 (Frequency)} = (\text{No. of failed tests / Total no. of tests}) * 100$$

$$\text{F3 (Excursion value)} = (\text{Failed test value}_i \text{ / Objective}) - 1$$

When the test value must not exceed the objective

Excursion = (Objective $_I$ */Failed Value* $_I$*) – 1*

For the cases in which the test value must not fall below the objective

The collective amount of excursions which are out of compliance is calculated according to the following formula:

Normalized sum excursion (nse) = Σ excursions / Total nmber of tests

Amplitude (F3) = nse / (0.01nse + 0.01) [16]

After determining F1, F2, and F3 through their individual formulas, CWQi is calculated using the formula.

$$CWQi = 100\text{-}\sqrt{(F1^2 + F2^2 + F3^2)}/\ 1.732^{12}$$

The CWQi thus calculated is then put into ranges and graded at different quality bands ranging from excellent to poor.

TABLE 12.6 Water Quality Assessment as Per CCME WQI Calculation.[12]

CCME WQI value	Quality of water
95–100	Excellent
80–94	Good
60–79	Fair
45–59	Marginal
0–44	Poor

c) Weighted Arithmetic Water Quality Index Method (WAQi): In this method, various physico-chemical parameters viz; (pH, BOD, COD, DO, P-PO4 3-, N-total, N-NO3-, N-NO2-, N-NH4+, SO4 2-, Cl-, Cr-total, Pb2+, Cd2+, Ni2+, Fe-total, Mn-total, Zn2+, As2+) are analyzed. WQI is calculated using the equation

$$WA\ WQI = \sum wiqi/\sum wi$$

q_i represents the relative value of the water quality.

where wi = *Weight of individual parameters.*

The following steps are done to arrive at the Weighted Average WQi index.[15]

Step 1: Collect data on various physio-chemical water quality parameters.

Step 2: Calculate Proportionality constant "K" value using formula

$$K = (1/(1/\sum Si))$$

where "Si" is the standard permissible for nth parameter.

Step 3: Calculate the quality rating for nth parameter ($q\ n$) where there are n parameters. This is calculated using the formula

q_n =100 { (v_n-v io)/(s_n-v io)}, where v_n = Estimated value of the nth parameter of the given sampling station.

v io = Ideal value of nth parameter in pure water.
And s_n = Standard permissible value of the nth parameter.

Step 4: Calculate the unit weight for the nth parameters. $Wn = (k/s\ n)$.

Step 5: Calculate WQI using formula,

$$WQI = ((\Sigma Wn\ X\ qn)/\Sigma\ Wn).$$

The indexes thus calculated are then graded as per ranges resulting in grading of quality of water from Excellent to Unsuitable for drinking (Table 12.7).

TABLE 12.7 WQI for Assessment of Water Quality.[16]

WQI	Assessment of quality
0–25	Excellent water quality
26–50	Good water quality
51–75	Poor water quality
76–100	Very poor water quality
>100	Unsuitable

Each WQI test method has its own advantages and disadvantages and has to be carefully selected depending on the following factors,

a. Number of parameters used for testing
b. Values of parameters experimented and calculated
c. Type of water body or sources
d. End objective—Whether the quality assessment is for general water quality indexing, assessing the extent of pollution, or assessing the potability of drinking water.

Some of the advantages and disadvantages of the WQIs are listed in Table 12.8.

TABLE 12.8 Advantages and Disadvantages of WQIs.

	NSF WQi	OWQi	CCME-WQi/CWQi	WAQi
Pros	• Fast and Objective • Index determines end water use	• Can determine the most impacted parameter • Sensitive to environmental factors	• High adaptability for different end objective • Low level of sensibility for missing data and hence can be used for automated sampling	• Suitable for surface water and groundwater • Importance of parameters can be completed • Easy to understand
Cons	• Does not use complex scale • Data loss during calculation	• Biological factors and impact not assessed	• No weightage for parameters • Biological parameters cannot be used	• Real quality of water is not assessed • Only determine the pollution level impact

12.3 CONTEXT OF USAGE OR ACCEPTANCE FOR WQI

As we discussed, WQi helps in the summarization of the water testing and placing a simple and effective way of water quality to the masses. However, as practitioners, we need to be sure about where and when to use these in context to the advantages and disadvantages of WQi indexes

- Advantages: WQi provides a lot of advantages in communicating the water quality.[15]

  a. Rationalization of the number of parameters required to compare water quality specific to the use of the index.
  b. Provision of a single number that represents overall water quality at a certain location and time.
  c. Provision of assurance on the safety of a water body for the specific use—habitat for aquatic life, irrigation water for agriculture and livestock, recreation activities, and drinking water supplies.
  d. An effective simple way for water quality monitoring is used by the general public and policymakers.
  e. Provides means of comparisons between different rivers and sampling sites.
  f. It simplifies a complex dataset into easily understandable and usable information.

- Disadvantages or Limitations: However, WQi usage has certain limitations and shortcomings which need to be known to the end users.[17]

  a. WQI is not an absolute measure of the degree of pollution or the actual water quality.
  b. Precision and accuracy are always a concern.
  c. Lack of a uniform method for measuring water pollution involving biological parameters.
  d. Inadequate to transfer complex environmental data into information.

12.4 CONCLUSION

The quality of water determines the healthy and sustainable life of society in an environment. Water pollution is a major cause of concern in various water user sectors viz; industrial, domestic, agriculture, aquatic life, etc. So, water quality analysis and thus mitigation of pollution of water resources is imperative for the socio-economic and environmental sustainability of the earth.

WQI gives a number that incorporates a comprehensive analysis of physico-chemical and biological parameters—pH, turbidity, conductivity, alkalinity, hardness, etc.—of water resources. According to the qualitative outcome, the water quality is divided into various indices like good, moderate, poor, etc. There are various methods involved in the calculation of water quality indices. This chapter deals with four different WQI calculation techniques—NSFWQI analysis, CCME WQI analysis WQI analysis, and weighted average WQI analysis.

12.4.1 LIMITATIONS IN USING WQI TO DETERMINE WATER QUALITY

Although there are many advantages in applying this technique, WQI may not be an absolute measure of the degree of pollution of the water body at a given time due to some reasons like,

- The WQI models are not generic as they are developed based on site-specific guidelines for a particular geographical region.[14]
- Despite the use of the same variables, the classification from each index differs, especially in the evaluation of drinking water. This difference is majorly due to the variation of anthropogenic activities in the particular geographical area chosen.[13]
- A single bad parameter value changes the overall assessment of WQI.

KEYWORDS

- **WQI**
- **quality rating scale**
- **index number**
- **average weightage**
- **relative importance**

REFERENCES

1. Chaturvedi, V.; Neog, K. K.; Basu, S.; Ghosh, A.; Gautam, S. K.; Jalan, I. *Reallocating Water for India's Growth: Sectoral Withdrawals, Water-efficient Agriculture, and Institutional Mechanisms*; Council on Energy, Environment and Water: New Delhi, **2020**.
2. Brown, R. M.; McCelland, N. I.; Deininger, R. A.; Tozer, R. G. A Water Quality Index-Do We Dare, **1970**.
3. World Health Organization. Water, sanitation and hygiene links to health [Online]. **2004**. www.who.int/wate sanitation-health/publications/fac ts2004/en/index.html
4. Lumb, A.; Sharma, T. C.; Bibeault, J. F. A Review of Genesis and Evolution of Water Quality Index (WQI) and Some Future Directions. *Water Qual. Expo. Health* **2011,** *3,* 11–24.
5. Madalina, P.; Breaban, L. G. In *Water Quality Index-An Instrument for Water Resource Management*, Conference: aerul şi apa componente ale mediuluiat: Cluj NapocaVolume, **2014**.
6. Ajoke Adelagun, R. O.; Etim, E. E.; Oko, G. E. Application of Water Quality Index for the Assessment of Water from Different Sources in Nigeria. **2021,** DOI: http://dx.doi.org/10.5772/intechopen.9869.
7. Alobaidy, A. H. M. J.; Maulood, B. K..; Kadhem, A J. Evaluating Raw and Treated Water Quality of Tigris River within Baghdad by Index Analysis. *J. Water Resour. Protect.* **2010,** *2,* 629–635.
8. Mukathe, S.; Wagh, V.; Panaskar, D.; Jacobs, J. A.; Sawanth, A. Development of New Integrated Water Quality Index (IWQI) Model to Evaluate the Drinking Suitability of Water. *Ecol. Indicat.* **2019,** *101,* 348–354.
9. Ram, A.; Tiwari, S. K.; Pandey, H. K.; Chaurasia, A. K.; Singh, S. ; Singh, Y. V. Groundwater Quality Assessment Using Water Quality Index (WQI) under GIS Framework. *Appl. Water Sci.* **2021,** *11,* 46. DOI: https://doi.org/10.1007/s13201-021-01376-7.
10. Banda, T. D.; Kumarasamy, M. Application of Multivariate Statistical Analysis in the Development of a Surrogate Water Quality Index (WQI) for South African Watersheds. *Water* **2020,** *12,* 1584. DOI: 10.3390/w12061584.
11. Uddin, M. G.; Moniruzzaman, M.; Khan, M.; Evaluation of Groundwater Quality Using CCME Water Quality Index in the Rooppur Nuclear Power Plant Area, Ishwardi, Pabna, Bangladesh. *Am. J. Environ. Prot.* **2017,** *5* (2), 33–43.
12. Călmuc, V.-A.; Călmuc, M.; Ţopa, M. C.; Timofti, M.; Iticescu, C.; Georgescu, L. P. Various Methods for Calculating the Water Quality Index, 2018, DOI: https://doi.org/10.35219/ann-ugal-math-phys-mec.2018.2.09. Annals of "Dunarea De Jos" University Of Galati Mathematics, Physics, Theoretical Mechanics Fascicle Ii, Year X (XLI) **2018**, No. 2, pp 171–178.
13. Uddin, M. G.; Stephen, N.; Agnieszka, O. I. A Review of Water Quality Index Models and their Use for Assessing Surface Water Quality. *Ecol. Indic.* **2021,** *122,* 107218.
14. Cude, C. G. Oregon Water Quality Index A Tool for Evaluating Water Quality Management Effectiveness, **2007**. DOI: https://doi.org/10.1111/j.1752-1688.2001.tb05480.x

15. Chandra, D. S.; Asadi, S. S.; Raju, M. V. S. Estimation Of Water Quality Index by Weighted Arithmetic Water Qualityindex Method: A Model Study. *Int. J. Civil Eng. Technol.* **2017,** *8* (4), 1215–1222.
16. Chatterjee, C.; Raziuddin, M. Determination of Water Quality Index of a Degraded River in Asanol Industrial Area, Raniganj, Burdwan, West Bengal. *Nat. Environ. Pollut. Technol.* **2002,** *1* (2), 181–189.
17. Akter, T.; Jhohura, F. T.; Akter, F.; Chowdhury, T. R.; Mistry, S. K.; Dey, D.; Barua, M. K. Islam, M. A.; Rahman, M. Water Quality Index for Measuring Drinking Water Quality in Rural Bangladesh: A Cross Sectional Study. *J. Health Popul. Nutr.* **2016,** *35,* 4–12.

CHAPTER 13

Environment Sustainability: Climate Change Mitigation, Renewable Resources, and Biodiversity Preservation Aided by IoT and AI

LINTA P. JOSEPH[1], LIYANA SHERIN[1], RIDHA ZAIN T.[1], SUNNY MATHEW[1], SARITHA K. JOSE[2], TINA SEBASTIAN[2], ARCHANA L. S.[3], VEENA R. K.[3], K. SREELATHA[4], and JINCEMON CYRIAC[1]

[1]*Marian College (Autonomous) Kuttikkanam, Kuttikkanam, Peermade, Idukki, India*

[2]*Department of Physics, Deva Matha College, Kuravilangad, Kottayam, Kerala, India*

[3]*Government College for Women, Vazhuthacaud, Thiruvananthapuram, Kerala, India*

[4]*Department of Physics, Sree Narayana College, Chengannur, Alappuzha, Kerala, India*

ABSTRACT

Environment sustainability is a necessary point of focus in this day and age. Resources are needed more and more quickly as the world's population grows by the second. Additionally, pollution and environmental degradation are also growing in scope. Carbon capture and storage is a technique for reducing greenhouse gas emissions caused by burning fossil fuels. Renewable energy sources are important for reducing dependence

The Food-Energy-Water Triangle for Sustainable Development: Integrative and Synergic Solutions.
Pedro Aguilar-Zárate, Mónica Chávez-González, Ann Rose Abraham, and A. K. Haghi (Eds.)

on fossil fuels. Hydro power, wind, geothermal, solar, and biomass power are the basis for sustainable energy. By replacing coal with solar energy, the threat of acid rain can be reduced. Renewable energy sources such as bioenergy, solar energy, wind energy, hydrogen, and geothermal energy offer sustainable solutions to the scarcity and harmful effects of nonrenewable sources. The integration of advanced technologies such as drones and deep learning algorithms can revolutionize wildlife monitoring and habitat preservation. By combining drone-borne thermal imaging with AI, researchers have made significant strides in locating ground nests of birds on agricultural land. This cross-pollination of conservation and agricultural practices contributes to wildlife preservation and ensures the long-term sustainability of food production. By harnessing these innovative approaches, we can navigate the complex challenges of conserving biodiversity in an ever-changing world, paving the way for a harmonious coexistence of humans and nature.

13.1 INTRODUCTION

Environmental sustainability has become one of the most pressing issues confronting mankind today. The increasing population, along with a massive increase in anthropogenic activity, has prompted various concerns about the sustainability of our earth's natural resources. The growing population and increase in average consumption have placed severe constraints on our planet's resources. Human activity and pollution have now affected every corner of the planet.[1]

According to estimates from a variety of organizations, including the United States Development Agency (USDA) and the Organization for Economic Cooperation and Development (OECD), the constantly growing emission of greenhouse gases has an immense impact on the earth and will end up resulting in a 2°C rise in temperature by the year 2050. The surface of the earth has altered and will keep on changing. As a result of global warming, polar ice caps and glaciers are melting at rates which are 2–3 times quicker than they were in the previous century. Estimates indicate that our planet is currently in one of the most severe periods of biodiversity loss, with unforeseen and unpredictable repercussions. According to the United Nations study, the occurrence of extinctions of species caused by human activity is hundreds of times higher than it was naturally, and if current trends continue, it might be thousands of times higher in the near future.

13.2 CLIMATIC CHANGES AND MITIGATION

Our heavy reliance on fossil fuels is a result of the extensive usage and utilization of petroleum derivatives, which, when coupled with the depletion of petroleum reserves, raises issues of environmental concern. There is unmistakable scientific proof that emissions of greenhouse gases (GHGs) such as carbon dioxide (CO_2), methane (CH_2), and nitrous oxide (N_2O), resulting from the burning of fossil fuels and changes in land use brought on by human activity, are causing climate change on Earth.

The term "greenhouse gases" (GHGs) refers to small gaseous elements with radiation characteristics capable of heating the atmosphere.[2] Some of the solar energy received in the upper layers of the atmosphere reaches the ground and heats it up. The amount of direct heating by solar energy relies on a variety of conditions, including surface reflectivity, cloud cover, and atmospheric dust. The interaction of the Earth's basic components, including the atmosphere, seas, biosphere (terrestrial and oceanic), and cryosphere (snow and ice), is also a key determinant of how energy is dispersed around the world.

The planet emits part of the solar energy it has absorbed back into space as infrared radiation to prevent overheating and continual energy absorption. GHGs absorb and reflect infrared radiation in certain bands of the infrared electromagnetic spectrum, including the Earth's atmosphere and surface. As a result, GHGs keep low-level air temperatures 35°C higher than what might otherwise be the situation. Without trace gases like carbon dioxide, the average temperature on Earth would be around –18°C. GHGs are thus life-sustaining; they account for less than 3% of the atmosphere's gaseous composition. In short, the gases that affect climate make up a relatively little fraction of the atmosphere, which is counterintuitive.

Since the onset of the industrial era, human activity has released massive quantities of GHGs into the atmosphere through industry, agriculture, energy generation, and transportation. Concerns have been raised that these emissions may be accidentally affecting the global climate by amplifying the natural greenhouse effect. Global mean temperatures could escalate by 1.5–5.8°C by the end of the coming century as a result of this increased radiative forcing, according to the Intergovernmental Panel on Climate Change (IPCC).[3,4] It should be highlighted that this warming is unusual in the last 10,000 years, even though it may seem small when compared to diurnal or seasonal levels of the temperature cycle. Concern among scientists is being raised not only by the magnitude of the change but also by the rate of global warming, particularly in light of how vulnerable these systems are and how they would react to climate change.

Changes in the distribution and pattern of precipitation and seasonalities will follow changes in the planet's surface temperatures. As the climate tends to change in the following decades, extreme climatic events may become more frequent and intense. Any significant and lasting shifts in the climate system will have an effect on human well-being, which is influenced by climate, food security, water quality, and environmental health. This is so because weather and climate have an impact on ecosystems, water accessibility and quality, agriculture, and the quality of the air. Burning of fossil fuels such as coal, gas, and oil results in the exothermic chemical process known as combustion.[5] Carbon dioxide (CO_2) is created as a chemical byproduct. Cars, trucks, power plants, and industrial facilities that consume these fossil fuels are the main sources of carbon emissions. Animals, including humans, create carbon dioxide, a colorless, odorless gas when they breathe. Photosynthesis, which keeps plants and animals alive, depends on carbon dioxide. However, as carbon dioxide builds up in the atmosphere, it can retain heat at the Earth's surface, causing warming. This is referred to as "the greenhouse effect." The majority of GHGs are CO_2.[6] Methane (CH_4), nitrous oxide (N_2O), hydrofluorocarbons (HFCs), perfluorocarbons (PCFs), and sulfur hexafluoride (SF_6) are additional GHGs.[7] Nevertheless, carbon dioxide is the main topic here.

The greenhouse effect is primarily caused by rising fossil fuel consumption, especially of coal and oil, which in turn causes rising carbon dioxide (CO_2) emissions. Global warming is caused by carbon emissions that prevent the sun's heat from escaping the atmosphere. Although people have always engaged in the process of burning, the onset of industrialization exponentially increased carbon emissions. Additionally, regardless of whether we live in a postindustrial civilization, emissions still rise as a result of rising energy demand as well as unsustainable energy production, distribution, and consumption.[8–11]

13.2.1 CARBON CAPTURE AND STORAGE

Carbon capture and storage (CCS) has the ability to play a significant role in accomplishing climate change goals by providing low-carbon power and heat, decarbonizing industry, and, more recently, allowing the overall elimination of atmospheric CO_2.

Carbon capture and storage, also known as CCS, is a technique for reducing greenhouse gas emissions caused by burning fossil fuels. The CCS process involves extracting CO_2 from significant energy-associated sources, capturing, transporting, and storing it using the many methods outlined below:

Storage options include:

- storage of liquid in the ocean
- solid storage (formed when CO_2 combines with metal oxides to generate stable carbonates)
- gas storage in various deep formations in the earth and environments (such as abandoned oil and gas fields, aquifers, etc.).

The removal of CO_2 from ambient air via geo-engineering technology is another name for CCS. Analysis of biological methods like biochar burial, which uses trees, plankton, etc. to take CO_2 from the air, has also been used in the process of carbon dioxide capture and storage. But for nonbiological processes, capturing carbon dioxide from burning at the source is preferable.[11–13]

Carbon dioxide can be collected after combustion or after the fuel has been treated before burning. The former is referred to as postcombustion capture, while the latter as precombustion capture. Nitrogen can be taken out of the air before combustion as well as fuel gases after combustion, in addition to the carbon capture aspect. CCS technology is available in a variety of forms, including chemical looping combustion and oxy-fuel combustion, depending on the need. In the oxy-fuel combustion process, pure oxygen is used to burn fossil fuels instead of air, resulting in a more thorough combustion. This produces an exhaust steam that contains water vapor, but condensation makes it simple to separate the water vapor from the carbon dioxide (CO_2). The key issue with this procedure is how to separate oxygen from air. Using the chemical looping type, an oxygen carrier is used to transfer oxygen from the air during combustion to the fuel. The gases produced by the oxidation of the fuel, such as carbon dioxide and water, escape the system as distinct steams because the fuel and combustion air are never mixed together.[14–19]

13.2.1.1 CHEMICAL LOOP COMBUSTION (CLC)

As CO_2 is naturally separated from the other fuel gas components, such as N_2 and unused O_2, and because no energy is required for the gas separation, chemical-looping combustion has become a popular choice for carbon dioxide capture. However, it is also important to consider how effective the existing technology is. If CLC were to be used to generate electricity from gaseous fuels without applying pressure, it would only be able to achieve a steam cycle's level of efficiency in place of a combined cycle, which is more efficient. This is typically not a problem, though, when the goal is to produce steam/heat, hydrogen, or if solid fuels are employed.

Two reactors—an air reactor and a fuel reactor—make up the CLC system. The process depends on the exchange of materials that carry oxygen between both the reactors. The oxygen carrier in the fuel reactor supplies the oxygen needed for fuel oxidation, while air in the air reactor deoxidizes the reduced oxygen carrier. At the fuel reactor exit, a flow of CO_2 and steam is thus obtained. The ability to carry oxygen and the reactivity of numerous synthetic materials, natural ores, and even some types of industrial waste have all been exploited as oxygen carriers in the past with varying degrees of success. The majority of the time, however, they have only been examined using a limited number of redox cycles, which is insufficient to cause and characterize any degradation. In order to increase the mechanical resistance of the particles, synthetic materials are typically composed of an active phase—a metal oxide that reacts with the fuel and is typically selected from Mn, Fe, Co, Ni, or Cu oxides—and a binding phase—an oxide that is stable in the conditions of the process, such as Al_2O_3 or SiO_2.

13.2.1.2 OXY-FUEL COMBUSTION

The idea behind oxygen fuel combustion technology is to burn the fuel with pure oxygen rather than air. The main components of the combustion process's flue gases include carbon dioxide and vaporized water. The high concentration of CO_2 in exhaust fumes makes it easier for it to be captured once water vapor condenses. Because nitrogen is not present, nitrogen oxide emissions are completely eliminated, barring leaks into the atmosphere or fuel impurities. Additionally, the absence of nitrogen causes the exhaust gases' high CO_2 concentrations. When compared to combustion using normal air, the absence of nitrogen in the oxidant mixture also results in reduced volumes in the combustion chamber and distinct combustion properties.

The availability of oxygen for burning is a significant issue. Currently, cryogenic methods are used to create oxygen from air. These traditional techniques are costly. Numerous researchers are looking into new, intriguing techniques that make use of membranes to distinguish oxygen from the air. Membranes can be created from ceramics or polymers. At high nitrogen concentrations at the permeate side, polymer membranes still offer substantial oxygen flow across the membrane. Ceramic membranes, on the other hand, produce extremely pure oxygen with a very low flux. Many researchers have been paying close attention to memory development in recent years. In the case of ceramic membranes, the goal is to obtain high purity of the membranes and high flow of oxygen.

The majority of oxycombustion research nowadays is compelled to use ceramic membranes with combined ionic and electronic conductivity properties. Additionally, significant scientific and industrial work have recently been concentrated on creating ceramic membranes for the partial oxidation of hydrocarbons to produce syngas. Particularly actively investigated is the process of producing syngas via partial oxidation of methane (POM).

13.2.1.3 CCS: ADVANTAGES AND DISADVANTAGES

According to the IPCC, 7000 industrial facilities (which include thermal power plants, cement factories, steel plants, refineries, and petrochemical plants) might adopt carbon capture and storage by 2050, potentially reducing carbon emissions by one-third. Additionally, it might significantly improve the ethical production of extra-heavy oil and sour gas. Capture (separation plus compression), transport, and storage (which involve measurement, monitoring, and verification) are the three main parts of a CCS system. These parts are offered in commerce in one shape or another. However, there is a limited commercial expertise in assembling all the parts to completely integrate CCS systems at a size that would probably apply to their implementation in the future. Carbon dioxide is able to be captured both during and after combustion, depending on how the fuel was prepared. Both precombustion capture and postcombustion capture are used to describe these processes. Along with the carbon capture component, nitrogen can also be taken out of the air prior to combustion as well as from exhaust gases following combustion. Depending on the requirement, CCS technology is offered in a range of configurations, such as oxy-fuel combustion and chemical looping combustion. When burning fossil fuels using oxy-fuels, pure oxygen is used in place of air, leading to more complete combustion. Condensation makes it straightforward for separating the water vapor and the carbon dioxide (CO_2) in the exhaust steam, which is produced as a result of this. Even though there is a remote chance of separation, it is preferable to store the CO_2 that has been caught in deep geological structures far from populous regions in order to reduce carbon emissions. There is a chance of leakage and subsequent rise in water acidity if it is kept in deep ocean masses. There are questions regarding the potential effects on marine ecosystems and how long CO_2 injected into the depths of the ocean will stay there. Additional study and development are required for this aspect. Moreover, because of CO_2 emissions into the atmosphere, the oceans are already becoming more acidic.[8,10]

Sources of renewable energy, when used to produce energy, help in reducing GHGs, which in turn aids in reducing pollution from energy sources based on fossil fuels, and helps to slow down climate change and also addresses issues for the environment and human health. Switching to renewable energy sources is the best course of action for halting climate change and supplying future generations' energy requirements.

13.3 RENEWABLE ENERGY SOURCES

For a population that is growing, renewable energy sources are crucial because nonrenewable resources will eventually run out. The most crucial factor is how utilizing them affects the environment. The best energy source should be chosen based on aspects including cost, high stability, economic benefits, and environmental effects. Even though they still constitute a threat to the environment and are responsible for many ecological problems, fossil fuels are still used to generate electricity. The usage of renewable resources must begin right away. It has a lower environmental impact than conventional fossil energy technologies.[20,21]

Renewable energy can be defined as energy sources that are plentiful in nature. The natural sources that provide renewable energy are the Sun, water, and wind. Fossil fuels are plentiful, yet consuming them will be terrible for the ecosystem nonetheless. Renewable energy sources are also referred to as green or clean energy because they are actually environmentally favorable. The use of fossil fuels can contribute to global warming by releasing large amounts of greenhouse gases. The pollutants that are formed in the environment can create many health issues as well as environmental issues.[21] We must use renewable, eco-friendly resources in daily life to protect the Earth. Hydro power, wind, geothermal, solar, and biomass power is the basis for the future of sustainable energy.[21,22]

13.3.1 BIOENERGY

The term "bioenergy" is derived from the word biomass. The use of biomass energy may minimize landfill waste, reliance on foreign petroleum, ozone-depleting substance emissions, and the greenhouse gas emissions. It additionally helps to support the small-scale agriculture and rural business initiatives. The paper mill waste, ambulatory plant byproduct, and municipal waste make up the three main biomass feedstocks for energy. The most well-known feedstocks for biomass power plants are soybean oil which can be used as biodiesel and maize grain for ethanol. The most beneficial use of biomass has some drawbacks but can also be seen as a part of the earth's carbon cycle. For instance, compared to energy sources derived from petroleum, biomass is a poor source of energy. The other application of biomass is biopower, which has the second position in power generation using renewable sources. This can be used as a replacement for coal. Additionally, by lowering sulfur emissions, the usage of biopower can lessen the possibility of acid rain.[23]

13.3.2 SOLAR ENERGY

When it comes to energy sources, the Sun is the largest source and it is present in the form of light and heat. Solar energy can be described as energy that is produced by the Sun. The origin of solar energy is the nuclear fission held in the Sun's atmosphere. There are many advantages to using solar energy. It does not emit any carbon dioxide like fossil fuels, so it does not cause global warming.[21,22] The amount of solar energy received by the earth is very high, so there will not be any shortage in availability. A very high amount of hydrogen is transformed into helium each second during the process, which is called the thermonuclear process. The solar energy generated has to have two parts: the collecting part and the storage part to generate power. The storage part will store extra energy generated during production and distribution. Solar energy can help solve the energy crisis in the world by using it in a systematic way. The world's oil reserves will only last up to 30 or 40 years; the use of this oil causes large amount of pollution. Solar energy does not cause any pollution, and there is no part that shifts. Due to the quantity of energy that may be generated by the solar panel, it will be considered a significant investment over a long period of time. Ancient people built their houses on south-facing cliff edges to absorb all the heat in the winter season and block maximum sunlight in the summer season, developing the use of passive solar lighting and solar heating. The passive solar design uses large, south-facing windows and brick flooring to absorb the heat of the Sun and release it back into the house at night. Energy-saving elements like adequate insulation, energy-efficient windows, and appliances can save a lot of money and energy.[21]

13.3.3 SOLAR ELECTRICITY

Using photovoltaic technology, sunlight can be converted directly into power. Since the initial days of the space programme, this technique has been the key source of power for spacecraft. For the last few decades, photovoltaic technology has also been used to run small electronics and agricultural applications. For powering urban residences and other buildings that are connected to the grid, a strong solar electric market has emerged during the past ten years. Modules, which convert sunlight into energy, convert that power into alternating current, and batteries that store extra electricity produced by the system are the three major components of solar electric systems.[24]

Solar cells are electrical appliances that use the photovoltaic effect to transform light energy into electricity. Solar cells serve as the fundamental elements of photovoltaic modules, and they also serve as light detectors, photodetectors, and light intensity meters. Three fundamental elements are important for a photovoltaic cell to function: its ability to absorb light, to separate charge carriers, and to separately extract those carriers for use in an external circuit. The intensity of the light will be determined by the voltage produced by the solar cell. The covalent bond formed between the semiconductor material's electrons and photons makes up the electromagnetic radiation, which energizes the electrons and causes them to emit. The term "photoelectrons" refers to the energetic electron, while the photoelectric effect refers to the phenomenon of electron emission, which is necessary for the photovoltaic cell to function. There are many studies held to increase the efficiency of solar cells with the minimum amount of pollution. The efficiency of solar cells is only 20 percent, which is very low, so there must be a new technique to increase the efficiency of solar cells. The product cost of solar cells is very low, and it is the best way to protect the environment.[24] These solar energy sources can be used in daily life in many ways, like as heaters, solar cookers, solar dryers, etc. The antisolar cell idea actually says that electricity can also be produced during the night when the sun is not there, which is actually the opposite of the traditional theory of solar cells. The same basic principle of traditional photovoltaic cells is used in thermo-radiative cells also, but the only difference is that here the current flows in the opposite direction and the resultant voltage has the opposite sign.[24]

13.3.4 SOLAR WATER HEATER

Solar water heater is the heating system that heats either water or a heat-transfer fluid, such as nontoxic antifreeze, with the use of a solar collector and water storage tank. Flat plates, evacuated tubes, and concentrated solar collectors are the three variations used to heat water. While active systems will use an electric pump to move water through the system, passive systems use normal water pressure to move water from a solar collector and into a storage tank. Although active systems are often more costly than passive systems, they are also more effective.[21–23]

13.3.5 WIND ENERGY

Wind energy has a very simple cycle that converts the wind's active energy, which is basically a movement, into mechanical energy, which is then used to produce electricity. Several advantages come with wind energy, including lower power costs and a decrease in ozone-harming material releases. Based on the axis of the blade rotation, there are two main categories of turbines. The most affordable renewable electricity technology for the bulk power market is wind energy, which is also a good fit for remote and scattered applications. The remote power industry is very interested in hybrid solutions, which combine a wind turbine with another renewable energy source and/or a traditional backup unit. In 1994, there were more than 1700 MW of installed wind turbines worldwide, with the majority of them situated in California.[25]

13.3.6 ENERGY FROM HYDROGEN

The most prevalent and most basic element in the universe is hydrogen. Through reforming, industry annually creates more than 4 trillion cubic feet of hydrogen. In order to create hydrogen from natural gas for proton exchange membrane fuel cells, researchers are creating highly efficient reformers. With the ability to store, transport, and transmit energy in a form that can be used, hydrogen may one day join electricity as an energy carrier.

Researchers eventually hope to use solar, wind, biomass, and biological technologies to directly create hydrogen from water.[20]

13.3.7 GEOTHERMAL ENERGY

Geothermal energy systems utilize the heat produced by the molten part of the earth as well as the radioactive material decaying in the crust. Currently, this energy is being used at an array of sites all over the world to generate power at a low cost and to deliver energy directly for space heating, food and industrial processing, aquaculture, and refrigeration. There are mainly three methods to create electricity: dry steam, flash steam, and binary conversion. Geothermal electricity might boost its present market share further with advances in conversion technology and methods for discovering and extracting geothermal energy. Low-temperature geothermal energy is used directly for space heating and the geothermal heat pump, which are examples of geothermal heating techniques.[22]

Climate change mitigation, energy use, and economic development are all going to benefit from renewable energy sources.[26] While distributing and emptying the world's resources, it is important to reduce any potential risk. Decrease in oil savings and an increase in ozone-harming compounds have resulted from the current consumption of petroleum gas and the other nonrenewable energy sources. It could be difficult for researchers to convince people to switch to renewable energy sources. In order to accomplish this, people must have access to accurate information about medical issues and environmental disasters that have impacted large urban areas. To encourage people to embrace sustainable power sources, they must create an environment where they are useful. In order to address the security issues that are related to petroleum products, such as explosions and collapsing coal mine shafts, environmentally friendly power can be created using the renewable sources.[21,22] Renewable energy sources reduce reliance on fossil fuels and minimize negative impacts on biodiversity, allowing us to mitigate climate change, reduce pollution, and protect biodiversity for future generations. In addition to this, newer technologies like IoT and AI technologies enable data-driven conservation to protect biodiversity.

13.4 BIODIVERSITY PRESERVATION

One thing that cannot be ignored while talking about environment sustainability is biodiversity. Biodiversity is a term that includes numerous components. From a tiny insect to large terrestrial animals like an elephant, biodiversity encompasses this huge variety of species, ecosystems, and genetic diversity. It essentially weaves a web of life laying a foundation for life-sustaining ecosystems and processes. From the moment we awaken in the morning to the moment we go to bed, everything we require comes from nature. It provides services such as nutrient cycling, climate regulation, and even air and water purification. The functionality and resilience of biodiversity is reliant on all its components, no matter how big or small. A common misconception, though, is that biodiversity is essentially just species richness but that is not entirely correct.

It is often customary to define biodiversity as "diversity" at three different scales: genetic, species, and ecosystem. Furthermore, the technical literature does not often make it obvious whether diversity at any of these levels is quantified by merely counting entities like species and ecosystems, or by a more complicated method (such as an information theoretic index). Again, "diversity" is nearly often interpreted as the quantity of things by both the general public and conservation biologists. But lately, structural and process diversity have been referred to by some scientists as essential elements of biodiversity (e.g., Franklin, 1988; Franklin et al., 1989) because in reality, species richness does not directly imply better forest health, in fact, sometimes ecosystem management can even decrease the species richness. So all these factors have to be taken into account rather than simply the number of species.[27]

The rapid growth of human population, as well as the growth of technology, demands availability of more and more resources. If we compare our current way of life to that of someone who lived a few centuries ago, we would be shocked to see the difference in amount of resources consumed. Technology, despite being useful in so many ways, can also drastically increase our requirement of raw materials from nature. Things like over consumption and capitalism, which is often poorly regulated, can directly and indirectly impose an enormous amount of stress on our natural resources. Habitat loss, which is mostly caused by deforestation, urbanization, and intensive agriculture, is another major problem. Ecological equilibrium is disrupted by habitat fragmentation and degradation, which results in species extinction and ecosystem collapse. By changing temperature and precipitation patterns, creating shifts in species' ranges, and interfering with crucial ecological processes, climate change makes these problems worse. Extreme weather and rising global temperatures can cause migratory patterns to be disrupted, keystone species to become extinct, and have a domino effect on other ecological processes.

The rapid rise in human population, in addition to other factors like habitat loss, pollution, poaching, over-exploitation, and climate change, poses a serious threat to the well-being of biodiversity. It is then undoubtedly necessary to protect or preserve biodiversity in order to sustain life on earth. Biodiversity preservation demands us to look at an ecosystem as a complex interconnected system of all its components rather than just, say, seeing a forest as simply a producer or timber.

Some of the methods to preserve biodiversity include establishing protected areas in the form of National parks, reserves, and sanctuaries, restoring habitats, increasing the use of sustainable practices, and spreading public awareness. In order to preserve biodiversity, we need to first learn about its current state and what the required changes should be. Extensive and rigorous monitoring is a key prerequisite for practically all studies needed to manage forests for biodiversity.

One thing that could be helpful in studying, and thus managing forests is the use of umbrella and indicator species. There is little research on these as management shortcuts, but their utility can be verified by extensive field study. However, as the indicator species may stop reflecting the status of other species at some point, management practices should not "evolve" toward management of these indicator species. Managing an umbrella species is not inherently contradictory, but since distinct umbrella species may protect various sets of species, management for one may be antagonistic to management for the other. The idea of keystone species may be helpful for managing forests. So we can say that, if management of one particular species results in effective management of other species in the area, then it can be useful to manage these particular 'keystone' species.[27] These keystone species have a greater impact on more species than what one would expect from their biomass or numbers.

13.4.1 IOT FOR DATA COLLECTION

Against the background of growing concerns over depleting natural resources and increasing human population, it is extremely useful, or even necessary, to look for more innovative and increasingly efficient methods to attain sustainable development goals. One such technology, which has a lot of potential, is the usage of IoT (Internet of Things) for collecting data.[38]

Internet of Things (IoT) revolutionizes data collection and its analysis by enabling the seamless linking of numerous devices, sensors, and systems. In order to collect data for environmental sustainability, we can use Internet of Things (IoT). By doing so, we can better understand ecosystems, encourage wise decision-making, and support successful conservation measures. By connecting devices and sensors to the Internet, we can directly collect and analyze data in real time.

IoT introduces an innovative model where connected objects, or "things," communicate with or without human intervention. IoT extends its data utilization beyond diagnostic data or data corresponding to consumption, in contrast to existing technologies like Automated Meter Reading (AMR) or condition monitoring systems. 'Big Data' is essential for integrating and utilizing the data from IoT devices, resulting in outcomes that are advantageous to all parties involved, including water utility owners and users. Easy data access and utilization are made possible by Internet connectivity via cellphones or PCs.[28]

"Big Data" is a cutting-edge technical concept for gathering enormous volumes of pertinent data from sensors deployed to keep track of structure utilizations, condition, and system performance. By implementing Internet of Things (IoT) technology throughout the entire system, this Big Data concept can be realized.[29] Big Data, being a quickly developing technical paradigm, has many definitions. These definitions typically include the gathering of enormous volumes of data, drawing insightful conclusions from previously unavailable material, and creating technology for data gathering. Big Data relies on the integration of information from various sources, such as sensors installed on infrastructure components that retrieve specific data like strain, temperature, pressure, flow, and corrosion, in order to collect enormous amounts of data. Big Data is built on this integration, also known as "data fusion."

IoT comprises several key elements which are as follows,

- A collection of physical components including sensors, actuators for the sensors, and communication boards integrated into a hardware system.
- An intermediate software layer that includes various data storage options such as cloud networks or built-in storage within Internet of Things (IoT) devices.
- A system for displaying information to users, facilitating their interaction, interpreting data, and identifying events, all within a user interface.

The future development of IoT represents a significant advancement in technology, integrating artificial intelligence and controlling various aspects of human well-being. This development aims to minimize human involvement, as algorithms derived from Big Data analysis will make most procedural decisions in regular processes. Initial network connections of things have already been utilized in various areas, such as installing pressure monitoring sensors in petroleum pipelines to prevent theft, vandalism, leaks, and ruptures. Pore-water pressure sensors have also been deployed to provide early warnings of potential landslides.[29] However, these early developments are limited to specific data collection.

Some of the ways in which IoT is useful in promoting sustainability are,

- IoT for more sustainable living arrangements (Smart Cities): A key component of sustainable smart cities is Internet of Things (IoT), which has the potential to improve environmental sustainability. The majority of recent research, however, has concentrated on the economic development and quality of life in smart cities, ignoring the contribution of the Internet of Things to improve or increase environmental sustainability. Urban planning, transportation, energy, infrastructure monitoring, and other areas of smart, sustainable cities can all benefit from the IoT and related big data applications. Internet of Things (IoT) makes it possible to gather and analyze urban data in order to improve energy efficiency and encourage environmentally sustainable development. The design and execution of smart cities, however, do not place enough emphasis on environmental sustainability, and the promise of cutting-edge technology like IoT is frequently disregarded. In response to the convergence of sustainability, urbanization, and ICT (Information and Communication Technology), the idea of smart sustainable cities was born. These cities make use of cutting-edge ICT to efficiently manage resources, enhance societal and economic outcomes, and satisfy the requirements of the present and future generations while respecting environmental constraints. Smart sustainable cities must integrate institutional, policy, urban design, scientific, technological, environmental, and technological advancements into complex socio-technical systems.
- IoT Data Collection System in Agriculture: Using IoT, we can create a system that includes a device to gather data and deliver it to cloud storage, where it is analyzed by machine learning algorithms to forecast environmental variables important for fungal diagnosis and

prevention. The IoT system gathers data on air temperature, relative air humidity, wind speed, and rainfall as these elements have a significant impact on the development and dispersal of fungal spores.[30] The data is processed by machine learning algorithms, which then forecast short-term environmental variables to help in the detection and treatment of fungal illnesses. The IoT system enables crop field managers to make informed decisions about disease management and prevention by delivering real-time information and predictions. By minimizing crop losses and reducing the need for chemical interventions, it supports resource efficiency and sustainable farming practices.[30]

13.4.2 ARTIFICIAL INTELLIGENCE AND MACHINE LEARNING FOR WILDLIFE PROTECTION

13.4.2.1 WHAT IS ARTIFICIAL INTELLIGENCE?

The science and engineering of building intelligent robots and computer programs is known as artificial intelligence (AI). It entails creating machines with goals and problem-solving abilities akin to human intelligence. But intelligence itself is a nuanced idea that differs between people, animals, and machines. AI does not just try to imitate human intelligence; it also studies the challenges that intelligence faces in the real world. After World War II, scholars like Alan Turing made significant advances to the field of artificial intelligence. According to Turing's test, a computer is said to be intelligent if it can persuade an informed observer that it is a human.[31] AI strives to develop intelligence comparable to that of humans, although viewpoints vary on whether this can be done just by writing programs or if new, basic concepts are required.

Although they are relevant to AI, computational complexity and computability theory do not deal with the core issues. AI's objective is to create algorithms that can solve problems as effectively as humans, especially if the problems fall within theoretically complex fields, despite the fact that these theories offer insights into the solvability and difficulty of specific problems.

The goal of the discipline of artificial intelligence (AI) is to build intelligent machines and software. Although it strives for human-level intelligence, the intellectual systems that make this possible are difficult to comprehend and replicate.

13.4.2.2 WHAT IS MACHINE LEARNING?

Machine learning (ML) is an evolving discipline of computational methods which aims to mimic human intelligence through the process of learning from its surroundings. It finds many applications across various different fields like pattern recognition, computer vision, aerospace engineering, finance, entertainment, computational biology, and biomedical applications.[32] ML has emerged as a key tool with the rise of big data.

ML algorithms can be divided into supervised and unsupervised learning types. The output that is wanted and the type of data being examined determine the algorithm to use. Algorithms are trained on labeled data in supervised learning to generate predictions or categorize data. On the other hand, unsupervised learning algorithms automatically find structures and patterns in unlabeled data.

Continuous advancement is possible in many sectors thanks to the ML algorithms' capacity to generalize from existing contexts and learn from previously untried tasks. It enables enterprises to automate difficult jobs, optimize processes, and produce more accurate predictions using massive volumes of data. The ability of ML to find hidden patterns and correlations in datasets has changed how we approach decision-making and problem-solving.

The field of machine learning is fast developing and has applications in many different sectors. ML keeps changing our environment, from improving software user experiences to revolutionizing healthcare analytics. As the big data era progresses, ML algorithms will become more and more crucial in converting data into insightful understandings and spurring innovation in a variety of industries.

13.4.2.3 HOW ARE THEY USEFUL IN WILDLIFE PROTECTION?

Forests are essential to the continuation of life on Earth because they support biodiversity, provide environmental services, and contribute to renewable energy and global food security. However, a number of issues, including deterioration, deforestation, and the effects of climate change, beset forests. Traditional methods of managing forests, such as those based on pen and paper, have drawbacks, including bias, sluggish data collecting and analysis, and lack of scalability.[34] Sectors like agriculture, in contrast, have embraced technological solutions more quickly.

The application of AI technology can address the shortcomings of conventional methods for collecting and analyzing forest data, resulting in more efficient forest management and conservation. Artificial intelligence

(AI) has the ability to close data collecting gaps, foster innovation, and speed up transformational change in the forest industry. New opportunities for the use of artificial intelligence (AI) in the forestry and wildlife sectors have emerged as a result of the development of data science, digital technology, and satellite technology. India can gain from the adoption of AI technology to efficiently monitor, manage, and maintain its biodiversity and forest resources because it is a biodiverse region with a sizable amount of forest cover.

13.4.2.4 UNMANNED AERIAL VEHICLES (UAVS) IN WILDLIFE MONITORING AND CONSERVATION

The monitoring and conservation of wildlife is being revolutionized by unmanned aerial vehicles (UAVs) and artificial intelligence (AI). Traditional ground-based monitoring methods, such as video traps and surveys, need a lot of resources and may not be accurate. UAVs combining AI and thermal imaging offer the possibility of conducting affordable surveys of big areas.[35] These systems have a video processing pipeline and infrared image collection for tracking and classifying animals in wide spaces and woodlands. They have also undergone successful testing, picking up all desired wildlife species. Because of their overhead perspective and capacity to overcome obstacles like wide geographic ranges and low population densities, UAVs have advantages when it comes to monitoring invasive and vulnerable species.

Some difficulties regarding UAVs include regulator restrictions and the requirement for sophisticated automatic image detection algorithms. UAVs have been used to track radio-tagged wildlife, monitor sea turtles, black bears, elephants, marine mammals, and birds. An essential area of study for increasing effectiveness is automating animal detection and counting in UAV footage. By providing precise population estimates and reducing disturbances, UAVs combined with AI have the potential to significantly improve wildlife monitoring and conservation efforts.

13.4.2.5 INTEGRATION OF DRONE-BORNE THERMAL IMAGING AND AI TO LOCATE BIRD NESTS ON AGRICULTURAL LAND

Drone-borne thermal imaging is a technology which has potential for being used to find bird nests on agricultural property. Traditional survey techniques for monitoring biodiversity are frequently time-consuming and constrained, highlighting the demand for effective and cutting-edge technologies. Sensor-equipped drones provide an affordable means of spotting and keeping an eye on wildlife.[36] By effectively identifying nests using technology, this novel strategy helps safeguard farmland birds and conserve wildlife. A scalable method for nest detection is provided by the incorporation of AI and thermal sensors in drone surveys, which reduces the negative effects of farming operations on bird populations.

13.5 CONCLUSION

Renewable energy sources, such as bioenergy, solar energy, wind energy, hydrogen, and geothermal energy, offer sustainable solutions to the scarcity and harmful effects of nonrenewable sources like fossil fuels. They emit little to no harmful gases, reduce greenhouse gas emissions, and have minimal environmental impact.

Bioenergy utilizes biomass to generate biofuels and reduce landfill waste. Solar energy is abundant, clean, and cost-effective, converting sunlight into electricity through solar cells. Wind energy harnesses the wind's kinetic energy to produce electricity, with continuous advancements in technology. Hydrogen, the most abundant element, can be produced from various sources and offers a clean energy carrier. Geothermal energy harnesses the Earth's heat for power and provides direct heat applications. By prioritizing renewable energy and educating the public about its benefits, we can mitigate climate change and pave the way for a greener future.

While exploring the potential of renewable energy sources to combat climate change and promote sustainability, it is just as important to utilize advanced technologies for effective conservation of biodiversity. While traditional survey methods are laborious and resource-intensive, advanced technologies like AI and ML can overcome these limitations. Automated identification of species and habitats using large-scale data collection can provide valuable insights for conservation planning and decision-making.

Further, we can say that the convergence of biodiversity preservation, Internet of Things (IoT), artificial intelligence (AI), and machine learning (ML) present promising opportunities for effective conservation of biodiversity. This is extremely necessary in this day and age because of rapid loss of biodiversity and the limited resources available for monitoring and conservation. The integration of advanced technologies like drones equipped with various sensors and the application of deep learning algorithms can revolutionize wildlife monitoring and habitat preservation. By combining drone-borne thermal imaging with AI, researchers have made significant strides in locating ground nests of birds on agricultural land.

The integration of AI and ML in precision agriculture, along with drone technology and sensor networks, opens up opportunities to protect ground nests and mitigate the impact of farming operations on wildlife. This cross-pollination of conservation and agricultural practices not only contributes to wildlife preservation but also ensures the long-term sustainability of food production.

By harnessing the potential of advanced technologies, such as drones and deep learning algorithms, we can enhance wildlife monitoring, habitat preservation, and cross-sector collaborations. Increasing the focus on these innovative approaches will enable us to navigate the complex challenges of conserving biodiversity in an ever-changing world, paving the way for a harmonious coexistence between humans and nature.

KEYWORDS

- **carbon capture**
- **storage**
- **renewable resources**
- **IoT**
- **artificial intelligence**

REFERENCES

1. Arora, N. K. Environmental Sustainability—Necessary for Survival. *Environ. Sustainability* **2018,** *1,* 1–2. DOI: https://doi.org/10.1007/s42398-018-0013-3.
2. Beniston, M. Climatic Change: Possible Impacts on Human Health. *Swiss. Med. Wkly.* **2002,** *132* (25–26), 332–337. DOI: https://doi: 10.4414/smw.2002.10041. PMID: 12422289.
3. IPCC; *Climate Change. The IPCC Second Assessment Report.* Cambridge and New York: Cambridge University Press; Volumes I (Science) II (Impacts) and III (Socio-economic implications); **1996**.
4. IPCC; *Climate Change. The IPCC Third Assessment Report.* Cambridge and New York: Cambridge University Press; Volumes I (The Scientific Basis) II (Impacts, Adaptation, and Vulnerability) and III (Mitigation); **2001**.

5. Butt, T. E.; Giddings, R. D.; Jones, Keith. Environmental Sustainability and Climate Change Mitigation—CCS Technology, Better Having it Than not Having it at all!. *Environ. Prog. Sustain. Energy* **2012,** *31*. DOI: https://doi.org/10.1002/ep.10590.
6. Total; *Ten Questions for Total—Environment and Society Report 2009*, Courbevoie, France: Total; **2009**.
7. HTC Purenergy; Glossary of Industry Terms, Glossary of Terms—CO_2, HTC Purenergy; **2009** [Online]. Available at: http://www.htcenergy.com/mediaCenter/glossary_CO2.htm.
8. Caldeira, K.; Wickett, M. E. Oceanography: Anthropogenic Carbon and Ocean pH. *Nature* **2003,** *425,* 365. DOI: https://doi.org/10.1038/425365a.
9. IPCC (Intergovernmental Panel on Climate Change). Carbon Dioxide Capture and Storage. In *The IPCC Special Report, Prepared by Working Group III of the Intergovernmental Panel on Climate Change*; Metz, B., Davidson, D., Coninck, H. C., Loos, M., Meyer, L. A., Eds.; Cambridge University Press: Cambridge, UK, **2005**.
10. Haszeldine, S. *Carbon Capture and Storage*; University of Edinburgh: CSC UK, **2009**.
11. Total. Energies No. 17 - France, Here and There, Total, **2010**.
12. POST (Parliamentary Office of Science and Technology). Carbon capture and storage (CCS), Postnote, Number 238, Parliamentary Copyright, **2005**, March [Online]. Available at: http://www.parliament.uk/briefing-papers/POST-PN-238.
13. Saether, A. K. *Bellona Report Calls for Legislation on Carbon Capture and Storage*, **2009** [Online]. Available at: http://www.bellona.org/articles/articles_2009.
14. Lyngfelt, A.; Leckner, B.; Mattisson, T. A Fluidized-bed Combustion Process with Inherent CO2 Separation, Application of Chemical-looping Combustion. *Chem. Eng. Sci.* **2001,** *56,* 3101–3113. DOI: https://doi.org/10.1016/S0009-2509(01)00007-0.
15. Cho, P.; Mattisson, T.; Lyngfelt, A. Comparison of Iron-, Nickel-, Copper- and Manganese-Based Oxygen Carriers for Chemical-looping Combustion. *Fuel* **2004,** *83,* 1215–1225. DOI: https://doi.org/10.1016/j.fuel.2003.11.013.
16. Buhre, B. J. P.; Elliott, L. K.; Sheng, C. D.; Gupta, R. P.; Wall, T. F. Oxy-fuel Combustion Technology for Coal-fired Power Generation. *Prog. Energy Comb. Sci.* **2005,** *31,* 283–307. DOI: https://doi.org/10.1016/j.pecs.2005.07.001.
17. Tan, Y.; Croiset, Eric; Douglas, Mark; Thambimuthu, Kelly. Combustion Characteristics of Coal in a Mixture of Oxygen and Recycled Flue Gas. *Fuel* **2006,** *85,* 507–512. DOI: https://10.1016/j.fuel.2005.08.010.
18. Syamlal, M.; Richards, G.; Plasynski, S. *Carbon Sequestration—Oxy-fuel Combustion, R&D Facts, R&D127.indd*; National Energy Technology Laboratory (NETOL), Morgantown, WV: US Department of Energy (DOE) Office of Fossil Energy; **2008**.
19. SCCS (Scottish Carbon Capture & Storage). Containment solutions to complement emissions reduction strategies—Oxy-Fuel Combustion Capture [Online]. Available at: http://www.geos.ed.ac.uk/sccs/capture/oxyfuel.html.
20. National Renewable Energy Lab. (NREL), Golden, CO (United States), **2001** [Online]. Available at: https://www.nrel.gov/docs/gen/fy01/29306.pdf.
21. Sagar, K.; Geodiversity And Impact On Environment Jnu-Eiacp. Importance of Renewable Energy and Sustainable Development in India, **2021** [Online]. Available at: https://www.researchgate.net/publication/350515490_Importance_of_Renewable_Energy_and_Sustainable_Development_in_India.
22. Ahmed, K. *Renewable Energy Technologies: A Review of the Status and Costs of Selected Technologies*; Wold Bank Technical Paper No. 240. The World Bank: Washington, DC, **1994**. DOI:https://doi.org/10.1596/0-8213-2744-5.

23. Capentieri, A. E.; Larson, E. D.; Woods, J. Future Biomass-Based Electricity Supply in Northeast Brazil. *Biomass Bioernergy* **1993,** *4,* 149–174.
24. Arachchige, U.; Weliwaththage, S. R. G. Solar Energy Technology, **2020** [Online]. https://www.researchgate.net/publication/342610877_Solar_Energy_Technology.
25. Grubb, M J.; Meyer, N. Wind Energy Resources, Systerns and Regional Strategies. In *Biomass for Energy: Supply Prospects*; Johansson, H. D. O., et al., Eds., **1993**.
26. Johansson, T. B.; et al. *Renewable Energy: Sources for Fuels and Electricity*; Island Press: Washington, DC, **1993**.
27. Simberloff, D. The Role of Science in the Preservation of Forest Biodiversity. *Forest Ecol. Manag.* **1999,** *115* (2–3), 101–111. DOI: https://doi.org/10.1016/s0378-1127(98)00391-0.
28. Bibri, S. E. The IoT for Smart Sustainable Cities of the Future: An Analytical Framework for Sensor-based Big Data Applications for Environmental Sustainability. *Sustainable Cities Soc.* **2018,** *38,* 230–253. DOI: https://doi.org/10.1016/j.scs.2017.12.034.
29. Koo, D.; Piratla, K.; Matthews, C. J. Towards Sustainable Water Supply: Schematic Development of Big Data Collection Using Internet of Things (IoT). *Procedia Eng.* **2015,** *118,* 489–497.
30. Truong, T.; Dinh, A.; Wahid, K. In *An IoT Environmental Data Collection System for Fungal Detection in Crop Fields*, **2017** IEEE 30th Canadian Conference on Electrical and Computer Engineering (CCECE), Windsor, ON, Canada, 2017; pp 1–4. DOI: 10.1109/CCECE.2017.7946787.
31. McCarthy, J. Hubert Dreyfus, What Computers Still Can't Do. *Artif. Intel.* **1996,** *80* (1), 143–150, DOI: https://doi.org/10.1016/0004-3702(95)00086-0.
32. Naqa, I. E.; Murphy, M. J. What Is Machine Learning? What Is Machine Learning? n.d. DOI: https://doi.org/10.1007/978-3-319-18305-3_1.
33. Carta, S.; Ed., Machine Learning and the City. In *Applications in Architecture and Urban Design*; Wiley-Blackwell, **2022**.
34. Shivaprakash, K. N.; Swami, N.; Mysorekar, S.; Arora, R.; Gangadharan, A.; Vohra, K.; Jadeyegowda, M.; Kiesecker, J. M. Potential for Artificial Intelligence (AI) and Machine Learning (ML) Applications in Biodiversity Conservation, Managing Forests, and Related Services in India. *Sustainability* **2022,** *14* (12), 7154. DOI: https://doi.org/10.3390/su14127154.
35. Gonzalez, L.; Montes, G.; Puig, E.; Johnson, S.; Mengersen, K.; Gaston, K. Unmanned Aerial Vehicles (UAVs) and Artificial Intelligence Revolutionizing Wildlife Monitoring and Conservation. *Sensors* **2016,** *16* (1), 97. DOI: https://doi.org/10.3390/s16010097.
36. Santangeli, A.; Chen, Y.; Kluen, E.; Chirumamilla, R.; Tiainen, J.; Loehr, J. Integrating Drone-borne Thermal Imaging with Artificial Intelligence to Locate Bird Nests on Agricultural Land. *Sci. Rep* **2020,** *10* (1). DOI: https://doi.org/10.1038/s41598-020-67898-3.
37. Kates, R. W.; Clark, W. C.; Corell, R.; Hall, J. M.; Jaeger, C. C.; Lowe, I.; McCarthy, J. J.; Schellnhuber, H. J.; Bolin, B.; Dickson, N. M.; Faucheux, S.; Gallopin, G. C.; Grübler, A.; Huntley, B.; Jäger, J.; Jodha, N. S.; Kasperson, R. E.; Mabogunje, A.; Matson, P.; … Svedin, U. Sustainability Science. *Science* **2001,** *292* (5517), 641–642.
38. Gubbi, J.; Buyya, R.; Marusic, S.; Palaniswami, M. Internet of Things (loT): A Vision, Architectural Elements, and Future Directions. *Future Generat. Comput. Syst.* **2012**. 29. DOI: https://doi.org/10.1016/j.future.2013.01.010.

Index

D

E

X

Y

For Product Safety Concerns and Information please contact our EU
representative GPSR@taylorandfrancis.com
Taylor & Francis Verlag GmbH, Kaufingerstraße 24, 80331 München, Germany

www.ingramcontent.com/pod-product-compliance
Ingram Content Group UK Ltd.
Pitfield, Milton Keynes, MK11 3LW, UK
UKHW022025061126
477531UK00002B/36

* 9 7 8 1 7 7 4 9 1 7 8 5 5 *